Geomorphology: An Earth Science Overview

Geomorphology: An Earth Science Overview

Edited by **Ken Shaw**

SYRAWOOD
PUBLISHING HOUSE

New York

Published by Syrawood Publishing House,
750 Third Avenue, 9th Floor,
New York, NY 10017, USA
www.syrawoodpublishinghouse.com

Geomorphology: An Earth Science Overview
Edited by Ken Shaw

International Standard Book Number: 978-1-68286-089-2 (Hardback)

Printed in the United States of America.

Contents

Permissions

List of Contributors

Preface

Every book is initially just a concept; it takes months of research and hard work to give it the final shape in which the readers receive it. In its early stages, this book also went through rigorous reviewing. The notable contributions made by experts from across the globe were first molded into patterned chapters and then arranged in a sensibly sequential manner to bring out the best results.

Geomorphology is an important field in the discipline of earth science that aims to analyze the formation of earth's topographic features by various physical and chemical interactions. Some of the chapters included in this book discuss significant concepts of geomorphology like data measurement and analysis, interaction between different spheres, remote sensing and modeling of earth surface processes, etc. It aims to shed light on some of the unexplored aspects of geomorphology and the recent researches in this field which will make this book an invaluable resource for academicians and professionals alike.

It has been my immense pleasure to be a part of this project and to contribute my years of learning in such a meaningful form. I would like to take this opportunity to thank all the people who have been associated with the completion of this book at any step.

Editor

Arctic–alpine blockfields in the northern Swedish Scandes: late Quaternary – not Neogene

B. W. Goodfellow[1,2], **A. P. Stroeven**[1], **D. Fabel**[3], **O. Fredin**[4,5], **M.-H. Derron**[5,6], **R. Bintanja**[7], and **M. W. Caffee**[8]

[1]Department of Physical Geography and Quaternary Geology, and Bolin Centre for Climate Research, Stockholm University, 10691 Stockholm, Sweden

[2]Department of Geology, Lund University, 22362 Lund, Sweden

[3]Department of Geographical and Earth Sciences, East Quadrangle, University Avenue, University of Glasgow, Glasgow G12 8QQ, UK

[4]Department of Geography, Norwegian University of Science and Technology (NTNU),7491, Trondheim, Norway

[5]Geological Survey of Norway, Leiv Eirikssons vei 39, 7491 Trondheim, Norway

[6]Institute of Geomatics and Risk Analysis, University of Lausanne, 1015 Lausanne, Switzerland

[7]Royal Netherlands Meteorological Institute, Wilhelminalaan 10, 3732 GK De Bilt, the Netherlands

[8]Department of Physics, Purdue University, West Lafayette, Indiana, USA

Correspondence to: B. W. Goodfellow (brad.goodfellow@natgeo.su.se)

Abstract. Autochthonous blockfield mantles may indicate alpine surfaces that have not been glacially eroded. These surfaces may therefore serve as markers against which to determine Quaternary erosion volumes in adjacent glacially eroded sectors. To explore these potential utilities, chemical weathering features, erosion rates, and regolith residence durations of mountain blockfields are investigated in the northern Swedish Scandes. This is done, firstly, by assessing the intensity of regolith chemical weathering along altitudinal transects descending from three blockfield-mantled summits. Clay / silt ratios, secondary mineral assemblages, and imaging of chemical etching of primary mineral grains in fine matrix are each used for this purpose. Secondly, erosion rates and regolith residence durations of two of the summits are inferred from concentrations of in situ-produced cosmogenic [10]Be and [26]Al in quartz at the blockfield surfaces. An interpretative model is adopted that includes temporal variations in nuclide production rates through surface burial by glacial ice and glacial isostasy-induced elevation changes of the blockfield surfaces. Together, our data indicate that these blockfields are not derived from remnants of intensely weathered Neogene weathering profiles, as is commonly considered. Evidence for this interpretation includes minor chemical weathering in each of the three examined blockfields, despite consistent variability according to slope position. In addition, average erosion rates of ~ 16.2 and $\sim 6.7\,\mathrm{mm\,ka^{-1}}$, calculated for the two blockfield-mantled summits, are low but of sufficient magnitude to remove present blockfield mantles, of up to a few metres in thickness, within a late Quaternary time frame. Hence, blockfield mantles appear to be replenished by regolith formation through, primarily physical, weathering processes that have operated during the Quaternary. The persistence of autochthonous blockfields over multiple glacial–interglacial cycles confirms their importance as key markers of surfaces that were not glacially eroded through, at least, the late Quaternary. However, presently blockfield-mantled surfaces may potentially be subjected to large spatial variations in erosion rates, and their Neogene regolith mantles may have been comprehensively eroded during the late Pliocene and early Pleistocene. Their role as markers by which to estimate glacial erosion volumes in surrounding landscape elements therefore remains uncertain.

1 Introduction

Autochthonous blockfields are diamicts comprised of clay- to boulder-sized regolith formed through in situ bedrock weathering (Potter and Moss, 1968; Nesje et al., 1988; Ballantyne, 1998; Boelhouwers, 2004). They are classically a feature of periglaciated landscapes, where they frequently mantle mountain summits and plateaus assumed to have undergone up to tens of metres of (non-glacial) erosion during the Quaternary (Dahl, 1966; Ives, 1966; Sugden, 1968, 1974; Nesje et al., 1988; Rea et al., 1996; Ballantyne, 1998; Small et al., 1999; Goodfellow et al., 2009; Rea, 2013). According to this interpretation, blockfields indicate surfaces that persisted as nunataks or were inundated by non-erosive cold-based ice during glacial periods. Blockfield-mantled surfaces may provide useful markers for quantifying Quaternary glacial erosion volumes in surrounding landscapes (Nesje and Whillans, 1994; Glasser and Hall, 1997; Kleman and Stroeven, 1997; Staiger et al., 2005; Goodfellow, 2007; Jansson et al., 2011). However, recent studies of landscape evolution and Quaternary sediment budgets along the Norwegian margin (Nielsen et al., 2009; Steer et al., 2012) imply that, rather than providing these markers, autochthonous blockfield-mantled surfaces have also undergone surface lowering of some hundreds of metres through the action of a Quaternary glacial and periglacial "buzz saw". The origins, ages, and erosion rates of blockfields remain enigmatic, so their utility for indicating non-glacially eroded surfaces and for estimating Quaternary erosion volumes is contentious. The weathering characteristics, erosion rates, and residence durations of present autochthonous blockfield regoliths in the periglaciated northern Scandinavian Mountains (Scandes) are investigated in this study.

Autochthonous blockfields in periglaciated landscapes are frequently hypothesized to be remnants of Neogene weathering profiles (Caine, 1968; Ives, 1974; Clapperton, 1975; Nesje et al., 1988; Rea et al., 1996; Boelhouwers et al., 2002; André, 2003; Marquette et al., 2004; Sumner and Meiklejohn, 2004; Fjellanger et al., 2006; Paasche et al., 2006; André et al., 2008; Strømsøe and Paasche, 2011). In this model, block production is initiated through chemical weathering of bedrock during the Neogene by a warmer-than-present climate. Regolith stripping occurred during the colder Quaternary, subaerially exposing rock made more porous by chemical weathering. Enhanced access by water permitted efficient frost shattering of this rock, which was periglacially reworked to produce blockfield mantles that armour these surfaces, making them resistant to further modification (Boelhouwers, 2004).

If chemical weathering depends upon a "warm" climate, certain characteristics of blockfields are incompatible with a Quaternary origin. These characteristics include the presence of saprolite (Caine, 1968) and/or secondary minerals, especially kaolinite and gibbsite (Rea et al., 1996; Fjellanger et al., 2006; André et al., 2008; Strømsøe and Paasche, 2011), and clay abundances exceeding about 10 % of the fine matrix (clay, silt, sand) by volume (Rea et al., 1996; Strømsøe and Paasche, 2011). An additional circumstantial argument is that there are apparently no actively forming blockfields (Boelhouwers, 2004), with the exception of those developing on highly frost susceptible limestone in the Canadian Arctic (Dredge, 1992). Also, blockfield-mantled surface remnants do not appear to have been glacially eroded (Sugden, 1968, 1974; Kleman and Stroeven, 1997; Fabel et al., 2002; Marquette et al., 2004; Stroeven et al., 2006; Goodfellow, 2007). Taken together, the evidence seemingly indicates that blockfields may pre-date the last glacial–interglacial cycle. Field observations in conjunction with geochemical features may indicate regolith residence durations extending back into the Neogene (Rea et al., 1996; Whalley et al., 2004; André et al., 2008; Strømsøe and Paasche, 2011).

The Neogene-origin model is not universally accepted, and some researchers conclude that blockfields in periglaciated landscapes are entirely Quaternary features (Dahl, 1966; Dredge, 1992; Ballantyne, 1998, 2010; Ballantyne et al., 1998; Goodfellow et al., 2009; Goodfellow, 2012). In this model, blockfields are produced through synergistic physical (e.g. through frost-cracking) and chemical weathering processes (Whalley et al., 2004) that operate independently of preconditioning by Neogene processes.

Key evidence supporting the Quaternary-origin model includes slow formation of clay-sized regolith and secondary minerals through chemical weathering. This is indicated, firstly, by a low ratio of clay to silt across a sample batch (clay$< \sim 0.5 \times$silt), compared with higher ratios (clay$> \sim 0.5 \times$silt) in regoliths located in non-periglaciated settings (Goodfellow, 2012). Secondly, low abundances of secondary minerals are mixed in with abundant primary minerals (Goodfellow, 2012). The secondary mineral assemblages may span a range of leaching intensities including low (minerals with interstratified primary and secondary layers), moderate (2 : 1 layer minerals such as vermiculite), high (1 : 1 minerals such as kaolinite), and extreme (Al- and Fe-oxides such as gibbsite and haematite). This may reflect the effect of hydrological heterogeneities on weathering intensity in blockfields and the varying susceptibility of different primary minerals to chemical weathering. These mineral assemblages differ from those occurring in subtropical and tropical regoliths, which are generally simpler and dominated by high volumes (i.e. > 30 % of the regolith) of kaolinite and Al- and Fe-oxides (Meunier et al., 2007; White et al., 1998; Goodfellow, 2012).

In the Quaternary-origin model, it is further argued that blocks form by frost cracking *within* the regolith, near the base of the permafrost active layer where liquid water accumulates and seasonally refreezes (Dahl, 1966; Anderson,

1998; Small et al., 1999; Hales and Roering, 2007; Goodfellow et al., 2009; Ballantyne, 2010). This mechanism might therefore explain the apparent absence of frost cracking of clasts comprising blockfield surfaces while further highlighting a possible key role of Quaternary, rather than Neogene, weathering processes in blockfield formation.

A critical problem with ascertaining blockfield ages and origins is that it has, until recently, been impossible to measure blockfield erosion rates or regolith residence durations. However, measurements of terrestrial cosmogenic nuclide (TCN) concentrations now offer some insight into these issues. Erosion rates of 1.1–12.0 mm ka^{-1} have been inferred for subaerially exposed bedrock within alpine blockfields (Small et al., 1997; Bierman et al., 1999; Staiger et al., 2005; Phillips et al., 2006). These rates may be lower than in the surrounding blockfields, because exposed bedrock sheds, rather than retains, water (Small et al., 1999; Cockburn and Summerfield, 2004; Phillips et al., 2006). However, regolith erosion rates in summit blockfields may remain low because of armouring of gently sloping surfaces by cobbles and boulders (Granger et al., 2001; Boelhouwers, 2004). For example, erosion rates of 13.4–14.0 mm ka^{-1} have been measured in plateau blockfields in the Wind River Range, Wyoming, which have not been inundated by glacial ice (Small et al., 1999). Where non-erosive cold-based ice has buried blockfields during glacial periods (Sugden and Watts, 1977; Kleman and Stroeven, 1997; Bierman et al., 1999; Hättestrand and Stroeven, 2002; Briner et al., 2003; Marquette et al., 2004), time-averaged erosion rates are further lowered. Subaerial exposure and burial durations of blockfield regoliths might then extend back in time to the early Quaternary or late Neogene. By combining measurements of TCN concentrations in bedrock or regolith with an inferred history of surface burial by ice sheets from benthic $\delta^{18}O$ records (Fabel et al., 2002; Stroeven et al., 2002; Li et al., 2008), it is possible to estimate minimum time spans over which present blockfield regoliths have mantled surfaces (i.e. minimum regolith residence durations).

In this study we test whether blockfields in the northern Swedish Scandes are remnants of intensely weathered Neogene regoliths or are formed solely by Quaternary weathering processes. We do this, firstly, by investigating the intensity of chemical weathering through grain size, X-ray diffraction, and scanning electron microscopy (SEM) analyses of blockfield fine matrix along three hillslope transects. Secondly, we examine regolith residence durations of two summit blockfields through the combination of apparent surface exposure durations, measured through TCN analyses, with burial durations, determined through an ice sheet model driven by benthic $\delta^{18}O$ records. Incorporating an elastic lithosphere, relaxed asthenosphere (ELRA) bedrock model, the ice sheet model is also used to study the effects of bedrock isostatic response to glacial loading and unloading on nuclide production rates (which vary with elevation above sea level) and subsequent regolith residence durations. Because uncertain-

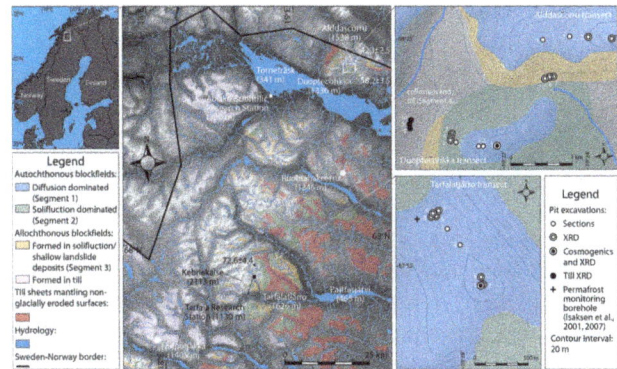

Figure 1. Map of the study areas in the northern Swedish Scandes. The map location and sample sites along three hillslope transects are shown in the adjoining panels. Autochthonous blockfields mantling low-gradient convex summits appear to be eroded by diffusive processes, such as regolith creep, and erosion of autochthonous blockfields on steep slopes appears dominated by solifluction. Colluvial boulder drapes provide evidence of shallow landsliding and form allochthonous blockfields on the steepest regolith-mantled slopes. These comprise slope segments 1, 2, and 3, respectively, and are the focus of regolith sampling in this study. Allochthonous blockfields also form in till sheets ($< \sim 1$ m thick) deposited on some glacially eroded summits (Goodfellow et al., 2008) and on some high-altitude non-glacially eroded surface remnants (Kleman and Stroeven, 1997; Fabel et al., 2002; Goodfellow et al., 2008). Pits along each transect are numbered according to those given in Tables 1 and 2 and S1 in the Supplement. Grey areas of the maps are cliff faces, talus slopes, or surfaces modified by glacial erosion or deposition (covered in $> \sim 1$ m thick tills on the transect maps). In the top right panel, mixed colluvium and till drape the landscape below Alddascorru and Duoptecohkka in the grey area west of these summits and comprises segment 4 for regolith sampling in this study. ^{10}Be bedrock exposure ages from sites close to those used in this study are reproduced from Fabel et al. (2002) and Stroeven et al. (2006).

ties associated with calculations of regolith residence durations are large, we confine our enquiry to an order of magnitude question: are regolith residence durations likely confined to the late Quaternary (< 1 Ma) or do they extend to the early Quaternary/late Neogene? A Neogene origin would imply low Quaternary-averaged surface erosion rates and a utility of autochthonous blockfield-mantled surfaces as markers from which to estimate glacial erosion depths in surrounding landscapes. In contrast, the implications of a Quaternary origin are more ambiguous. These could not exclude tens of metres, to perhaps more than one hundred metres, of surface lowering during the Plio-Pleistocene transition of currently blockfield-mantled surfaces, resulting in a lowered utility of these surfaces as markers from which to estimate glacial erosion depths in surrounding landscape sectors.

2 Study area

Blockfields were examined along hillslope transects descending from three summits in the northern Swedish Scandes (Fig. 1): Alddasčorru (68°25′ N, 19°24′ E; 1538 m above sea level (a.s.l.)), Duoptečohkka (68°24′ N, 19°22′ E; 1336 m a.s.l.), and Tarfalatjårro (67°55′ N, 18°39′ E; 1626 m a.s.l.). The transects intersect slope segments shaped by contrasting assemblages of surface processes. Diffusive processes, such as regolith creep, have apparently shaped gently convex summits and solifluction has dominated on higher gradient downslope segments. On the steepest, lowermost slopes imbricated blocks and boulder sheets indicate that shallow landsliding and boulder tumbling have been active in addition to solifluction. Further deposition of transported material has occurred at the concave bases of these slopes where scattered boulders are embedded in, and rest upon, a fine-matrix-rich regolith. The abundance of the fine matrix here and numerous boulders, some of non-local lithology, resting on the ground surface also indicate an additional contribution of glacial till. The Tarfalatjårro transect terminates 73 m below the summit in an autochthonous blockfield-mantled saddle and intersects only the diffusion-dominated segment. In contrast, the Alddasčorru transect terminates on the steep mass-wasting segment 278 m below the summit, whereas the Duoptečohkka transect intersects each of these segments and the solifluction-dominated segment and terminates 270 m below the summit on the mixed colluvium and till segment (Fig. 1).

Each blockfield is developed on amphibolite. Lithological variations were not observed along either of the Alddasčorru or Duoptečohkka transects except for some granitic glacial erratics. In contrast, plagioclase-porphyritic and highly schistic amphibolites were observed along the Tarfalatjårro transect, which terminates at its lower end on metapsammite. Glacial erratics also occur occasionally on Tarfalatjårro but were not observed along the profile.

The blockfields along each transect form areally continuous mantles, and bedrock outcrops are generally absent. An exception occurs on the narrow, ridge-like summit of Duoptečohkka, where bedrock is frequently exposed. The blockfield surfaces are dominated by cobbles and boulders (> 90 % area) with patches of fine matrix visible only in the centres of interspersed periglacially sorted circles (mean diameters of 1.5–2.0 m) on Alddasčorru, Tarfalatjårro, and on the upper flanks of Duoptečohkka. The Duoptečohkka summit blockfield is not periglacially sorted, and fine matrix is absent from the surface. Ventifacted boulders and loess deposits are absent from each transect.

The study area is located where the Arctic maritime climate of Norway converges with the continental climate of northern Sweden. The mean annual air temperature (MAAT) on Tarfalatjårro during 1946–1995, as inferred from records of nearby Tarfala Research Station at 1130 m a.s.l. (Fig. 1), was approximately −6 °C and mean annual precipitation

(MAP) was about 500 mm (Grudd and Schneider, 1996). More recent data from a permafrost monitoring borehole on Tarfalatjårro indicate a MAAT at 2 m above the ground of −4.3 °C and mean annual ground temperatures of −2.8 and −3.0 °C, at 0.2 and 2.5 m depth respectively, over 2003–2005 (Isaksen et al., 2007). The closest meteorological station to Alddasčorru and Duoptečohkka is located at 380 m a.s.l. at the Abisko Scientific Research Station. (Fig. 1). It has recorded a MAAT of −0.9 °C and MAP of about 320 mm (Eriksson, 1982), which is a warmer and drier climate than occurs on Tarfalatjårro. Based on this MAAT we infer that MAATs are also below zero on Alddasčorru and Duoptečohkka. Permafrost is present on each of the three summits, with the monitoring borehole on Tarfalatjårro indicating a distinct warming trend and a present active layer thickness of 1.4–1.6 m (Isaksen et al., 2001, 2007). Snow covers Tarfalatjårro from about October to May, although strong winds limit the maximum snow depth to about 0.3 m (Isaksen et al., 2001). Similar temperature, snow, and permafrost conditions are expected and assumed for Alddasčorru and Duoptečohkka. Vegetation along each transect is restricted to lichens, mosses, and occasional grasses, except for the base of the Duoptečohkka transect, which is well grassed. Stable lichen-covered surface clasts indicate that, although they have occurred in the past, large-scale periglacial sorting and gelifluction processes appear to be now largely inactive. However, upfreezing of pebbles and creep and gelifluction processes over a few tens of centimetres remain active.

The northern Swedish Scandes have been repeatedly glaciated during the Quaternary, with cirque glaciation inferred to have been dominant before 2.0 million years ago (Ma), mountain ice sheets dominant between 2.0 and 0.7 Ma, and Fennoscandian ice sheets developing over the last 0.7 Ma (Kleman and Stroeven, 1997; Kleman et al., 2008). Current glaciation in the region is confined to small icecaps and small cirque and valley glaciers. During glacial periods, relatively high-altitude surfaces such as Tarfalatjårro, Alddasčorru, and Duoptečohkka were either exposed as nunataks or covered by cold-based ice sheets (Stroeven et al., 2006). The occasional erratics on Tarfalatjårro and abundant granitic erratics on Alddasčorru and the flanks of Duoptečohkka confirm former ice sheet coverage as late as 12 ka (Fabel et al., 2002; Stroeven et al., 2006). However, the presence of autochthonous blockfields and the absence of till sheets and glacial erosion features, such as striated bedrock outcrops, indicate extremely minor glacial modification of Tarfalatjårro and Alddasčorru. Clear evidence of glacial processes is also absent from Duoptečohkka. We therefore consider the presently thin autochthonous blockfield and outcropping bedrock to be attributable to slope transport processes operating across this narrow summit, although some glacial entrainment of blocks cannot be entirely discounted.

3 Methods

3.1 Field techniques

To determine the composition of blockfield regoliths, a total of 26 pits were hand excavated along the three hillslope transects during August 2004–2006: 7 pits on Alddasčorru, 10 pits on Duoptečohkka, and 9 pits on Tarfalatjårro (Fig. 1). Blockfield sections were examined in 15 of these pits, which were excavated across sorted circles, from clast-dominated rings to fine matrix-rich circle centres, or into clast-rich solifluction lobes. The pit excavated into the summit of Duoptečohkka was an exception because the thin regolith (< 0.5 m) and frequent bedrock exposures prevented periglacial sorting. The pits were excavated either until large amphibolite slabs prohibited sampling of deeper sections, the water table was intersected, or bedrock was reached. Fine matrix samples were taken from 16 blockfield pits for grain size, XRD, and SEM analyses. For replication purposes, at least three fine matrix samples were taken from each of the four surface process segments: (1) low-gradient diffusive, (2) solifluction slope, (3) steep mass-wasting, and (4) concave depositional, if and where they occur on the three transects. Segment 3 was only sampled on the Alddasčorru transect, and only one sample was analysed from segment 2 on the Alddasčorru transect. Because we previously found that only minor variations occur in fine matrix granulometry and secondary mineralogy with depth beneath the ground surface (Goodfellow et al., 2009), only one sample was analysed from each pit. An exception occurred for the Alddasčorru summit pit, from which surface, 0.16 m, and 0.60 m depth samples were processed. For comparative purposes, fine matrix samples were taken for grain size, XRD, and SEM analyses from tills covering Ruohtahakčorru (Fig. 1; 68°09′ N, 19°20′ E; 1342–1346 m a.s.l.; three samples from 0.5, 0.9, and 1.2 m depth) and Nulpotjåkka (Fig. 1; 67°48′ N, 18°01′ E; 1405 m a.s.l.; one sample from 0.9 m depth).

Two quartz clasts were collected from the summit surfaces of Duoptečohkka and Tarfalatjårro for measurements of in situ-produced ^{10}Be and ^{26}Al concentrations. Sampling was undertaken on summits to eliminate the possibility of these clasts having been transported and buried by slope processes, which would complicate estimates of regolith residence durations from measurements of ^{10}Be and ^{26}Al concentrations. This constraint, coupled with the scarcity of summit vein quartz, limited our sampling for TCN analyses to these two sites. Zero vertical mixing was assumed for the vein quartz clast (4 cm thick) sampled from the Duoptečohkka summit (Fig. 2a, b). This is because this long clast (0.19 m) resided on the surface of a thin regolith (0.3 m depth), and periglacially sorted circles were absent from this site. Because of the absence of glacial erratics from Duoptečohkka and the presence of quartz veins in these blockfields, we considered the sampled clast to be locally derived. On Tarfalatjårro, the clast (3–4 cm thick) was taken from a shattered

Figure 2. Sample sites for surface quartz clasts on the summits of **(a, b)** Duoptečohkka and Tarfalatjårro **(c, d)**. The quartz sampled on Duoptečohkka (arrow in **b**) was an isolated mass attached to an amphibolite block whereas the quartz clast sampled on Tarfalatjårro (arrow in **d**) was part of a frost-shattered quartz vein that extended ~ 8 m across the blockfield surface (arrows in **c**).

quartz vein (~ 8 m in length) that extended ~ 8 m across the blockfield surface (Fig. 2c, d). Zero vertical mixing of the sampled clast through the regolith profile was again assumed because the quartz vein was clearly expressed at the blockfield surface, periglacially sorted circles did not intersect the quartz vein, and the fine matrix required for periglacial sorting (Ballantyne and Harris, 1994, pp. 85–96) was sparse, resulting in limited vertical and lateral sorting of clasts. While both samples were taken from autochthonous amphibolite blockfields, the narrow, high curvature summit of Duoptečohkka, which also displays frequent bedrock outcrops, contrasts with the broad, comparatively low curvature, comprehensively regolith-mantled summit of Tarfalatjårro. A higher erosion rate and shorter regolith residence duration was therefore expected for Duoptečohkka.

To correct for topographic shielding the surface geometries of the sampled blockfields and surrounding summits were measured with a clinometer and compass. Sample locations were recorded with a handheld GPS and on a 1:50 000 topographic map. Three amphibolite clasts and three fine matrix samples extracted in a cylinder of known volume were collected for regolith clast and matrix density measurements.

3.2 Fine matrix analyses

To determine the chemical weathering characteristics of blockfields, grain size, SEM, and XRD analyses were completed. Grain sizes were determined on dried samples with a Coulter LS particle size analyser. SEM analyses of chemical etching were performed on surface bulk fine matrix samples and semi-quantitative analyses of grain chemistry and mineralogy were completed according to energy dispersive spectrometer techniques (Goldstein et al., 2003). Thin sections for mineralogical interpretation of parent material were prepared from two Alddasčorru rock samples.

For XRD analysis of clay mineralogies, the $< 2\,\mu m$ size fraction of each sample was separated by settling, Mg-saturated, and purified with a ceramic filter to produce oriented samples. An initial XRD scan was performed at 2–69° 2θ with a scan speed of 0.02° $2\theta\,s^{-1}$ and a step size of 0.04° 2θ. Second and third scans were performed following ethylglycol saturation and heating of the samples to 550 °C, respectively. These scans were performed at 2–35° 2θ with a scan speed of 0.0067° $2\theta\,s^{-1}$ and a step size of 0.04° 2θ. Diffraction peaks were analysed with peak search software and manually reviewed using Brindley and Brown (1980) and Moore and Reynolds (1997).

3.3 Cosmogenic radionuclide analyses and ice sheet modelling

Concentrations of ^{10}Be and ^{26}Al in samples of vein quartz were measured to estimate erosion rates and residence durations of blockfield regoliths. Clean quartz separates were processed for cosmogenic nuclide analyses through methods adapted from Kohl and Nishiizumi (1992) and Child et al. (2000). Accelerator mass spectrometry (AMS) measurement of the Tarfalatjårro sample was completed at PRIME Lab, Purdue University, USA, and AMS measurement of the Duoptečohkka sample was completed at the SUERC AMS Laboratory, East Kilbride, UK. Measured TCN concentrations were corrected by full chemistry procedural blanks and normalized using the NIST ^{10}Be standard (SRM4325) with a ^{10}Be / ^{9}Be ratio of $(2.79 \pm 0.03) \times 10^{-11}$ and using a ^{10}Be half-life of 1.36×10^6 a (Nishiizumi et al., 2007) and the PRIME Lab ^{26}Al standard (Z92-0222) with a nominal ^{26}Al / ^{27}Al ratio of 4.11×10^{-11} and using an ^{26}Al half-life of 7.05×10^5 a (Nishiizumi, 2004). Errors in nuclide concentrations include the quadrature sum of analytical uncertainty calculated from AMS counting statistics and procedural errors.

Apparent exposure ages were calculated from ^{10}Be and ^{26}Al concentrations using the CRONUS-Earth exposure age calculator (version 2.2; Balco et al., 2008) assuming zero erosion and the Lal–Stone time-independent ^{10}Be production rate model (Lal, 1991; Stone, 2000). The time-independent Lal–Stone scaling was used here because we also used it for modelling regolith residence durations for reasons described below. We do though cite in our results the full age ranges given for all production rate models incorporated into the CRONUS-Earth exposure age calculator. Corrections were applied for topographic shielding (scaling factors > 0.9998) and for sample thickness using a clast density of 2.65 g cm^{-3}. Apparent exposure ages are not corrected for snow shielding because of high uncertainties, and a correction for vegetation shielding is not required. Quoted exposure age uncertainties (1σ external) include nuclide production rate uncertainties and the concentration errors described above. The apparent exposure ages for these sites are minimum durations to which the effects on nuclide production rates of surface burial by snow and glacial ice, and elevation changes attributable to glacial isostasy are subsequently added.

Regolith residence durations, incorporating periods of subaerial exposure and burial by glacial ice, of the Duoptečohkka and Tarfalatjårro summit blockfields were calculated from measured ^{10}Be concentrations accordingly (rearranged from Lal, 1991):

$$\frac{N_i}{N} = [1 - \beta]\exp\{(\lambda + \alpha)t\} + \beta, \tag{1}$$

where N is nuclide concentration, with the subscript i indicating one step back in time; λ the nuclide half-life; t the time step, and where

$$\alpha = \frac{E\rho}{\Lambda}, \tag{2}$$

where E is the erosion rate (cm a^{-1}), ρ the regolith density, and Λ the attenuation mean free path, and

$$\beta = \frac{P}{N(\lambda + \alpha)}, \tag{3}$$

where P is nuclide production rate. During periods of surface burial by ice sheets, Eq. (1) is simplified to

$$\frac{N_i}{N} = \exp\{\lambda t\}. \tag{4}$$

Regolith residence durations are obtained when zero nuclide concentrations are reached.

The source code for the CRONUS-Earth exposure age calculator (version 2.2; Balco et al., 2008) was not used for calculating these regolith residence durations because of the complexities introduced by accounting for depth- and time-averaged ^{10}Be production rates. Rather, a sea-level high-latitude ($> 60°$) ^{10}Be production rate of 4.59 ± 0.28 atoms g^{-1} a^{-1}, from the Nishiizumi et al. (2007) ^{10}Be half-life of 1.36×10^6 a, was used in these calculations. Production rates were scaled to latitude and altitude using Stone (2000) and a sea surface temperature of 5 °C. The errors in regolith residence durations attributable to our use of simplified ^{10}Be production rates are minor compared with uncertainties attributable to surface burial by snow and ice and isostatic responses to ice sheet loading and unloading.

A constant blockfield density of $2.60\,\mathrm{g\,cm^{-3}}$ was assumed based on a blockfield containing 15 % fine matrix, with a density of $2.10\,\mathrm{g\,cm^{-3}}$ (mean of three samples, $1\sigma = 0.09$), and 85 % amphibolite, with a density of $2.79\,\mathrm{g\,cm^{-3}}$ (mean of three samples, $1\sigma = 0.02$). The effects of regolith dissolution were ignored because regolith residence durations that include multiple periods of surface exposure and burial by ice sheets preclude erosion rate calculations directly from nuclide concentrations. Furthermore, chemical weathering in these blockfields is likely to be minor (Goodfellow et al., 2009). Corrections were applied for shielding and for sample thickness using an attenuation mean free path of $160\,\mathrm{g\,cm^{-2}}$ and a quartzite density of $2.65\,\mathrm{g\,cm^{-3}}$. A correction for snow burial was also incorporated assuming 0.3 m of snow (Isaksen et al., 2007), with a density of $0.3\,\mathrm{g\,cm^{-3}}$, for a duration of 7 months per year. The resulting annual shielding by snow of only $5.25\,\mathrm{g\,cm^{-2}}$ is assumed to be representative of all ice-free periods. Associated uncertainties are, however, unknown but, because they are high, a sensitivity analysis was performed by increasing the depth of burial from 0.30 to 0.50 m and increasing duration of burial from 7 to 10 months a year in a calculation of regolith residence duration.

To incorporate periods of surface burial by ice sheets into calculations of regolith residence durations, and to explore the effects of glacial isostasy on these calculations, we used a 3-dimensional ice-dynamical model forced by the Lisiecki and Raymo (2005) stack of global benthic $\delta^{18}\mathrm{O}$ records and an ELRA bedrock model. Full details of the ice-dynamical model can be found in Bintanja et al. (2002, 2005). An ELRA model offers the best glacial isostasy approximation among the group of simple models, with its primary weakness being that it incorporates only one time constant (Le Meur and Huybrechts, 1996). Whereas self-gravitating visco-elastic spherical Earth models are the most accurate, they are much more complex and require greater computational power and time (Le Meur and Huybrechts, 1996). The ice sheet model was run at 40 km resolution, with a 100-year time step over the last 1.07 Ma. Although spatial resolution is coarse, using a 20 km grid provides minimal change in bedrock topography and calculations on a 50 m digital elevation model of the northern Swedish mountains produced the same mean elevations for 40×40 km squares centred on the relevant model grid points. Furthermore, the wavelength of glacial isostasy is much longer than the topographic wavelength (Le Meur and Huybrechts, 1996). Regional ice sheet thickness is subsequently more important than local ice sheet thicknesses for determining isostatic response, and a grid size of some tens of kilometres appears justified. A limitation of our models' treatment of isostasy is that it lacks an erosion component. However, the consequence of this on isostasy over successive late Quaternary glacial cycles is likely to be less important in this landscape of apparent selective linear erosion than in other alpine locations where glacial erosion was more aerially extensive. For example, Staiger et al. (2005) estimate limited net regional glacial erosion and

a low glacial erosion efficiency for the Torngat Mountains in Labrador, Canada, which consist of similar lithologies to the Scandes and which were also subjected to selective linear glacial erosion during the Quaternary. Our model may also underestimate surface burial durations because smaller ice masses may have formed on summits in between the periods of regional ice sheet coverage that are captured in our model. However, ice sheet models that have the required resolution to account for the formation of these small ice masses are not run over multiple glacial cycles because of the high computational burden. Furthermore, the two summits used in our study are perhaps unlikely locations for small ice masses to form, because they are subjected to strong winds, or to persist well after larger ice masses have retreated, because of their locations on the crests of high ridges. We therefore consider our model to offer a reasonable approximation (with estimated ±20 % error margins) of ice sheet burial durations and glacial isostasy.

4 Results

4.1 Blockfield structure

Blockfield vertical sections along the Alddasčorru, Duoptečohkka, and Tarfalatjårro altitudinal transects display a number of common features (Fig. 3). Surface exposures of cobbles and boulders comprise the outer rings of periglacially sorted circles on low-gradient surfaces and delineate solifluction lobes on slopes. In addition to lichen covers, subaerially exposed clast surfaces on sorted circles display rounding through granular disintegration, which further confirms presently limited regolith mixing on low-gradient surfaces. In contrast, freshly exposed fine matrix and loose clasts indicate that some solifluction lobes remain active. In each setting, surface cobbles and boulders are underlain by a layer of gravel and larger clasts to average depths of 0.9–1.0 m beneath the ground surface. In the centres of fine matrix-rich sorted circle centres, surface layers of cobbles and boulders are usually underlain by fine matrix, granules and gravel, in which cobbles and boulders are embedded, to depths up to 0.7 m. Excavation depths generally ranged between 0.6 and 1.3 m, and the bottom of each pit typically consisted of boulders embedded in a fine matrix that ranged from damp to water saturated (Table S1 in the Supplement). The Duoptečohkka summit pit was excavated to bedrock, which was reached at only 0.30 m. None of the pits revealed soil horizons or saprolite.

Although only about 10 % of the ground surface consists of fine matrix, it appears to comprise about 10–20 % of the subsurface regolith (Table S1 in the Supplement). Surface fine matrix is most abundant on the section of the plateau into which Alddasčorru pit 4 was dug (~ 50 % of the surface area) and in the saddle into which Tarfalatjårro pits 6–9 were dug (~ 30 % of the surface area). Sub-surface fine matrix appears most limited on the summits of Alddasčorru and

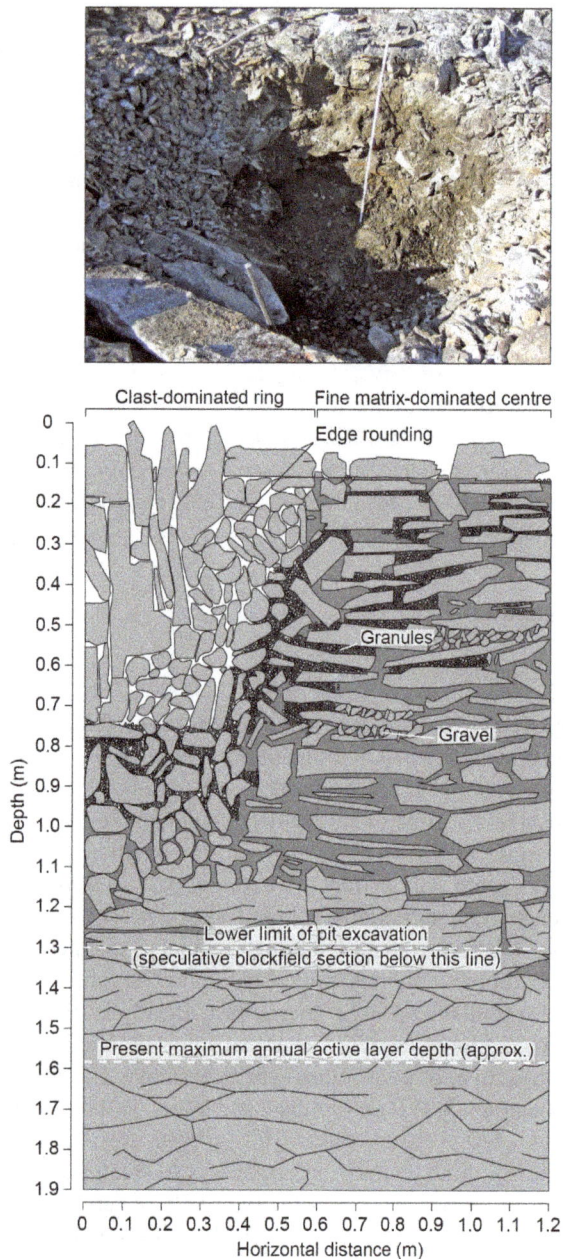

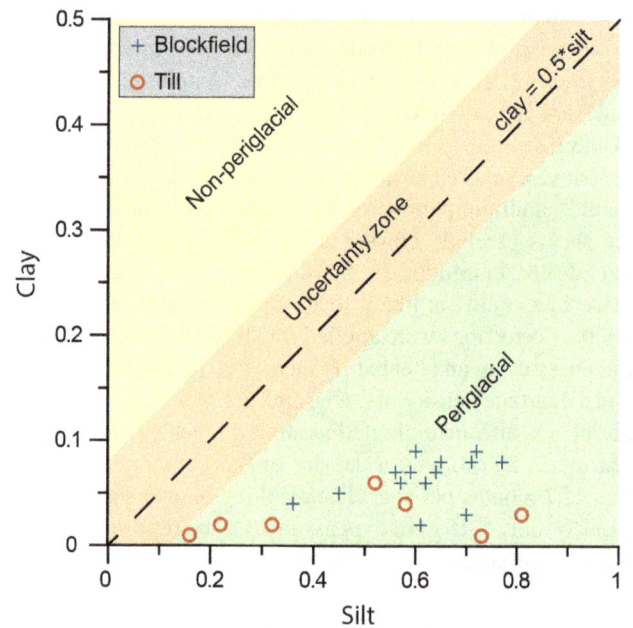

Figure 4. Quantities of clay and silt in fine matrix samples collected from blockfields (blue crosses) and till (red circles). All samples contain clay / silt ratios less than 0.5, which indicates fine matrix production under conditions that have been, at least seasonally, periglacial (Goodfellow, 2012). Fine matrix samples falling in the uncertainty zone may have been exposed to periglacial conditions during their formation (Goodfellow, 2012). All data have been divided by 65 to fit on 0–1 scales.

Tarfalatjårro (5–15 %), and in the solifluction lobes on the slope of Duoptečohkka (pits 4 and 5, 5–10 %; Table S1 in the Supplement). Here, cobbles and boulders were embedded in gravel throughout most of the subsurface, with only small accumulations of fine matrix on boulder tops or in poorly defined sorted circle centres.

In summary, blockfield sections indicate present surface stability as well as regolith sorting and transport, particularly during former periods of colder climate where blockfield-mantled surfaces also remained free of glacial ice cover. Chemical weathering rates have been insufficient to produce soil horizons or saprolite.

4.2 Fine matrix granulometry

All blockfield fine matrix samples are sandy loams (Table 1; US Department of Agriculture, 1993, p. 138). However, minor variations occur in the distribution of sand ($1\sigma = 7.5\%$), silt ($1\sigma = 6.9\%$), and clay ($1\sigma = 1.3\%$) according to sampling depth and the subsurface distribution of boulders, between which granules, pebbles, and gravel accumulate. Till samples vary between loamy sand, sandy loam, and silt loam, and have lower mean clay quantities (1.8%, $1\sigma = 1.1\%$) than the blockfield fine matrix (4.2%, $1\sigma = 1.3\%$). Although clay abundances are higher in blockfields than in till,

Figure 3. Representative vertical sections of autochthonous blockfields. The photograph shows an excavation across one of the sparsely distributed sorted circles developed in the summit blockfield of Tarfalatjårro, and the line drawing summarizes the general features of autochthonous blockfields in a vertical section cut across a periglacially sorted circle. The ruler in the photograph is 1 m in length. Angular cobbles and boulders are embedded in fine matrix (clay, silt, sand) in sorted circle centres. Granules and gravel accumulate between clasts distributed vertically through the section that have subhorizontally oriented long axes. Conversely, the outer ring of the sorted circle is comprised of gravel, cobbles, and boulders, whereas granules accumulate near the base of the section. Clast surfaces are sub-rounded where they are subaerially exposed. The pit bases generally intersect large rock slabs and are wet.

Table 1. Particle size distribution and secondary minerals in fine matrix from Alddasčorru, Duoptečohkka, and Tarfalatjårro transects, and from Ruohtahakčorru and Nulpotjåkka summit tills.

Transect and pit	Sample[a]	Location[b] (slope segment)	Elevation (m a.s.l.)	Depth (m)	Clay (%)	Silt (%)	Sand (%)	Clay / silt	Clay minerals[c]
Alddasčorru 1	BG-05-11	Summit (1)	1538	Surface	4.9	42.2	52.9	0.12	C, I, A, P, Q
Alddasčorru 1	BG-05-04	Summit (1)	1538	0.16	3.4	29.3	67.3	0.12	C, I, A, P, Q
Alddasčorru 1	BG-05-16	Summit (1)	1538	0.60	2.7	23.6	73.7	0.11	C, I, A, P, Q
Alddasčorru 2	BG-05-26	Slope (1)	1500	0.60	4.5	36.2	59.4	0.12	C, I, A, G?, P, Q
Alddasčorru 5	BG-05-67	Slope base (3)	1260	0.40	5.8	46.5	47.7	0.12	C, V, I, A, G, P, Q
Alddasčorru 6	BG-05-68	Slope base (3)	1260	0.40	4.1	37.0	58.8	0.11	C, V, I, A, G, P, Q
Alddasčorru 7	BG-05-69	Slope base (3)	1260	0.40	3.2	29.2	67.6	0.11	C, V, I, A, G, P, Q
Duoptečohkka 5	BG-05-64	Slope (2)	1200	0.40	4.0	40.6	55.4	0.10	C, V, I, A, P, Q
Duoptečohkka 6	BG-05-65	Slope (2)	1200	0.40	4.1	40.3	55.6	0.10	C, V, I, A, G, P, Q
Duoptečohkka 7	BG-05-66	Slope (2)	1200	0.40	4.4	38.4	57.3	0.11	C, V, I, A, G?, P, Q
Duoptečohkka 8	BG-05-70	Colluvium/till (4)	1060	0.40	1.1	14.1	84.8	0.08	C, V, I, A, P, Q
Duoptečohkka 9	BG-05-71	Colluvium/till (4)	1060	0.40	3.6	34.1	62.3	0.11	C, V, I, A, G, P, Q
Duoptečohkka 10	BG-05-72	Colluvium/till (4)	1060	0.40	1.5	20.8	77.7	0.07	C, V, I, A, G?, P, Q
Tarfalatjårro 1	BG-04-25	Summit (1)	1626	1.25	1.5	39.5	59.0	0.04	C, I, A, P, Q
Tarfalatjårro 2	BG-06-22	Summit (1)	1626	Surface	5.6	39.3	55.2	0.14	C, I, A, P, Q
Tarfalatjårro 3	BG-04-26	Summit/slope (1)	1623	0.80	2.1	45.5	52.4	0.05	C, I, A, P, Q
Tarfalatjårro 6	BG-05-83	Saddle (1)	1553	0.50	4.7	41.3	54.0	0.11	C, V, I, A, G, P, Q
Tarfalatjårro 7	BG-05-84	Saddle (1)	1553	0.20	5.3	49.9	44.8	0.11	C, V, I, A, G, P, Q
Tarfalatjårro 8	BG-05-85	Saddle (1)	1553	0.20	5.4	49.8	44.8	0.11	C, V, I, A, G, P, Q
Tarfalatjårro 9	BG-06-07	Saddle (1)	1553	0.70	5.4	46.1	48.5	0.12	C, V, I, A, G, P, Q
Ruohtahakčorru	BG-04-22	Summit till	1342	0.50	2.2	52.8	45.0	0.04	C, V, I, A, P, Q
Ruohtahakčorru	BG-04-23	Summit till	1346	0.90	0.7	10.3	89.0	0.07	C, V, I, A, P, Q
Ruohtahakčorru	BG-04-24	Summit till	1343	1.20	2.6	37.7	59.7	0.07	C, V, I, A, P, Q
Nulpotjåkka	BG-04-27	Summit till	1405	0.90	0.7	47.2	52.1	0.01	C, V, I, A, P, Q

[a] BG-06-22 and BG-06-07 are representative of eight samples from Tarfalatjårro pit 2 and seven samples from Tarfalatjårro pit 9, respectively (Goodfellow et al., 2009). [b] Slope segments are numbered as follows: (1) diffusion-dominated summit, (2) solifluction-dominated slope, (3) steep mass wasting, (4) concave depositional, where regolith comprises colluvium and till. Till is indicated by the presence of clasts of different lithologies and a high abundance of fine matrix. [c] C – chlorite; V – vermiculite; I – illite; A – amphibole; G – gibbsite (? indicates uncertain); P – plagioclase (dominant feldspar); Q – quartz.

clay comprises only 1.5 to 5.8 % of the fine matrix volume, which remains at the low end of the range (1–30 %) previously reported for other blockfields (e.g. Caine, 1968; Rea et al., 1996; Dredge, 2000; Marquette et al., 2004; Paasche et al., 2006; Table A.1 in Goodfellow, 2012). Clay / silt ratios are all ≤ 0.14 (Table 1; Fig. 4). These indicate a low intensity of chemical weathering that is typical for regolith formation under, at least seasonal, periglacial conditions (Goodfellow, 2012).

4.3 Fine matrix mineralogy

XRD analyses of the clay-sized fraction of the regolith indicate the presence of primary and secondary minerals in all samples (Table 1, Fig. 5). Primary minerals are abundant and include chlorite, amphibole, and feldspar. The presence of these minerals, along with epidote, was also confirmed using thin sections. In addition to these primary minerals, small quantities of poorly crystallized Al- and Fe-oxyhydroxides are identifiable by XRD in the Alddasčorru and Tarfalatjårro summit samples. In contrast, vermiculite, gibbsite, and larger quantities of poorly crystallized oxyhydroxides are also identifiable in concave locations, such as at the base of Alddasčorru and in the Tarfalatjårro saddle. Gibbsite generally occurs together with poorly crystallized Al- and Fe-oxyhydroxides and vermiculite. A sample from the upper slope of Alddasčorru (BG-05-26; Fig. 5b) forms a possible exception, where poorly crystallized oxyhydroxides and gibbsite appear to be the only secondary minerals present. Up to two samples from the solifluction-dominated slope segment on the Duoptečohkka transect are also gibbsite-bearing. All till samples contain poorly crystallized oxyhydroxides and vermiculite, but up to two samples of mixed colluvium and till at the base of the Duoptečohkka transect also contain gibbsite. We are unable to distinguish kaolinite according to the standard XRD techniques we used because of the ubiquitous presence of chlorite (Moore and Reynolds, 1997, p. 234). However, we believe that kaolinite may be present in our samples in small quantities. Quartz and muscovite are also commonly present. Because of the scarcity of quartz in the amphibolitic parent rock, these likely represent aeolian additions to blockfields and/or are till components.

Chemical weathering was observed on only two albite grains under SEM: one through surface etching and a second through more general disintegration (Fig. 6). It was otherwise absent, even on easily weathered minerals such as amphibolite and epidote, in addition to most albite grains. This observation of sparsely weathered silt- and sand-sized grains complements the mixed primary and secondary mineralogy of clay-sized grains. Together, they provide an overall impression of generally minor chemical weathering.

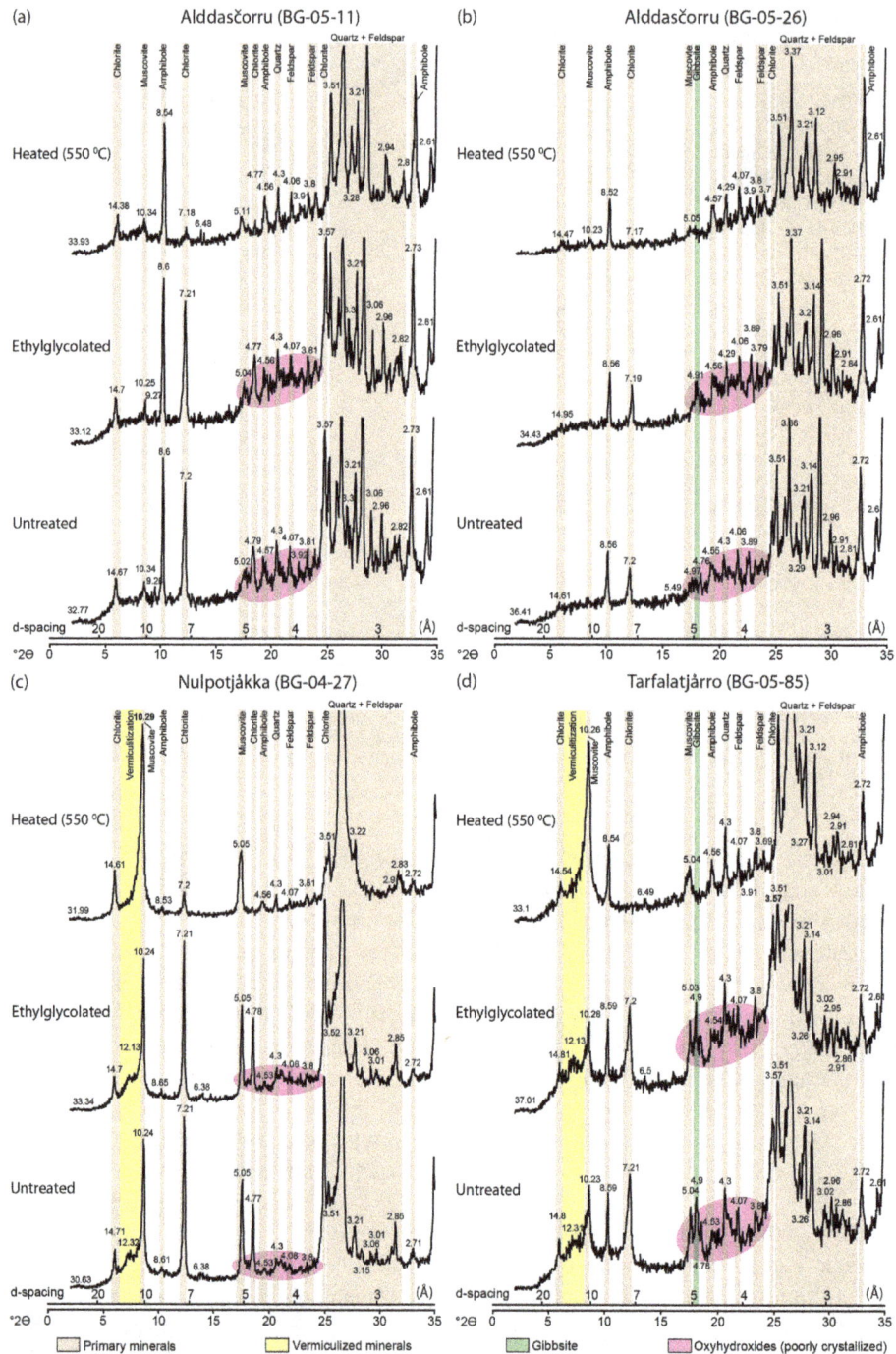

Figure 5. Four representative X-ray diffractograms of the clay-sized fraction ($< 2\,\mu$m) of fine matrix samples from (**a**) Alddasčorru, (**b**) Ald-dasčorru, (**c**) Nulpotjåkka, and (**d**) Tarfalatjårro. For each sample three diffractograms are shown. In the bottom diffractograms the samples are untreated, in the middle diffractograms the samples are ethylglycolated, and in the top diffractograms the samples are heated to 550 °C. These diffractograms illustrate the range of minerals present (labelled) in the Alddasčorru, Duoptečohkka, and Tarfalatjårro blockfields, and in the till samples. Poorly crystallized oxyhydroxides produce a rise in the diffractogram baseline, which disappears on heating, at d spacings between 5 and 3.5 Å (pink-filled circles). Vermiculization of chlorite and/or mica is shown by peaks in the ~ 10–14 Å (yellow) area that collapse to 10 Å on heating. Gibbsite is shown by peaks at 4.9 Å (green), which also collapse on heating.

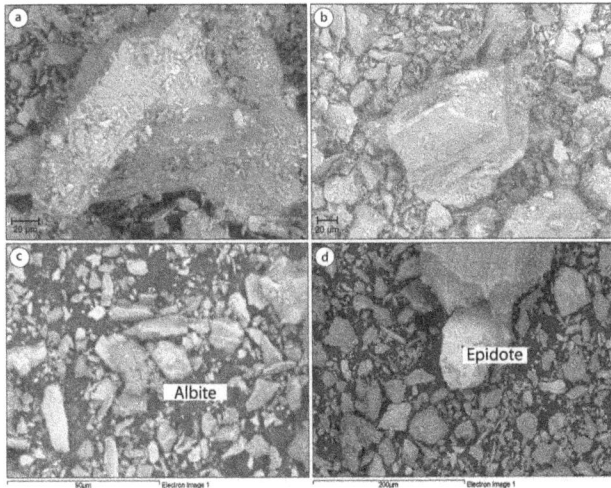

Figure 6. SEM images indicating only slight chemical weathering of fine matrix. (**a**) Albite, with a chemically etched surface; the only etched grain identified (BG-05-84, Tarfalatjårro pit 7). (**b**) Chemically unaltered amphibole, typical of all SEM images of amphibole (BG-05-26, Alddasčorru pit 2). (**c**) Disintegrating albite, possibly through chemical processes (BG-05-11, Alddasčorru pit 1). (**d**) Chemically unaltered epidote, typical of all SEM images of epidote (BG-05-04, Alddasčorru pit 1).

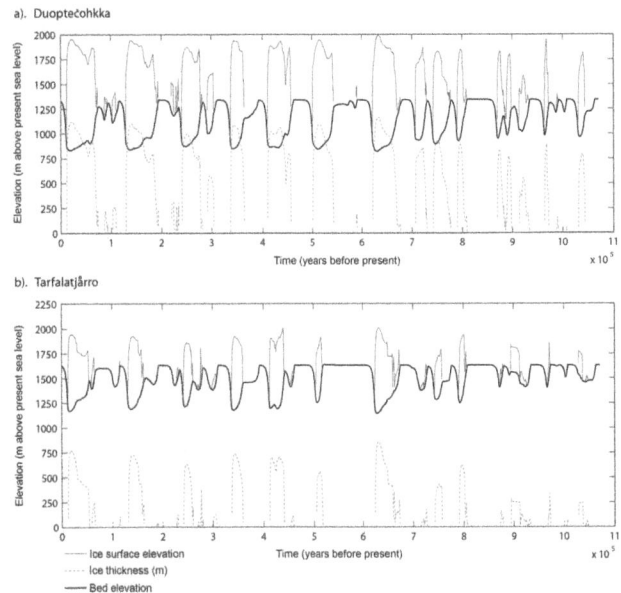

Figure 7. Modelled ice sheet surface elevation, ice thickness, and bedrock response to ice sheet loading and unloading for (**a**) Duoptečohkka and (**b**) Tarfalatjårro over the last 1.07 Ma. These data were generated using a 3-dimensional ice-dynamical model forced by the Lisiecki and Raymo (2005) stack of global benthic $\delta^{18}O$ records and the ELRA bedrock model (Bintanja et al., 2002, 2005).

In summary, XRD and SEM analyses indicate chemical weathering in blockfield and till samples, albeit in limited quantities. Samples from concave blockfield sites appear most chemically weathered, and summit blockfield fine matrix appears the least chemically weathered. Chemical weathering of till samples displays an intermediate intensity.

4.4 Regolith residence durations

Apparent ^{10}Be surface exposure durations for two surface quartzite clasts on Duoptečohkka and Tarfalatjårro are 33.5 ± 3.2 ka and 81.8 ± 7.8 ka, respectively (Table 2). These ages are based on the time-invariant spallogenic production rate model (Lal, 1991; Stone, 2000; Balco et al., 2008). For Duoptečohkka, this provides a younger age than any of the time-varying ^{10}Be spallogenic production rate models included on the CRONUS-Earth exposure age calculator (version 2.2; Balco et al., 2008). The full apparent surface exposure age range is 33.5 ± 3.2 ka to 35.6 ± 4.4 ka. For Tarfalatjårro, the apparent surface exposure age range from these production rate models is 80.4 ± 8.5 ka to 86.2 ± 10.8 ka.

The ^{26}Al / ^{10}Be ratio of 6.57 ± 0.43 for the quartz sample of the Duoptečohkka summit (Table 2) indicates no apparent burial (within error) by the Fennoscandian Ice Sheet. However, even a full exposure nuclide ratio does not exclude a complex exposure history including short intermittent periods of surface burial beneath an ice sheet. In contrast, the lower ratio of 5.92 ± 0.41 for the quartz sample of the Tar-

falatjårro surface requires some previous period of burial, and, by inference, indicates periods of burial by glacial ice.

Model results for ice sheet surface elevations, ice sheet thicknesses, and glacial isostasy over multiple glacial cycles indicate that the Duoptečohkka and Tarfalatjårro summits have been repeatedly covered by ice sheets (Fig. 7). According to the model, thicker ice has formed over Duoptečohkka (a maximum of 1595 m compared with 1150 m for Tarfalatjårro) and burial durations have been longer on this summit. These data are consistent with what might be expected for the lower elevation of Duoptečohkka (1336 m a.s.l. versus 1626 m a.s.l. for Tarfalatjårro). However, they seemingly contrast with inferences from the ^{26}Al / ^{10}Be ratio for Duoptečohkka (Table 2) of either no glacial burial of this summit or surface burial during short periods relative to intermittent full-exposure durations. The thickness and duration of ice cover may therefore be overestimated in our model, and a comparison with data from other models supports this possibility. Firstly, ice sheet thicknesses produced by our model are either similar to those indicated for our study areas by other ice sheet models (Fjeldskaar et al., 2000; Milne et al., 2004; Peltier, 2004; Steffen and Kauffman, 2005) or exceed them (Peltier, 1994; Kauffman et al., 2000; Lambeck et al., 2006; Steffen et al., 2006; Charbit et al., 2007). Secondly, the magnitude of isostatic rebound following the last glaciation is 497 m for Duoptečohkka and 457 m for Tarfalatjårro. These values exceed those indicated by the isostatic rebound

Table 2. Cosmogenic nuclide data, apparent exposure ages, and nuclide ratios.

Sample[a]	Location (°N/°E)	Elevation (m a.s.l.)	Thickness[b] (cm)	Shielding factor	Quartz (g)	Be carrier (mg)	^{10}Be/^{9}Be[c] (×10^{-13})	[^{10}Be][c,d,e,f] (10^6 atoms g^{-1})	Al carrier (mg)	^{26}Al/^{27}Al[c] (×10^{-13})	[^{26}Al][c,f,g,h] (10^6 atoms g^{-1})	^{10}Be apparent age[c,i,j] (ka) Age ± 1σ (int)	±1σ (ext)	^{26}Al apparent age[c,i,j] (ka) Age ± 1σ (int)	±1σ (ext)	^{26}Al/^{10}Be[c] ratio
Duo 1	68.42/19.37	1330	3	1.0000	41.8211	0.3094	11.40±0.30	0.51±0.02	0.7222	50.10±1.05	3.52±0.19	33.5±1.2	±3.2	32.7±1.8	±3.4	6.57±0.43
Tar 1	67.61/18.52	1626	4	0.9998	54.2814	0.2741	44.70±1.20	1.55±0.05	0.9028	170.00±6.00	9.59±0.59	81.8±2.8	±7.8	72.6±4.6	±8.0	5.92±0.41

[a] Duo 1 = surface sample from Duoptečohkka pit 1; Tar 1 = surface sample from Tarfalatjårro pit 1. [b] A quartzite density of 2.65 g cm^{-3} was used for thickness corrections. [c] Uncertainties are reported at the 1σ confidence level. [d] Measured ^{10}Be concentrations were normalized to NIST SRM 4325, with a ^{10}Be/^{9}Be ratio of 2.79 ± 0.03 × 10^{-11} and using a ^{10}Be half-life of 1.36 × 10^6 a (Nishiizumi et al., 2007). [e] Blank values of 11 5436 ± 37 556 ^{10}Be atoms ([^{10}Be/^{9}Be = 6.6 × 10^{-15} ± 1.6 × 10^{-15}) and 56 776 ± 39 917 ^{10}Be atoms ([^{10}Be/^{9}Be = 4.1 × 10^{-15} ± 2.0 × 10^{-15}) were used to correct for background in Duoptečohkka 1 and Tarfalatjårro 1, respectively. [f] Propagated uncertainties include error in the blank, carrier mass (1%), and counting statistics. [g] Measured ^{26}Al concentrations were normalized to PRIME standard Z92-0222 with a nominal ^{26}Al/^{27}Al ratio of 4.11 × 10^{-11} and using an ^{26}Al half-life of 7.05 × 10^5 a (Nishiizumi, 2004). [h] Blank values of 337 056 ± 99 378 ^{26}Al atoms (^{26}Al/^{27}Al = 11.4 × 10^{-15} ± 3.3 × 10^{-15}) and 183 701 ± 275 552 ^{26}Al atoms (^{26}Al/^{27}Al = 6.0 × 10^{-15} ± 9.0 × 10^{-15}) were used to correct for background in Duoptečohkka 1 and Tarfalatjårro 1, respectively. [i] Apparent exposure ages were calculated using the CRONUS-Earth calculator (version 2.2; Balco et al., 2008). Constant (time-invariant) ^{10}Be and ^{26}Al spallogenic production rate models (Lal, 1991; Stone, 2000) were used. Muogenic production was also incorporated into the production rate models giving total ^{10}Be production rates of 16.138 atoms g^{-1} a^{-1} for Duoptečohkka 1 and 20.196 atoms g^{-1} a^{-1} for Tarfalatjårro 1. Total ^{26}Al production rates are 109.368 atoms g^{-1} a^{-1} for Duoptečohkka 1 and 136.788 atoms g^{-1} a^{-1} for Tarfalatjårro 1. [j] (int) – internal (analytical) uncertainties; (ext) – propagated external uncertainties (Balco et al., 2008).

map in the National Atlas of Sweden (Fredén, 2002, p. 101) by 150–250 m, which again indicates a possible overestimation of ice sheet thicknesses and durations of ice coverage by our model. The key consequences of this for our subsequent analysis of regolith residence durations are that the lengths of the ice-free periods during which cosmogenic nuclides accumulate are likely underestimated, whereas nuclide decay periods during ice sheet burial are likely overestimated. If nuclides have accumulated in surface regolith more quickly than provided for in our model and nuclide decay has been less, inferred maximum erosion rates will be underestimated and regolith residence durations, for a given erosion rate, will be overestimated in our analyses. We consider the regolith residence duration calculations to remain valid for our purposes, however, because we are interested in an order of magnitude question (i.e. whether or not regolith residence durations are confined to the late Quaternary) and to be conservative in our interpretations prefer to err on the side of overestimating regolith residence durations.

Modelled regolith residence durations for Duoptečohkka and Tarfalatjårro are shown in Fig. 8. Steps in these regolith residence duration curves indicate periods of surface burial by glacial ice. The timing of these steps in each modelled scenario varies according to erosion rate and durations of surface burial by glacial ice. The primary model output, which considers the effects of both ice sheet burial and bedrock isostasy on cosmogenic nuclide accumulation (labelled "burial and isostasy"), indicates a maximum surface erosion rate of ∼ 16.2 mm ka^{-1} for Duoptečohkka and ∼ 6.7 mm ka^{-1} for Tarfalatjårro. As maximum surface erosion rates are asymptotically approached, maximum surface ages become infinite. However, the regolith residence durations of Duoptečohkka and Tarfalatjårro become asymptotic above cut-off values of ∼ 380 and ∼ 490 ka before present, respectively. This indicates that the late Quaternary has likely offered sufficient time for the present regolith mantles on both summits to gain their respective ^{10}Be inventories.

Four additional regolith residence duration scenarios help define the sensitivity of derived maximum erosion rates to durations of surface burial by snow and glacial ice and to the magnitude of glacial isostasy (Fig. 8). As expected, maximum erosion rates are highest in the absence of former glaciation ("0 burial, 0 isostasy" lines in Fig. 8). These rates are ∼ 18.2 mm ka^{-1} for Duoptečohkka and ∼ 7.3 mm ka^{-1} for Tarfalatjårro. Accordingly, regolith residence durations

for these simple exposure conditions are also lowest for this scenario. These "simple exposure" ages are ∼ 34–170 ka for Duoptečohkka and ∼ 82–375 ka for Tarfalatjårro for the range of erosion rates up to where the ages become asymptotic (Table 2; Fig. 8).

When intermittent surface burial by glacial ice is introduced, maximum erosion rates decrease to ∼ 17.5 mm ka^{-1} and ∼ 7.1 mm ka^{-1} for Duoptečohkka and Tarfalatjårro, respectively ("burial, 0 isostasy" lines in Fig. 8). Regolith residence durations also increase for a given erosion rate up to the erosion rate where the ages become asymptotic. These ages vary from 110 to ∼ 370 ka and 166 to ∼ 450 ka for Duoptečohkka and Tarfalatjårro, respectively.

Increasing by 10 % the duration of each period of surface burial by glacial ice has negligible impact on maximum erosion rates for either summit ("10 % more burial, isostasy" in Fig. 8). However, long burial periods are reached on Duoptečohkka at lower surface erosion rates than otherwise occur, resulting in longer regolith residence durations at these erosion rates. For example, at an erosion rate of 8 mm ka^{-1}, the regolith residence duration on Duoptečohkka increases from 123 to 199 ka. A similar effect is induced by increasing the duration of snow cover from 7 to 10 months a year and the depth of snow from 30 to 50 cm ("burial, isostasy, more snow" lines in Fig. 8). Increasing the duration and depth of snow cover decreases maximum erosion rates. These values are now ∼ 15.7 mm ka^{-1} for Duoptečohkka and ∼ 6.3 mm ka^{-1} for Tarfalatjårro.

In summary, the summits of both Duoptečohkka and Tarfalatjårro appear to have been repeatedly inundated by glacial ice over the past 1.07 Ma. The durations of burial and depths of glacial isostatic depression have had notable impacts on regolith residence durations for each summit. It remains likely, however, that the residence durations of regolith mantles on both summits are confined to the late Quaternary. Modelled maximum erosion rates are ∼ 16.2 mm ka^{-1} and ∼ 6.7 mm ka^{-1} for Duoptečohkka and Tarfalatjårro, respectively.

5 Discussion

Minimal chemical weathering of blockfields in the northern Swedish mountains is indicated by the following fine-matrix characteristics: clay / silt ratios ≤ 0.14 in all samples ($n = 16$), the presence of mixed primary and secondary minerals

a).

b).

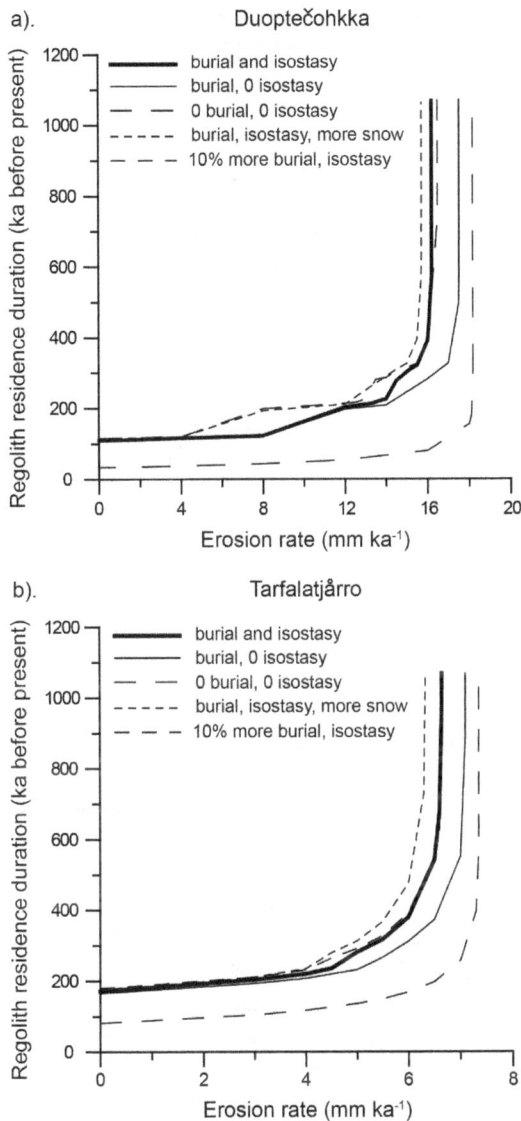

Figure 8. Regolith residence durations for **(a)** Duoptečohkka and **(b)** Tarfalatjårro plotted against surface erosion rates. These are modelled using ^{10}Be concentrations in regolith surface quartz clasts and incorporate periods of burial by ice sheets and changes in ^{10}Be production rates attributable to glacial isostasy. Seasonal burial of ground surfaces by 30 cm of snow for 7 months of the year is included. Five different scenarios were modelled: (i) ice sheet burial duration and glacial isostasy (marked "burial and isostasy" in the plots), which is our primary model output; (ii) isostasy is removed ("burial, 0 isostasy"); (iii) a simple surface exposure history, from which burial and glacial isostasy are excluded ("0 burial, 0 isostasy"); (iv) the first scenario is replicated, but snow depth and snow cover duration are increased to 50 cm and 10 months of the year, respectively ("burial, isostasy, more snow"); (v) each burial period is extended by 10 % and exposure periods commensurately shortened ("10 % more burial, isostasy").

in clay-sized regolith ($n = 16$), and a scarcity of chemically etched grains in bulk fine matrix (Table 1; Figs. 2–5). In addition, soil horizons and saprolite are absent from all blockfield sections (Table S1 in the Supplement). These findings support those from Goodfellow et al. (2009) and their model of blockfield formation, primarily through physical weathering processes. Conversely, the data do not support blockfield initiation through intense chemical weathering under a warm, non-periglacial climate.

Chemical weathering intensity, although generally low, varies predictably along hillslope transects. Convex summit areas are the least chemically weathered, as indicated by the absence of well-crystallized secondary minerals (Table 1). This is possibly because these areas are drier (Table S1 in the Supplement) or because fine matrix may not be resident on summits long enough, before being transported downslope, for secondary minerals to become well-crystallized. Concave locations, such as at the Alddasčorru slope base and the Tarfalatjårro saddle, exhibit the highest chemical weathering intensity, as indicated by the formation of vermiculite and gibbsite. This may be attributable to longer residence durations within the blockfields of fine matrix that has been transported downslope, wetter conditions, and/or changes in bedrock mineralogy (Table S1 in the Supplement). The relative paucity of summit fine matrix (Table S1 in the Supplement), which also displays lowest chemical weathering intensity (Table 1; Fig. 5), might therefore indicate erosion through, for example, surface creep and subsurface water flow. Altitudinal differences along the transects are generally insufficient for weathering variations to be clearly related to temperature changes, particularly along the 73 m Tarfalatjårro transect (Fig. 1). However, a slightly milder temperature regime and an extensive grass cover may enhance chemical weathering of the colluvium–till mixture at the base of the Duoptečohkka transect (270 m below the summit), which contains vermiculized minerals and gibbsite (Table 1). Secondary mineral assemblages are consistent with contemporary climatic conditions and hillslope position, rather than indicating palaeoregoliths.

The presence of gibbsite in some fine matrix samples does not indicate a Neogene deep-weathering origin of blockfields. Although it is an end product of chemical weathering and can be abundant in intensely weathered regoliths, formation of limited quantities of gibbsite is apparently also favoured by the generally low temperatures and spatially and temporally variable hydrologic conditions that occur in alpine regoliths. The spatial distribution of gibbsite in the blockfields of the northern Swedish Scandes indicates that its precipitation may be favoured by wetter regolith conditions found, for example, in concave locations (Table 1). Seasonally abundant liquid water may result in some leaching along spatially discrete flow paths in these blockfield regoliths (Meunier et al., 2007; Goodfellow, 2012). Together with low temperatures, which inhibit chemical reactions, this may maintain the low concentrations of H_4SiO_4 along these

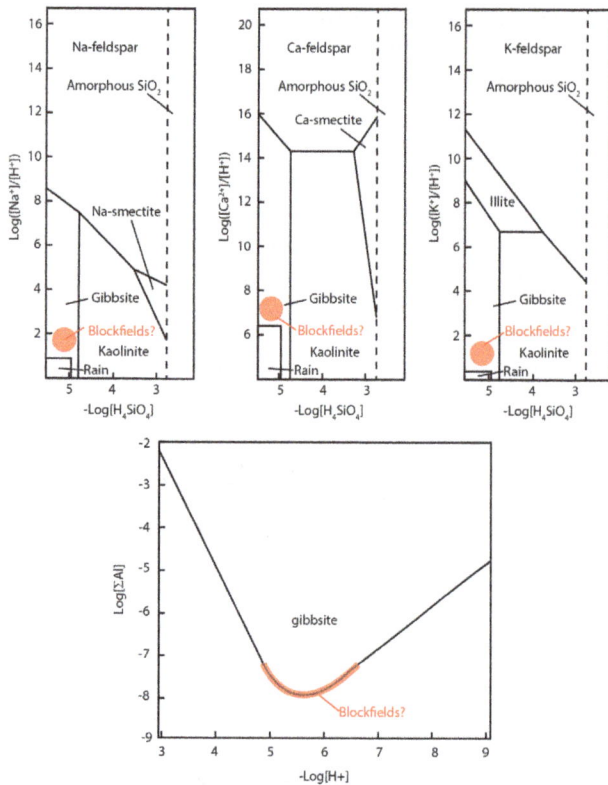

Figure 9. Thermodynamic stability relations between feldspars, secondary minerals, and weathering solutions (top panels) and the relationship between gibbsite solubility and the pH of regolith water (bottom panel). Relatively low Na^+ Ca^{2+}, and K^+ abundances and high H^+ abundances favour secondary mineral formation. Solute concentrations are usually most favourable to kaolinite precipitation, but gibbsite forms where regolith waters have very dilute silica concentrations, similar to rainwater (adapted from Stumm and Morgan, 1981, p. 547, and Nesbitt and Young, 1984, p. 1524, with permission of Wiley and Elsevier, respectively). Gibbsite is least soluble where pH is 5–6, and its solubility increases as pH is elevated or lowered from this value (adapted from Wesolowski and Palmer, 1994). It is expected that these key conditions of low silica concentration and slight acidity occur in Arctic–alpine blockfields, which favours gibbsite precipitation in these regoliths.

flow paths that are favourable to the precipitation of gibbsite rather than kaolinite (Fig. 9; Nesbitt and Young, 1984). Gibbsite may persist in blockfields because of the general absence of macro-vegetation and associated organic acids. This permits porewaters to remain slightly acidic, under which conditions gibbsite is least soluble (Fig. 9; Reynolds, 1971; May et al., 1979; Gardner, 1992; Wesolowski and Palmer, 1994). Marquette et al. (2004) measured a mean pH of 5.7 in alpine blockfield porewaters, which coincides with minimum gibbsite solubility (Fig. 9). As expected, gibbsite is present in those blockfields (located in NE Canada). The low temperature constraint on weathering reactions may further ensure the persistence of gibbsite rather than it being converted to kaolinite through resilication (Watanabe et al., 2010). The

presence of gibbsite in alpine regoliths is well documented (Green and Eden, 1971; Reynolds, 1971; McKeague et al., 1983; Bain et al., 1994; Rea et al., 1996; Dahlgren et al., 1997; Ballantyne, 1998; Ballantyne et al., 1998; Burkins et al., 1999; Marquette et al., 2004; Paasche et al., 2006; Watanabe et al., 2010; Hopkinson and Ballantyne, 2014), and it should not be used to indicate that regoliths have necessarily originated under a warmer-than-present pre-Quaternary climate.

Late Quaternary surface erosion rates on the Duoptečohkka and Tarfalatjårro summits are low. These values are $16.2\,mm\,ka^{-1}$ (with a range of $15.7–18.2\,mm\,ka^{-1}$ for the modelled scenarios) and $6.7\,mm\,ka^{-1}$ (with a range of $6.3–7.3\,mm\,ka^{-1}$ for the modelled scenarios) for Duoptečohkka and Tarfalatjårro, respectively (Fig. 8). The higher erosion rate for Duoptečohkka likely reflects enhanced regolith transport across the narrow, steeply sided, summit ridgeline. This is perhaps indicated by a patchy regolith only a few tens of centimetres thick and by older bedrock apparent surface exposure durations from relict non-glacial surfaces on a nearby part of Alddasčorru (1380 m a.s.l.) and Olmáčohkka (1355 m a.s.l.) of 42.1 ± 2.5 and $58.2 \pm 3.5\,ka$ (analytical errors only), respectively (Fig. 1; Fabel et al., 2002; Stroeven et al., 2006). The lower erosion rate for the broad, low gradient summit of Tarfalatjårro likely approaches the minimum limit for blockfield-mantled surfaces in this landscape, and is based on an apparent exposure duration which is similar to one derived from bedrock on nearby Dárfalcohkka (1790 m a.s.l.) of $72.6 \pm 4.4\,ka$ (analytical error only; Fig. 1; Stroeven et al., 2006). Blockfields therefore appear to represent end-stage landforms (Granger et al., 2001; Hättestrand and Stroeven, 2002; Boelhouwers, 2004) that effectively armour low-gradient surfaces, making them resistant to erosion and limiting further modification of surface morphology and regolith composition and thickness.

While average erosion rates of blockfield-mantled summits are low, they are of sufficient magnitude to remove ~ 1–2 m thick regolith profiles within a late Quaternary time frame, even accounting for periods of surface protection during burial by cold-based glacial ice (Figs. 7–8). If our conclusion that blockfields represent end-stage landforms resistant to further modification is correct, then it is reasonable to also assume steady-state conditions, where the rate of regolith production through weathering processes equals the erosion rate. Together, erosion and regolith production rates of $6–18\,mm\,ka^{-1}$ and geochemical features including mixtures of primary and secondary minerals in clay-sized regolith, low clay / silt ratios, and sparsely weathered primary mineral grains are consistent with blockfield formation under, at least, seasonal periglacial conditions during the late Quaternary.

The contrast between low rates of regolith production and erosion in blockfield mantles and the apparent comprehensive removal of presumably more-intensely weathered

Neogene regoliths from the northern Swedish Scandes is intriguing. One possible explanation for this contrast is that erosion rates may have greatly outpaced regolith production rates during the onset of cold Quaternary climatic conditions. This is because protective vegetative covers may have been lost during this period, and the intensity of periglaciation may have greatly increased. In Canadian Arctic plateau landscapes similar to the Scandinavian Mountains, there is evidence of extensive removal of Neogene regolith during the early Quaternary (Refsnider and Miller, 2013). Different conditions, including perhaps less intense Quaternary periglaciation, have allowed the persistence of pre-Quaternary saprolite remnants in low-altitude locations in Scandinavia (Lidmar-Bergström, 1997). While our data indicate residence durations of present regoliths, they do not address the question of how long blockfields have mantled portions of the Scandes. Furthermore, they provide only minimum estimates of the ages of surfaces on which the blockfields reside. However, because the Plio-Pleistocene transition may have been a period of marked disequilibrium in surface processes, blockfields perhaps formed during or after this period.

Our data indicate relative stability of blockfield-mantled summits during the late Quaternary. They are therefore incompatible with the operation on these summits of a periglacial or glacial "buzz saw" during this period. However, they discount neither a "buzz saw" effect on topographic relief during earlier periods, as has been previously suggested (Pedersen and Egholm, 2013), nor more rapid late Quaternary erosion rates on higher-gradient slopes. Field observations of extensive solifluction on slopes coupled with a marked absence of remnant glacial erosion features across a range of spatial scales, such as striae, roches moutonnées, rock drumlins, whalebacks, and crag and tails, are compatible with lowering of relief on blockfield-mantled surfaces through periglacial, rather than glacial, processes. We also speculate that formation of autochthonous blockfields on glacially eroded bedrock surfaces would be greatly inhibited. This is because of low water retention against, and low infiltration into, subaerially exposed, glacially polished, and generally convex rock surfaces (André, 2002; Ericson, 2004; Hall and Phillips, 2006), which are essential for regolith production through chemical processes and frost action (Walder and Hallet, 1985; Anderson, 1998; Whalley et al., 2004; Dixon and Thorn, 2005; Goodfellow et al., 2009). Glacially scoured bedrock surfaces might therefore be resistant to regolith formation over timescales of 10^5–10^6 years (or longer), and the sum of ice-free periods during the Quaternary may have been insufficient for weathering and erosive processes to have completely removed evidence of any early Quaternary glacial erosion from the landscape. We do consider though that (early) Quaternary periglacial processes may have modified presently blockfield-mantled surfaces to a greater extent than can be easily recognized (Anderson, 2002; Goodfellow, 2007; Berthling and Etzelmuller, 2011).

The utility of blockfield-mantled surfaces as markers from which to estimate Quaternary glacial erosion volumes in surrounding landscape elements remains uncertain. This is because potentially large spatial variations in rates of non-glacial erosion on blockfield-mantled topography may have lowered local relief and altered the topography from its preglacial configuration (Anderson, 2002; Goodfellow, 2007). In addition, presently blockfield-mantled surfaces may have undergone vertical erosion exceeding some tens of metres, as well as topographic reconfiguration, during the Plio-Pleistocene transition to a colder climate. Confirmation that these slowly eroding landforms persist over multiple glacial–interglacial cycles does, however, further demonstrate the utility of these landforms as key markers of nunataks and/or cold-based ice coverage during (at least) late Quaternary glacial periods.

6   Conclusions

Blockfields on three mountains in the northern Swedish Scandes were examined, and none of them appear to be remnants of thick, intensely weathered Neogene weathering profiles, which has been the prevailing opinion for these regoliths. Minor chemical weathering is indicated in each of the three examined blockfields, with predictable differences according to slope position. Average erosion rates of ~ 16.2 and ~ 6.7 mm ka^{-1} are calculated for two blockfield-mantled summits, from concentrations of in situ-produced cosmogenic ^{10}Be in surface quartz clasts that were inferred not to have been vertically mixed through the regolith. Although low, these erosion rates are of sufficiently high magnitude to remove present blockfield mantles, which appear to be commonly $< \sim 2$ m thick, within a late Quaternary time frame. This finding remains valid even when accounting for temporal variations in ^{10}Be production rates attributable to glacial isostasy and burial of ground surfaces by snow and cold-based glacial ice. Blockfield mantles appear to be replenished by regolith formation through, primarily physical, weathering processes that operated during the Quaternary.

The persistence of autochthonous blockfields over multiple glacial–interglacial cycles confirms their importance as key markers of surfaces that were not glacially eroded through, at least, the late Quaternary. However, presently blockfield-mantled surfaces may undergo potentially large spatial variations in erosion rates, and their regolith mantles may have been comprehensively eroded during the late Pliocene and early Pleistocene. Their role as markers by which to estimate glacial erosion volumes in surrounding landscape elements therefore remains uncertain.

Acknowledgements. We thank Ulf Jonsell, Ben Kirk, Damian Waldron, Mark Wareing, and staff at the Tarfala Research Station and Abisko Scientific Research Station for field support, Ann E. Karlsen and Wieslawa Koziel (Geological Survey of Norway, Trondheim) for completing the grain size and XRD analyses, Bjørn Willemoes-Wissing (Geological Survey of Norway, Trondheim) for assisting with SEM, Maria Miguens-Rodriguez (SUERC, East Kilbride) for guiding Goodfellow through sample preparation for AMS, and Stewart Freeman for running AMS on sample "Duo 1". Henriette Linge, an anonymous reviewer, and journal editor David Egholm are thanked for insightful comments that improved the manuscript. John Gosse, three anonymous journal reviewers and an associate editor are also thanked for helpful comments on an earlier version. The Swedish Society for Anthropology and Geography Andréefonden, Axel Lagrelius' Fond, Hans and Lillemor Ahlmanns Fond, C. F. Liljevalchs Fond, Carl Mannerfelts Fond, and Lars Hiertas minne provided funding to Goodfellow, and Swedish Research Council grants 621-2001-2331 and 621-2005-4972, and a PRIME Lab SEED grant provided funding to Stroeven. Figure 9 is adapted from p. 547 in Stumm, W., Morgan, J. J., 1981, Aquatic Chemistry, 2nd Edn., John Wiley and Sons, New York, and from Fig. 1 in Nesbitt, H. W., Young, G. M., 1984, Prediction of some weathering trends of plutonic and volcanic rocks based on thermodynamic and kinetic considerations, Geochimica et Cosmochimica Acta 48, 1523–1534, with permission of Wiley and Elsevier, respectively.

Edited by: D. Lundbek Egholm

References

Anderson, R. S.: Near-surface thermal profiles in alpine bedrock: implications for the frost-weathering of rock, Arct. Alpine Res., 30, 362–372, 1998.

Anderson, R. S.: Modeling the tor-dotted crests, bedrock edges, and parabolic profiles of high alpine surfaces of the Wind River Range, Wyoming, Geomorphol., 46, 35–58, 2002.

André, M. F.: Rates of postglacial rock weathering on glacially scoured outcrops (Abisko-Riksgränsen area, 68° N), Geograf. Annal., 84A, 139–150, 2002.

André, M. F.: Do periglacial landscapes evolve under periglacial conditions?, Geomorphology, 52, 149–1164, 2003.

André, M. F., Hall, K., Bertran, P., and Arocena, J.: Stone runs in the Falkland Islands: Periglacial or tropical?, Geomorphol., 95, 524–543, 2008.

Bain, D. C., Mellor, A., Wilson, M. J., and Duthie, D. M. L.: Chemical and mineralogical weathering rates and processes in an upland granitic till catchment in Scotland. Water, Air Soil Pollut., 73, 11–27, 1994.

Balco, G., Stone, J. O., Lifton, N. A., and Dunai, T. J.: A complete and easily accessible means of calculating surface exposure ages or erosion rates from ^{10}Be and ^{26}Al measurements, Quaternary Geochronol., 3, 174–195, 2008.

Ballantyne, C. K.: Age and significance of mountain-top detritus, Permafrost Perigl. Process., 9, 327–345, 1998.

Ballantyne, C. K.: A general model of autochthonous blockfield evolution, Permafrost Perigl. Process. 21, 289–300, 2010.

Ballantyne, C. K. and Harris, C.: The Periglaciation of Great Britain, Cambridge University Press, Cambridge, 1994.

Ballantyne, C. K., McCarroll, D., Nesje, A., Dahl, S. O., and Stone, J. O.: The last ice sheet in north-west Scotland: Reconstruction and implications, Quaternary Sci. Rev. 17, 1149–1184, 1998.

Berthling, I. and Etzelmüller, B.: The concept of cryo-conditioning in landscape evolution, Quaternary Res., 75, 378–384, 2011.

Bierman, P. R., Marsella, K. A., Patterson, C., Davis, P. T., and Caffee, M.: Mid-Pleistocene cosmogenic minimum-age limits for pre-Wisconsinan glacial surfaces in southwestern Minnesota and southern Baffin Island: a multiple nuclide approach, Geomorphology, 27, 25–39, 1999.

Bintanja, R., van de Wal, R. S. W., and Oerlemans, J.: Global ice volume variations through the last glacial cycle simulated by a 3-D ice-dynamical model, Quaternary Int., 95–96, 11–23, 2002.

Bintanja, R., van de Wal, R. S. W., and Oerlemans, J.: Modelled atmospheric temperatures and global sea levels over the past million years, Nature, 437, 125–128, 2005.

Boelhouwers, J. C.: New perspectives on autochthonous blockfield development, Polar Geogr., 28, 133–146, 2004.

Boelhouwers, J., Holness, S. Meiklejohn, I., and Sumner, P.: Observations on a blockstream in the vicinity of Sani Pass, Lesotho Highlands, southern Africa, Permafrost Perigl. Process., 13, 251–257, 2002.

Brindley, G. W. and Brown, G.: Crystal Structures of Clay Minerals and their X-ray Identification, Mineralogical Society, London, 495 pp., 1980.

Briner, J. P., Miller, G. H., Davis, P. T., Bierman, P. R., and Caffee, M.: Last Glacial Maximum ice sheet dynamics in Arctic Canada inferred from young erratics perched on ancient tors, Quaternary Sci. Rev., 22, 437–444, 2003.

Burkins, D. L., Blum, J. D., Brown, K., Reynolds, R. C., and Erel, Y.: Chemistry and mineralogy of a granitic, glacial soil chronosequence, Sierra Nevada Mountains, California, Chem. Geol., 162, 1–14, 1999.

Caine, N.: The Blockfields of Northeastern Tasmania. Publication G/6, Department of Geography, Research School of Pacific Studies, The Australian National University, Canberra, 127 pp., 1968.

Charbit, S., Ritz, C., Philippon, G., Peyaud, V., and Kageyama, M.: Numerical reconstructions of the Northern Hemisphere ice sheets through the last glacial-interglacial cycle, Clim. Past, 3, 15–37, doi:10.5194/cp-3-15-2007, 2007.

Child, D., Elliott, G., Mifsud, C., Smith, A. M., and Fink, D.: Sample processing for earth science studies at ANTARES. Nuclear Instruments and Methods in Physics Research Section B, Beam Interactions with Materials and Atoms, 172, 856–860, 2000.

Clapperton, C. M.: Further observations on the stone runs of the Falkland Islands, Biuletyn Peryglacjalny, 24, 211–217, 1975.

Cockburn, H. A. P. and Summerfield, M. A.: Geomorphological applications of cosmogenic isotope analysis, Progr. Phys. Geogr., 28, 1–42, 2004.

Dahl, R.: Block fields, weathering pits and tor-like forms in the Narvik mountains, Nordland, Norway, Geograf. Annal., 48A, 55–85, 1966.

Dahlgren, R. A., Boettinger, J. L., Huntington, G. L., and Amundson, R. G.: Soil development along an elevational transect in the western Sierra Nevada, California, Geoderma, 78, 207–236, 1997.

Dixon, J. C. and Thorn, C. E.: Chemical weathering and landscape development in mid-latitude alpine environments, Geomorphology, 67, 127–145, 2005.

Dredge, L.: Breakup of limestone bedrock by frost shattering and chemical weathering, Eastern Canadian Arctic, Arctic Alpine Res., 24, 314–323, 1992.

Dredge, L. A.: Age and origin of upland block fields on Melville Peninsula, eastern Canadian Arctic, Geograf. Annal., 82A, 443–454, 2000.

Ericson, K.: Geomorphological surfaces of different age and origin in granite landscapes: An evaluation of the Schmidt hammer test, Earth Surf. Process. Landform., 29, 495–509, 2004.

Eriksson, B.: Data rörandeSveriges temperaturklimat (Data concerning the air temperature of Sweden). SMHI Reports, Meteorology and Climatology, RMK 39, 1982.

Fabel, D., Stroeven, A. P., Harbor, J., Kleman, J., Elmore, D., and Fink, D.: Landscape preservation under Fennoscandian ice sheets determined from in situ produced [10]Be and [26]Al, Earth Planet. Sci. Lett., 201, 397–406, 2002.

Fjeldskaar, W., Lindholm, C., Dehls, J. F., and Fjeldskaar, I.: Postglacial uplift, neotectonics and seismicity in Fennoscandia, Quaternary Sci. Rev., 19, 1413–1422, 2000.

Fjellanger, J., Sørbel, L., Linge, H., Brook, E. J., Raisbeck, G. M., and Yiou, F.: Glacial survival of blockfields on the Varanger Peninsula, northern Norway, Geomorphology, 82, 255–272, 2006.

Fredén, C. (Ed): National Atlas of Sweden: Land and soils, Sveriges Nationalatlas, Vällingby, 208 pp., 2002.

Gardner, L. R.: Long-term isovolumetric leaching of aluminum from rocks during weathering: implications for the genesis of saprolite, Catena, 19, 521–537, 1992.

Glasser, N. and Hall, A.: Calculating Quaternary glacial erosion rates in northeast Scotland, Geomorphology, 20, 29–48, 1997.

Goldstein, J., Newbury, D. E., Joy, D. C., Lyman, C. E., Echlin, P., Lifshin, E., Sawyer, L. C., and Michael, J. R.: Scanning Electron Microscopy and X-ray Microanalysis, 3rd Edn., Kluwer Academic/Plenum Publishers, New York, 689 pp., 2003.

Goodfellow, B. W.: Relict non-glacial surfaces in formerly glaciated landscapes, Earth-Sci. Rev., 80, 47–73, 2007.

Goodfellow, B. W.: A granulometry and secondary mineral fingerprint of chemical weathering in periglacial landscapes and its application to blockfield origins, Quaternary Sci. Rev., 57, 121–135, 2012.

Goodfellow, B. W., Stroeven, A. P., Hättestrand, C., Kleman, J., and Jansson, K. N.: Deciphering a non-glacial/glacial landscape mosaic in the northern Swedish mountains, Geomorphology, 93, 213–232, 2008.

Goodfellow, B. W., Fredin, O., Derron, M.-H., and Stroeven, A. P.: Weathering processes and Quaternary origin of an alpine blockfield in Arctic Sweden, Boreas, 38, 379–398, 2009.

Granger, D. E., Riebe, C. S., Kirchner, J. W., and Finkel, R. C.: Modulation of erosion on steep granitic slopes by boulder armoring, as revealed by cosmogenic [26]Al and [10]Be, Earth Planet. Sci. Lett., 86, 269–281, 2001.

Green, C. P. and Eden, M. J.: Gibbsite in the weathered Dartmoor granite, Geoderma, 6, 315–317, 1971.

Grudd, H. and Schneider, T.: Air temperature at Tarfala Research Station 1946–1995, Geograf. Annal., 78A, 115–119, 1996.

Hales, T. C. and Roering, J. J.: Climatic controls on frost cracking and implications for the evolution of bedrock landscapes, J. Geophys. Res., 112, F02033, doi:10.1029/2006JF000616, 2007.

Hall, A. M. and Phillips, W. M.: Weathering pits as indicators of the relative age of granite surfaces in the Cairngorm Mountains, Scotland, Geograf. Annal., 88A, 135–150, 2006.

Hättestrand, C. and Stroeven, A. P.: A relict landscape in the centre of Fennoscandian glaciation; geomorphological evidence of minimal Quaternary glacial erosion, Geomorphology, 44, 127–143, 2002.

Hopkinson, C. and Ballantyne, C. K.: Age and Origin of Blockfields on Scottish Mountains, Scottish Geogr. J., 130, 116–141, 2014.

Isaksen, K., Holmlund, P., Sollid, J. L., and Harris, C.: Three deep alpine-permafrost boreholes in Svalbard and Scandinavia, Permafrost Perigl. Process., 12, 13–25, 2001.

Isaksen, K., Sollid, J. L., Holmlund, P., and Harris, C.: Recent warming of mountain permafrost in Svalbard and Scandinavia, J. Geophys. Res., 112, F02S04, doi:10.1029/2006JF000522, 2007.

Ives, J. D.: Block fields, associated weathering forms on mountain tops and the Nunatak hypothesis, Geograf. Annal., 4, 220–223, 1966.

Ives, J. D.: Biological refugia and the nunatak hypothesis, in: Arctic and Alpine Environments, edited by: Ives, J. D. and Barry, R. G., Methuen, London, 605–636, 1974.

Jansson, K. N., Stroeven, A. P., Alm, G., Dahlgren, K. I. T., Glasser, N. F., and Goodfellow, B. W.: Using a GIS filtering approach to replicate patterns of glacial erosion, Earth Surf. Process Landf., 36, 408–418, 2011.

Kauffman, G., Wu, P., and Li, G.: Glacial isostatic adjustment in Fennoscandia for a laterally heterogenous earth, Geophys. J. Int., 143, 262–273, 2000.

Kleman, J. and Stroeven, A. P.: Preglacial surface remnants and Quaternary glacial regimes in northwestern Sweden, Geomorphology, 19, 35–54, 1997.

Kleman, J., Stroeven, A. P., and Lundqvist, J.: Patterns of Quaternary ice sheet erosion and deposition in Fennoscandia and a theoretical framework for explanation, Geomorphology, 97, 73–90, 2008.

Kohl, C. P. and Nishiizumi, K.: Chemical isolation of quartz for measurement of in situ-produced cosmogenic nuclides, Geochim. Cosmochim. Acta, 56, 3586–3587, 1992.

Lal, D.: Cosmic ray labeling of erosion surfaces: in situ nuclide production rates and erosion models, Earth Planet. Sci. Lett., 104, 424–439, 1991.

Lambeck, K., Purcell, A., Funder, S., Kjær, K. H., Larsen, E., and Möller, P.: Constraints on the Late Saalian to early Middle Weichselian ice sheet of Eurasia from field data and rebound modeling, Boreas, 35, 539–575, 2006.

Le Meur, E. and Huybrechts, P.: A comparison of different ways of dealing with isostasy: examples from modelling the Antarctic ice sheet during the last glacial cycle, Ann. Glaciol., 23, 309–317, 1996.

Li, Y., Fabel, D., Stroeven, A. P., and Harbor, J.: Unraveling complex exposure-burial histories of bedrock surfaces under ice sheets by integrating cosmogenic nuclide concentrations with climate proxy records, Geomorphology, 99, 139–149, 2008.

Lidmar-Bergström, K.: A long-term perspective on glacial erosion, Earth Surf. Process. Landf., 22, 297–306, 1997.

Lisiecki, L. E. and Raymo, M. E.: A Pliocene-Pleistocene stack of 57 globally distributed benthic $\delta^{18}O$ records, Paleooceanography, 20, PA1003, doi:10.1029/2004PA001071, 2005.

Marquette, G. C., Gray, J. T., Gosse, J. C., Courchesne, F., Stockli, L., MacPherson, G., and Finkel, R.: Felsenmeer persistence under non-erosive ice in the Torngat and Kaumajet mountains, Quebec and Labrador, as determined by soil weathering and cosmogenic nuclide exposure dating, Can. J. Earth Sci., 41, 19–38, 2004.

May, H. M., Helmke, P. A., and Jackson, M. L.: Gibbsite solubility and thermodynamic properties of hydroxy-aluminum ions in aqueous solution at 25 °C, Geochim. Cosmochim. Acta, 43, 861–868, 1979.

McKeague, J. A., Grant, D. R., Kodama, H., Beke, G. J., and Wang, C.: Properties and genesis of a soil and the underlying gibbsite-bearing saprolite, Cape Breton Island, Canada, Can. J. Earth Sci., 20, 37–48, 1983.

Meunier, A., Sardini, P., Robinet, J. C., and Prêt, D.: The petrography of weathering processes: facts and outlooks, Clay Min., 42, 415–435, 2007.

Milne, G. A., Mitrovica, J. X., Scherneck, H.-G., Davis, J. L., Johansson, J. M., Koivula, H., and Vermeer, M.:. Continuous GPS measurements of postglacial adjustment in Fennoscandia: 2. Modelling results, J. Geophys. Res., 109, 1–18, 2004.

Moore, D. M. and Reynolds, R. C.: X-ray Diffraction and the Identification and Analysis of Clay Minerals, 2nd Edn., University Press, Oxford, 378 pp., 1997.

Nesbitt, H. W. and Young, G. M.: Prediction of some weathering trends of plutonic and volcanic rocks based on thermodynamic and kinetic considerations, Geochimi. Cosmochim. Acta, 48, 1523–1534, 1984.

Nesje, A. and Whillans, I. M.: Erosion of Sognefjord, Norway, Geomorphology, 9, 33–45, 1994.

Nesje, A., Dahl, S. O., Anda, E., and Rye, N.: Block fields in southern Norway: Significance for the Late Weichselian ice sheet, Norsk Geologisk Tidsskrift, 68, 149–169, 1988.

Nielsen, S. B., Gallagher, K., Leighton, C., Balling, N., Svenningsen, L., Jacobsen, B. H., Thomsen, E., Nielsen, O. B., Heilmann-Clausen, C., Egholm, D. L., Summerfield, M. A., Clausen, O. R., Piotrowski, J. A., Thorsen, M. R., Huuse, M., Abrahamsen, N., King, C., and Holger Lykke-Andersen, H.: The evolution of western Scandinavian topography: a review of Neogene uplift versus the ICE (isostasy-climate-erosion) hypothesis, J. Geodynam., 47, 72–95, 2009.

Nishiizumi, K.: Preparation of ^{26}Al AMS standards, Nuclear Instrume. Methods Phys. Res. B, 223–224, 388–392, 2004.

Nishiizumi, K., Imamura, M., Caffee, M. W., Southern, J. R., Finkel, R. C., and McAninch, J.: Absolute calibration of ^{10}Be AMS standards, Nuclear Instrum. Methods Phys. Res. B, 258, 403–413, 2007.

Paasche, Ø., Strømsøe, J. R., Dahl, S. O., and Linge, H.: Weathering characteristics of arctic islands in northern Norway, Geomorphology, 82, 430–452, 2006.

Pedersen, V. K. and Egholm, D. L.: Glaciations in response to climate variations preconditioned by evolving topography, Nature, 493, 206–2010, 2013.

Peltier, W. R.: Ice age paleotopography, Science, 265, 195–201, 1994.

Peltier, W. R.: Global glacial isostasy and the surface of the ice-age Earth: the ICE-5G (VM2) Model and GRACE, Ann. Rev. Earth Planet. Sci., 32, 111–149, 2004.

Phillips, W. M., Hall, A. M., Mottram, R., Fifield, L. K., and Sugden, D.: Cosmogenic ^{10}Be and ^{26}Al exposure ages of tors and erratics, Cairngorm Mountains, Scotland: Timescales for the development of a classic landscape of selective linear glacial erosion, Geomorphology, 73, 224–245, 2006.

Potter, N. and Moss, J. H.: Origin of the Blue Rocks block field and adjacent deposits, Berks County, Pennsylvania, Geol. Soc. Am. Bull., 79, 255–262, 1968.

Rea, B. R.: Blockfields (Felsenmeer), in: The Encyclopedia of Quaternary Science, edited by: Elias, S. A., Vol. 3, Elsevier, Amsterdam, 523–534, 2013.

Rea, B. R., Whalley, B., Rainey, M. M., and Gordon, J. E.: Blockfields, old or new? Evidence and implications from some plateaus in northern Norway, Geomorphology, 15, 109–121, 1996.

Refsnider, K. A. and Miller, G. H.: Ice-sheet erosion and the stripping of Tertiary regolith from Baffin Island, eastern Canadian Arctic, Quaternary Sci. Rev., 67, 176–189, 2013.

Reynolds, R. C.: Clay mineral formation in an alpine environment, Clays Clay Mineral., 19, 361–374, 1971.

Small, E. E., Anderson, R. S., Repka, J. L., and Finkel, R.: Erosion rates of alpine bedrock summit surfaces deduced from in situ ^{10}Be and ^{26}Al, Earth Planet. Sci. Lett., 150, 413–425, 1997.

Small, E. E., Anderson, R. S., and Hancock, G. S.: Estimates of the rate of regolith production using ^{10}Be and ^{26}Al from an alpine hillslope, Geomorphology, 27, 131–150, 1999.

Staiger, J. K. W., Gosse, J. C., Johnson, J. V., Fastook, J., Gray, J. T., Stockli, D. F., Stockli, L., and Finkel, R.: Quaternary relief generation by polythermal glacier ice, Earth Surf. Process. Landform., 30, 1145–1159, 2005.

Steer, P., Huismans, R. S., Valla, P. G., Gac, S., and Herman, F.: Bimodal Plio–Quaternary glacial erosion of fjords and low-relief surfaces in Scandinavia, Nat. Geosci., 5, 635–639, 2012.

Steffen, H. and Kauffman, G.: Glacial isostatic adjustment of Scandinavia and northwestern Europe and the radial viscosity structure of the Earth's mantle, Geophys. J. In., 163, 801–812, 2005.

Steffen, H., Kauffman, G., and Wu, P.: Three-dimensional finite-element modeling of the glacial isostatic adjustment in Fennoscandia, Earth Planet. Sci. Lett., 250, 358–375, 2006.

Stone, J. O.: Air pressure and cosmogenic isotope production, J. Geophys. Res., 105, 23753–23759, 2000.

Stroeven, A. P., Fabel, D., Hattestrand, C., and Harbor, J.: A relict landscape in the centre of Fennoscandian glaciation; cosmogenic radionuclide evidence of tors preserved through multiple glacial cycles, Geomorphology, 44, 145–154, 2002.

Stroeven, A. P., Harbor, J., Fabel, D., Kleman, J., Hättestrand, C. Elmore, D., Fink, D., and Fredin, O.: Slow, patchy landscape evolution in northern Sweden despite repeated ice-sheet glaciation, Geol. Soc. Am. Special Paper, 398, 387–396, 2006.

Strømsøe, J. R. and Paasche, Ø.: Weathering patterns in high-latitude regolith, J. Geophys. Res., 116, F03021, doi:10.1029/2010JF001954, 2011.

Stumm, W. and Morgan, J. J.: Aquatic Chemistry, 2nd Edn., John Wiley and Sons, New York, 1981.

Sugden, D. E.: The selectivity of glacial erosion in the Cairngorm Mountains, Scotland, Trans. Institute of British Geographers, 45, 79–92, 1968.

Sugden, D. E.: Landscapes of glacial erosion in Greenland and their relationship to ice, topographic and bedrock conditions, in: Progress in Geomorphology: Papers in honour of D.L. Linton, edited by: Brown, E. H. and Waters, R. S., Institute of British Geographers Special Publication, Vol. 7, 177–195, 1974.

Sugden, D. E. and Watts, S. H.: Tors, felsenmeer, and glaciation in northern Cumberland Peninsula, Baffin Island, Can. J. Earth Sci., 14, 2817–2823, 1977.

Sumner, P. D. and Meiklejohn, K. I.: On the development of autochthonous blockfields in the grey basalts of sub-Antarctic Marion Island, Polar Geogr., 28, 120–132, 2004.

US Department of Agriculture: Soil Survey Manual, Agricultural Handbook No. 18. Government Printing Office, Washington D.C., 532 pp., 1993.

Walder, J. S. and Hallet, B.: A theoretical model of the fracture of rock during freezing, Geol. Soc. Am. Bull., 96, 336–346, 1985.

Watanabe, T., Funakawa, S., and Kosaki, T.: Distribution and formation conditions of gibbsite in the upland soils of humid Asia: Japan, Thailand and Indonesia, 19th World Congress of Soil Science, Soil Solutions for a Changing World, 1–6 August 2010, Brisbane, Australia, 17–20, 2010.

Wesolowski, D. J. and Palmer, D. A.: Aluminum speciation and equilibria in aqueous solution: V. Gibbsite solubility at 50 °C and pH 3–9 in 0.1 molal NaCl solutions (a general model for aluminum speciation; analytical methods), Geochim. Cosmochim. Acta, 58, 2947–2969, 1994.

Whalley, W. B., Rea, B. R., and Rainey, M.: Weathering, blockfields, and fracture systems and the implications for long-term landscape formation: some evidence from Lyngen and Øksfordjøkelen areas in north Norway, Polar Geogr., 28, 93–119, 2004.

White, A. F., Blum, A. E., Schulz, M. S., Vivit, D. V., Stonestrom, D. A., Larsen, M., Murphy, S. F., and Eberl, D.: Chemical weathering in a tropical watershed, Luquillo Mountains, Puerto Rico: I. Long-term versus short-term weathering fluxes, Geochim. Cosmochim. Acta, 62, 209–226, 1998.

A geomorphology-based approach for digital elevation model fusion – case study in Danang city, Vietnam

T. A. Tran[1], V. Raghavan[1], S. Masumoto[2], P. Vinayaraj[1], and G. Yonezawa[1]

[1]Graduate School for Creative Cities, Osaka City University, Osaka, Japan
[2]Graduate School of Science, Osaka City University, Osaka, Japan

Correspondence to: T. A. Tran (tranthian.gis@gmail.com)

Abstract. Global digital elevation models (DEM) are considered a source of vital spatial information and find wide use in several applications. The Advanced Spaceborne Thermal Emission and Reflection Radiometer (ASTER) Global DEM (GDEM) and Shuttle Radar Topographic Mission (SRTM) DEM offer almost global coverage and provide elevation data for geospatial analysis. However, GDEM and SRTM still contain some height errors that affect the quality of elevation data significantly. This study aims to examine methods to improve the resolution as well as accuracy of available free DEMs by data fusion techniques and evaluating the results with a high-quality reference DEM. The DEM fusion method is based on the accuracy assessment of each global DEM and geomorphological characteristics of the study area. Land cover units were also considered to correct the elevation of GDEM and SRTM with respect to the bare-earth surface. The weighted averaging method was used to fuse the input DEMs based on a landform classification map. According to the landform types, the different weights were used for GDEM and SRTM. Finally, a denoising algorithm (Sun et al., 2007) was applied to filter the output-fused DEM. This fused DEM shows excellent correlation to the reference DEM, having a correlation coefficient $R^2 = 0.9986$, and the accuracy was also improved from a root mean square error (RMSE) of 14.9 m in GDEM and 14.8 m in SRTM to 11.6 m in the fused DEM. The results of terrain-related parameters extracted from this fused DEM such as slope, curvature, terrain roughness index and normal vector of topographic surface are also very comparable to reference data.

1 Introduction

A digital elevation model (DEM) is a digital model representing a surface which is presently used in many applications such as hydrology, geomorphology, geology and disaster risk mitigation. It is one of the essential inputs in modeling or simulating landscapes as well as dynamic natural phenomena such as flooding, soil erosion and landslides. Due to the important role of DEMs in terrain-related research and applications, it is necessary to create high-quality DEMs at various levels of details. DEM can be generated using photogrammetry, interferometry, ground and laser surveying and other techniques (Mukherjee et al., 2013a). Usually, aerial photos, high-resolution satellite data or field-surveyed spot height and light detection and ranging (lidar) data are used as inputs to generate high-resolution/high-quality DEMs. Surveying

data collections is not only time consuming but also expensive. Even though a good number of aerial photos, high-resolution synthetic aperture radar (SAR) and optical remote-sensing data are available, it is not always easy and affordable to generate a DEM over large areas.

Recently, global, free DEMs, including the Advanced Spaceborne Thermal Emission and Reflection Radiometer (ASTER) Global DEM (GDEM) and Shuttle Radar Topographic Mission (SRTM) DEM, have been offering almost global coverage and easily accessible data. These DEMs have been used in many applications, especially in geomorphology and hydrology (Zandbergen, 2008). However, GDEM and SRTM display some height errors, which affect the quality of elevation data significantly. Therefore, there have been several attempts to develop methodologies for enhancing quality of these global, free DEMs.

Several authors (e.g., Li et al., 2013; Ravibabu et al., 2010; Zhao et al., 2011; Suwandana et al., 2012; Mukherjee et al., 2013a; Czubski et al., 2013) have evaluated the accuracy of GDEM as well as SRTM and carried out comparative evaluation of two DEMs. Results from these studies indicated that, due to the inherent difficulties in acquiring satellite data both with the optical stereoscopic and the interferometric synthetic aperture radar (InSAR) technologies, global DEMs are not complete in and of themselves (Yang and Moon, 2003). Some authors (e.g., Reuter et al., 2007; Mukherjee et al., 2013a; Czubski et al., 2013; Fuss, 2013) have also evaluated the accuracy of global DEMs based on terrain characteristic. The vertical accuracy of these quasi-global DEMs vary depending on the terrain and land cover (Czubski et al., 2013). The main purpose of these studies was to verify the quality of global DEMs. However the unique characteristics and different factors affecting the vertical accuracy of optical stereoscopy and InSAR provide an opportunity for DEM fusion (Kaab, 2005).

This study proposes a geomorphological approach for DEM fusion based on evaluation of the accuracy of GDEM and SRTM in mountain slopes, valleys and flat areas. This approach was used to combine DEMs from different sources with appropriate weights to generate a fused elevation data. This could be an effective method to enhance the quality of global DEMs that have not been attempted in previous studies on DEM fusion (e.g., Corsetto and Crippa, 1998; Kaab, 2005; Karkee et al., 2008; Papasaika et al., 2011; Lucca, 2011; Fuss, 2013)

2 Study area

This study was conducted in Danang city in the middle of central Vietnam (Fig. 1). The test site of 950 km^2 covers the inland area of Danang city and is characterized by elevation ranging from 0 to 1664 m a.m.s.l. Danang city is located on the Eastern Sea coast, extending from 15°55′ N to 16°14′ N and 107°18′ E to 108°20′ E. The topography of this area has great variation from flat to mountainous regions. Due to varying of topography and geomorphology, the optical stereoscopy technique used to generate GDEM as well as the InSAR technique used in SRTM show different representation on DEM data, and contain inherent anomalies that need to be detected and minimized.

There are few studies in this area using global, free DEMs such as GDEM or SRTM. Ho and Umitsu (2011) and Ho et al. (2013) developed a landform classification method and flood hazard assessment of the Thu Bon alluvial plain, central Vietnam. In their study, the authors used SRTM as an input DEM source and applied bias elimination method to correct surface elevation data to the height of bare-earth surface. However, SRTM with low resolution (90 m) may not give sufficient terrain information. Also, the InSAR technique used in SRTM may fail to provide reliable estimate of elevation if images contain layovers, nonlinear distortion of the images due to slanted geometry of the radar sensing and shadows, or suffer from temporal decorrelation and changes in atmospheric conditions between two acquisitions (Karkee et al., 2008). Although Ho et al. (2013) already masked the high and upland areas and focused only on a low-lying alluvial plain, their research did not discuss methods to enhance accuracy of free DEMs, especially in the areas that have high topographic relief.

3 DEM data sets

The global, free DEMs used in this study include GDEM Version 2 (http://earthexplorer.usgs.gov) and SRTM Version 4.1 (http://www.cgiar-csi.org). GDEM Version 2 released in October 2011 has the resolution of 30 m. GDEM data were compiled from over 1.2 million scene-based DEMs covering land surface between 83° N and 83° S latitudes. GDEM was generated from ASTER optical satellite images using stereoscopy technique with differing sensor look angles. The Terra spacecraft used in ASTER GDEM is capable of collecting in-track stereo using nadir- and aft-looking near-infrared cameras (ASTER GDEM Validation Team, 2011). DEM from such optical satellite images as GDEM usually contains some height errors because of cloud coverage. ASTER GDEM Version 2 was improved with respect to Version 1 (released on June 2009) due to a better data-processing algorithm and additional data used during the processing. However, the revised version still contains anomalies and artifacts which need to be corrected before being used in any application, especially on a local scale (ASTER GDEM Validation Team, 2011).

SRTM Version 4.1 has been obtained from the Consortium for Spatial Information (CGIAR-CSI; http://www.cgiar-csi.org). The DEM data were derived from the 11-day Shuttle Radar Topographic Mission that flew in February 2000, and have provided publicly available elevation surface data for approximately 80 % (from 60° N to 56° S) of the world's land surface area (Reuter et al., 2007). The SRTM elevation data are derived from X-band and C-band InSAR sensor. The first release of SRTM was provided in 1° DEM tiles in 2003. When the data were processed by NASA and the USGS, they were made available at 1 arcsec resolution (approximately 30 m) for the United States, and 3 arcsec resolution (approximately 90 m) for the rest of the world. The Consortium for Spatial Information of the CGIAR (CGIAR-CSI) is offering post-processed 3 arcsec DEM data for the globe. The original SRTM has been subjected to a number of processing steps to provide seamless and complete elevation surface for the globe. In its original release, SRTM data contained regions of no data, specifically over water bodies (lakes and rivers) and in areas where insufficient textural detail was available

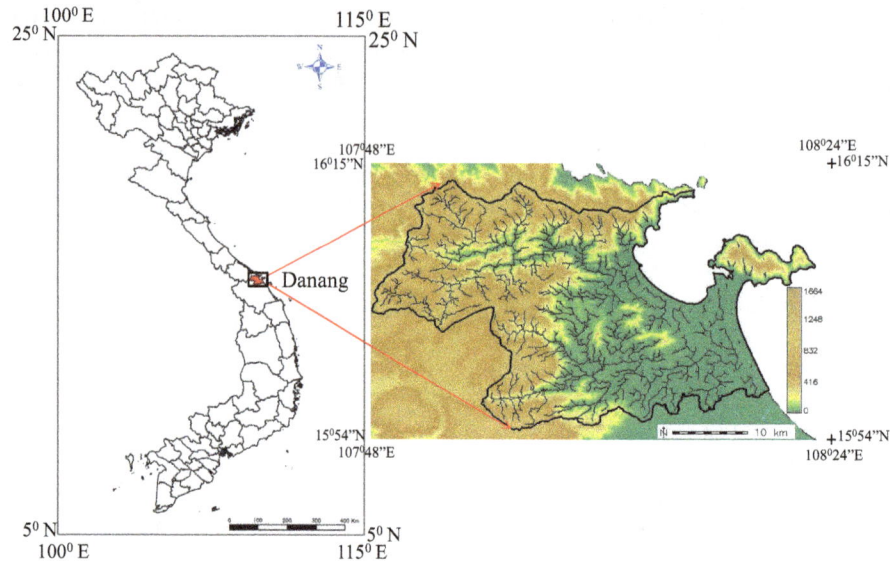

Figure 1. Location of study area and topographic overview.

Table 1. General information on the global DEMs and reference DEM (all the negative values were filled by neighboring pixels). Unit: m

	Min	Max	Mean	SD
GDEM	0	8016	271.8	319
SRTM	0	1634	277.5	304.6
Reference DEM	0	1664	268.1	302.6

in the original radar images to produce three-dimensional elevation data (http://www.cgiar-csi.org). Presently, the latest version of SRTM released by CGIAR-CSI is SRTM Version 4.1. SRTM V4.1 has some advantages over previous versions, such as filling void areas and masking water bodies. SRTM used in this study has the resolution of 90m. Although SRTM has lower resolution than GDEM, it offers coverage in all weather conditions since it uses the InSAR technique. However, because of the limitation of resolution and vertical error in some areas, SRTM needs to be edited before being used in any application. Both GDEM and SRTM are in a geographic coordinate system, with the World Geodetic System 1984 (WGS84) horizontal datum and the Earth Gravitational Model 1996 (EGM96) vertical datum.

The reference elevation data used in this study are from a DEM generated from the 1:10 000 topographic map of Danang city published in 2010, including contour lines with 5 m intervals and spot height elevation data developed by the Department of Natural Resources and Environment (DONRE), Danang city, Vietnam. Contour lines were derived from aerial photos of Danang city captured on 2003, and additionally surveyed and modified during 2009. Spot height elevation data were surveyed in 2009. The data are

projected in a Vietnamese projection named VN2000. In this study, the DEM generated from contour and spot height elevation is referred to as the "reference" DEM. Firstly a DEM was generated from the contour map using the regularized spline with tension (RST) algorithm. The RST interpolation is considered as one of the effective interpolation methods available for elevation data (Hofierka et al., 2002). The RST method is based on the assumption that the approximation function should pass as closely as possible to the given data and should be as smooth as possible (Mitasova et al., 1995). RST interpolation was carried out in GRASS GIS open-source software (http://grass.osgeo.org). However, the contour lines do not cover the whole area of Danang city. In flat areas with elevation less than 10 m, there are no contour lines. A large number of spot height data are available for flat areas (more than 190 000 elevation points), and inverse distance weighting (IDW) interpolation was applied to generate the DEM where contour data are not available and merged with DEM generated using RST with contour data for hilly areas. This reference DEM was also generated at a resolution of 30 m. The RMSE of the reference DEM comparing to spot height data is 1.66 m. Some statistical data on the global DEMs and reference DEM are shown in Table 1. The mean elevation and standard deviation (SD) in GDEM and SRTM are analogous to the reference DEM. Due to some artifacts located in GDEM, the maximum elevation value of GDEM (8016 m) shows significant dissimilarity. Compared to GDEM, the SD of SRTM (304.6 m) is almost similar to the reference DEM (302.6 m).

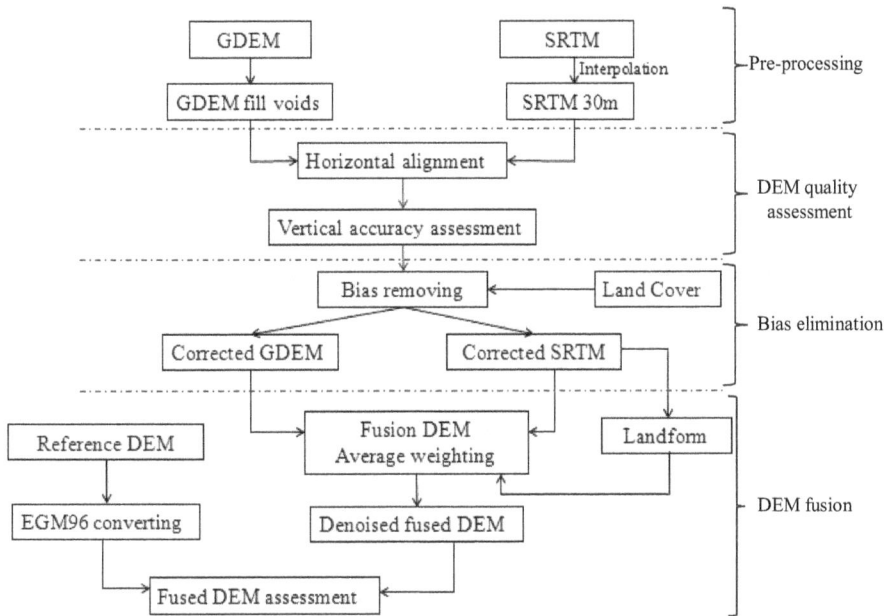

Figure 2. Flowchart of data processing.

4 Methodology

SRTM was interpolated from 90 to 30 m resolution in order to compare with other DEM sources. The artifacts in GDEM were eliminated using the fill and feather method (Dowding et al., 2004). DEM alignment was also carried out in order to co-register GDEM and interpolated SRTM with respect to the reference DEM. Next, both GDEM and SRTM were evaluated in terms of vertical and horizontal accuracy. The quality of each DEM was also assessed according to different topographic conditions. The result of evaluation has been used to devise an appropriate DEM fusion method considering various factors responsible for degradation of data quality. Basically, there is a difference between the digital surface model (DSM) like GDEM, SRTM and the digital terrain model (DTM) that refers to the bare-earth surface. The overestimations as well as underestimated elevation values in GDEM and SRTM need to be detected and corrected by comparing these elevation data to the reference DEM on the basis of the geomorphology and land cover map. In the case of land cover category, the offsets were calculated by taking mean values of the difference in elevation between the global DEMs and reference DEM. The corrected GDEM and SRTM were used as input data for the DEM fusion process. The landform classification map was generated from SRTM to determine the area suitable for different fusion methods. The algorithm used in the DEM fusion process is weighted averaging based on geomorphologic characteristics. In relatively flat areas, the higher weight was used for SRTM and lower weights for GDEM. In the mountainous areas, SRTM and GDEM were weighted equally. The higher weight was applied for GDEM in the valley areas, because of the limitation

of SRTM in those areas. The output-fused DEM was filtered using a denoising algorithm according to Sun et al. (2007). Finally, the fused DEM was compared to the reference DEM to assess the efficiency of the DEM fusion method.

The data processing described above is shown in Fig. 2. The data fusion workflow includes four main steps, namely, pre-processing, DEM quality assessment, bias elimination and DEM fusion.

4.1 Pre-processing

It is observed that SRTM has anomalies in the coastal area and some small areas inland with negative values; 377 pixels show negative values and cover about 0.34 km². These pixels were filled by averaging elevation of 3×3 neighboring pixels. SRTM and GDEM have been converted from geographic coordinates to UTM_WGS84_zone 49N projection. The reference DEM was also converted from VN2000 to UTM_WGS84_zone 49N projection. The vertical datums used in the global DEMs and reference DEM are different. The global DEMs use EGM96 vertical datum, while the reference DEM uses the Vietnamese vertical datum named Hon Dau–Hai Phong, which is related to m.s.l. in Hon Dau Island, Hai Phong province, Vietnam. An offset 1.5 m downwards was applied to convert the global DEMs from EGM96 to Hon Dau–Hai Phong vertical datum.

SRTM was interpolated from 90 to 30 m using the RST algorithm, which is available in GRASS GIS as the *r.resamp.rst* function. RST interpolation not only re-samples the DEM to higher resolution but also reduces the staircase effect in the original SRTM and smoothens the DEM surface. Figure 5a and b show the profile of SRTM compared to

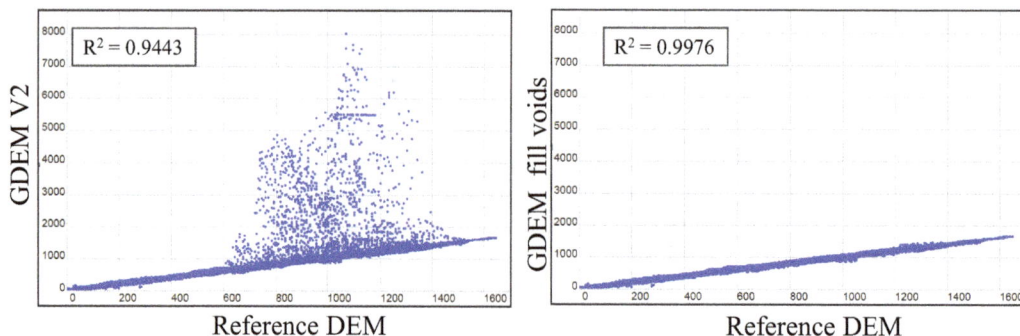

Figure 3. Correlation between GDEM and the reference DEM before (left panel) and after (right panel) filling voids.

Table 2. Results of GDEM after filling artifacts and shifting.

	RMSE (m)			Correlation
	Mountain	Flat	Whole area	coefficient (R^2)
Original GDEM	91.2	4.2	75.6	0.9443
GDEM filled voids	17.8	4.2	14.9	0.9976
GDEM after shifting	15.4	4.1	13.0	0.9983

Table 3. SRTM before and after interpolation to 30 m.

	RMSE (m)			Correlation
	Mountain	Flat	Whole area	coefficient (R^2)
Original SRTM	17.6	3.3	14.8	0.9979
Interpolated SRTM (30 m)	15.0	3.2	12.6	0.9986

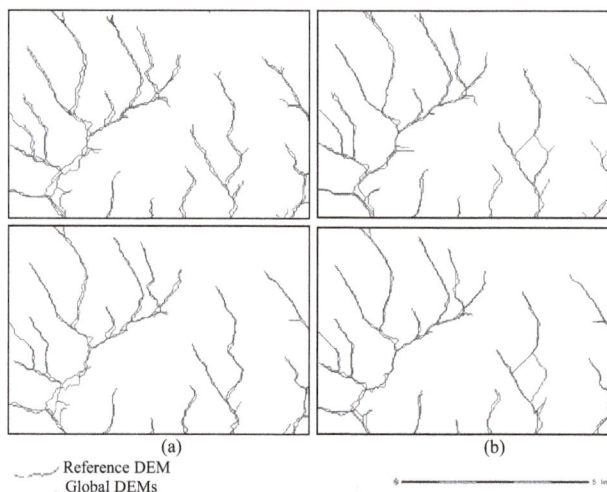

Reference DEM
Global DEMs

(a)   (b)

Figure 4. Comparing stream networks of the global DEMs and reference DEM before (top panels) and after (bottom panels) shifting DEM: **(a)** GDEM, **(b)** SRTM.

the reference DEM before and after interpolation. The interpolated SRTM also has better RMSE and correlation to the reference DEM than the original 90 m data (Table 3).

GDEM has some artifacts in the western mountain part of Danang city, with extreme high-elevation values. These artifacts may be caused due to cloud coverage that is very common in optical satellite data. These artifacts are the main reason for high RMSE (75.6 m) observed in raw GDEM (Table 2). The artifacts in GDEM need to be eliminated before further processing. Several algorithms for void filling have been proposed, such as kriging, spline, IDW (Reuter et al., 2007), moving window (Karkee et al., 2008), fill and feather (Dowding et al., 2004) and delta surface fill (Grohman et al., 2006). All the void-filling algorithms can be categorized into three groups, namely, interpolation, moving window and fill and feather (F & F). The F & F method proposed by Dowding et al. (2004) was applied in this study to fill artifacts in GDEM. In the F & F approach, an artifact is replaced with the most accurate digital elevation source available with the void-specific perimeter bias removed (Grohman et al., 2006).

The artifacts were detected by overlaying the slope map of GDEM and the difference elevation map between GDEM and the reference DEM, and digitizing from the anomalies that can be visualized from the overlaying display. SRTM was chosen as auxiliary data to fill the artifacts for GDEM. After filling these artifacts, the surface is feathered to mitigate any abrupt change (Grohman et al., 2006). In this case study, DEM surface will be feathered in the final step of data processing using a filtering algorithm. As the result, GDEM after filling artifacts has a RMSE of only 14.9 m. The scatter plot of GDEM after applying F & F also shows a good correlation to the reference DEM, while the original one has several outliers (Fig. 3). Comparing to original GDEM, it can also be seen that most of the artifacts were eliminated.

4.2 DEM quality assessment

The horizontal accuracy of the global DEMs was evaluated by comparing the extracted stream networks (Fig. 4). Stream networks extracted from the reference DEM, GDEM and

SRTM indicate that SRTM has a horizontal difference of about 15 m, and GDEM has a difference of around 30 m with respect to the reference DEM. Therefore, GDEM was shifted one pixel to the east, and SRTM was shifted half a pixel to the west, in order to align all input DEMs before fusion process. Figure 5 compares the profiles of GDEM, SRTM and the reference DEM before and after shifting. The ridge lines as well as canyon bottoms in GDEM and SRTM become more similar to the reference DEM. In Table 2, GDEM after shifting shows better RMSE and correlation with the reference DEM as compared to before shifting.

In this study area, the RMSE of GDEM and SRTM with respect to the reference DEM was observed as 14.9 and 14.8 m, respectively (Tables 2 and 3). The correlation coefficient (R^2) of GDEM in the whole area is 0.9976, while this value in the original SRTM is 0.9979. The accuracy of the individual DEM should be considered based on the different topographic condition. Figure 6 shows the correlation coefficients of each global DEM in flat and mountain areas. In mountain areas, GDEM and SRTM have a similar correlation with the reference DEM (0.9966 and 0.9969, Fig. 6b). However, in some specific areas, especially in the steep valleys, GDEM provides better accuracy than SRTM. The circled areas in Fig. 5 show that GDEM preserves the considerable details of topography in the valley areas, while SRTM is ineffective in those areas. In such valley areas, SRTM seems to suffer from layover and shadow effects. In the case of a very steep slope, targets in the valley have a larger slant range than related mountain tops; consequently the fore-slopes are "reversed" in the slant range image. This is referred to as layover effect when the ordering of surface elements on the radar image is the opposite of the ordering on the ground (European Space Agency, https://earth.esa.int/applications/data_util/SARDOCS/spaceborne/RadarCourses/Radar_CourseIII/layover.htm). Radar shadow is caused when a slope is away from the radar illumination with an angle that is steeper than the sensor depression angle (European Space Agency, https://earth.esa.int/applications/data_util/SARDOCS/spaceborne/RadarCourses/Radar_CourseIII/shadow.htm). In such areas, SRTM may not provide sufficient information, compared to GDEM or other DEM sources. In relatively flat areas, the correlation coefficient between SRTM and the reference DEM ($R^2 = 0.8504$) is better than GDEM ($R^2 = 0.5578$) (Fig. 6a). This is because degradation of the elevation estimate of GDEM in the area has low topographic relief. In the profile of Fig. 7, it can be seen that GDEM has many spikes and unstable elevation values in this flat area, while SRTM shows similar trends to the reference DEM.

The difference elevation maps of the global DEMs were also generated by subtracting GDEM and SRTM values from the reference DEM. Both GDEM and SRTM show high vertical error in mountain areas, and lower vertical error in flat areas (Fig. 8). These errors occur because of the forest cover in mountain areas and due to some limitations of the sensing

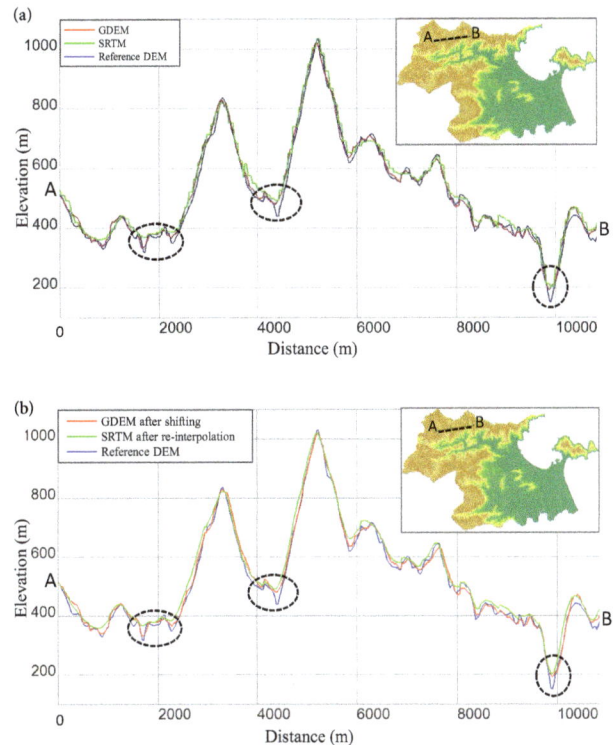

Figure 5. Comparing GDEM and SRTM to the reference DEM: **(a)** before re-interpolation SRTM and shifting data, **(b)** after re-interpolation SRTM and shifting data.

techniques used to generate DEM in high-relief areas. The profile of SRTM from the difference elevation map in flat areas is closer to 0 m line (Fig. 8), while GDEM contains higher difference and spikes that affect the quality of GDEM significantly.

4.3 Minimizing DEM bias effect

The topographic height variation between the global DEMs and reference DEM is caused due to the differences in vertical datum used and in primary data collection methods. Vertical datum is one of the reasons for difference in elevation between the global DEMs and reference DEM. In addition, both GDEM and SRTM, which were generated from satellite data, are DSMs, while the reference DEM is considered a bare-earth DTM; this difference also introduces the bias offsets depending on the land cover.

Firstly, the global DEMs were converted to the Hon Dau–Hai Phong vertical datum. According to the Vietnam Land Administration, the global EGM96 model is almost similar to the Vietnamese vertical datum; 97 % of data shows the height difference around 1.5 m, while only 3 % of data shows higher than 1.5 m (Nguyen and Le, 2002). Therefore, an offset of 1.5 m was subtracted from the global DEMs, considering height difference between EGM96 and Vietnamese vertical datum.

a. Flat area

b. Mountain area

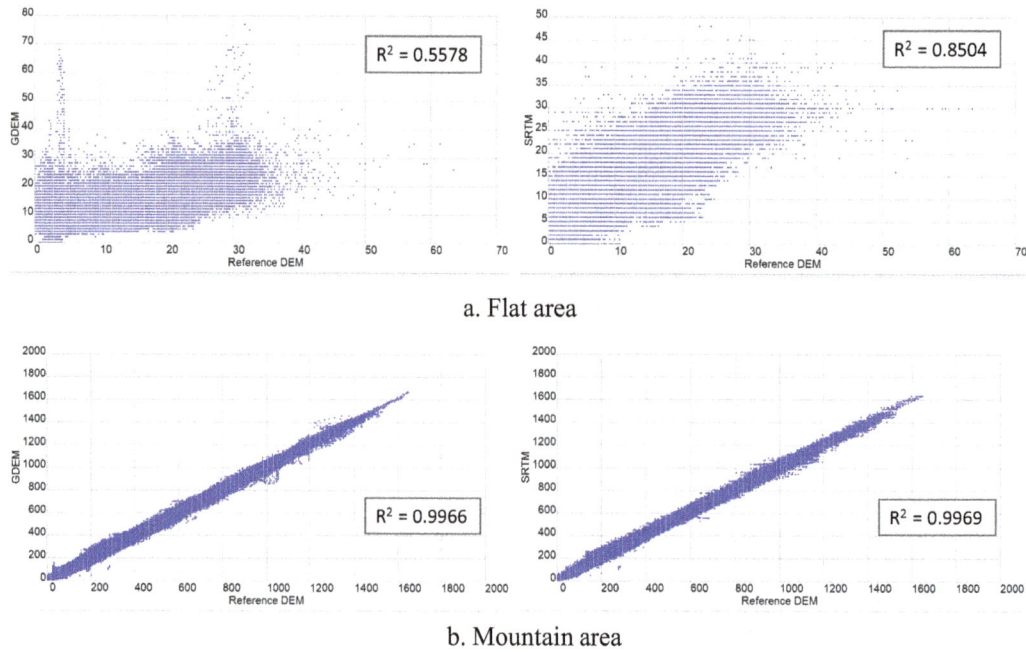

Figure 6. Correlation of GDEM and SRTM in flat (**a**) and mountainous (**b**) area.

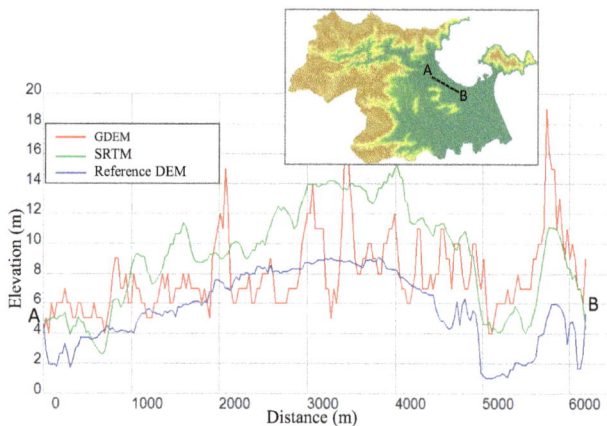

Figure 7. A profile of GDEM and SRTM compared to the reference DEM in flat areas.

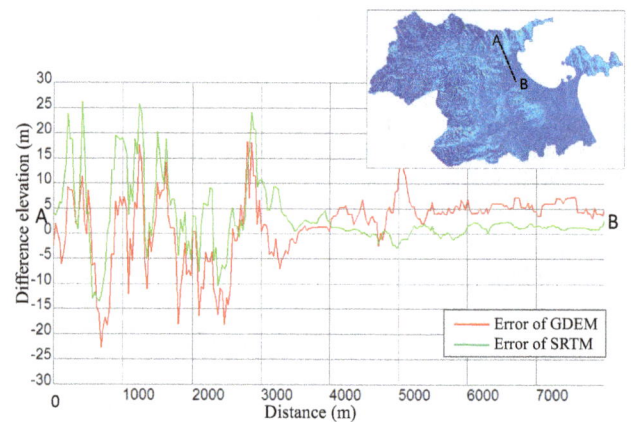

Figure 8. Difference elevation of GDEM and SRTM with respect to the reference DEM from mountain to flat areas.

Secondly, the height offsets of the global DEMs were determined based on the land cover map. Because the SRTM data were derived in 2000 and GDEM data were collected from millions of ASTER images from 1999 to 2009, a land cover map of Danang city in 2001 was used to calculate the height offsets for the global DEMs. These offsets were calculated based on the difference elevation maps of GDEM and SRTM with respect to the reference DEM considering land cover. This was done using the *r.statistics* function in GRASS GIS. The mean elevation differences on each land cover type were calculated, and used as offsets to verify elevation for GDEM and SRTM (Table 4). As the result, GDEM has the highest difference in the water body (4 m). This error is

common in GDEM because water surfaces give very low reflectance values in optical satellite data. The elevation value of GDEM in bare land is underestimated (−2 m), on average 2 m lower than the reference DEM. These bare-land surfaces are located in flat areas where the topographic relief is inadequate for the optical stereoscopy technique. GDEM in such areas can, therefore, not provide reliable elevation information. In SRTM, the highest error is observed in the forest land cover type (6.3 m), which mostly covers mountainous areas. SRTM in mountainous areas revealed relatively higher errors, because layovers and shadows affect the quality of radar data. The significant error in SRTM is also observed in bare-land areas (3.8 m). The back scatter from bare land

Table 4. The mean errors of GDEM and SRTM according to the land cover map. Unit: m.

	Agriculture	Forest	Built-up	Bare land	Water
GDEM	0.7	1.0	1.1	−2.0	4.0
SRTM	1.9	6.3	2.5	3.8	0.4

is too small to create a radar image. From global assessment of the SRTM data, voids were found to be very common in mountainous areas, as well as in very flat areas especially in deserts (Zandbergen, 2008). SRTM V4 used in this study already dealt with the water body problem using a number of interpolation techniques and void-filling algorithms (Reuter et al., 2007). Therefore, the error of SRTM in water bodies currently is only 0.4 m (Table 4).

Based on the above investigations, the elevations for GDEM and SRTM with respect to the reference DEM were corrected by subtracting GDEM and SRTM from the elevation offsets for each land cover type (Table 4). The calculation was executed by the *r.mapcalc* function in GRASS GIS software using the land cover map as the base. The corrected GDEM and SRTM were used as input data for DEM fusion processing.

After removing the offsets, GDEM and SRTM were compared to the reference DEM again to make better input for DEM fusion processing. The mean value of GDEM and SRTM with respect to each elevation value in the reference DEM was calculated. Figure 9a shows the behavior of the global DEMs with respect to the reference DEM, from flat to mountainous areas. In the A and C area (Fig. 9b and d), the mean elevation of SRTM is closer to the reference DEM, while the profile of GDEM shows higher error. In the case of the B area (Fig. 9c), both SRTM and GDEM show the good correlation to the reference DEM. In Fig. 9e, the profile of GDEM is comparable to the reference DEM in this mountainous area. From this analysis, it is evident that using a global data fusion for the whole area is not a good solution. Appropriate weights for the DEM fusion process need to be considered depending upon the topographic context, and is used as the basis for DEM fusion in this study.

4.4 DEM fusion algorithm

Both GDEM and SRTM contain intrinsic errors due to primary data acquisition technology and processing methodology in relation with a particular terrain and land cover type (Mukherjee et al., 2013a). The optical stereoscopy technique used in GDEM is limited by the cloud coverage, radiometric variation and low levels of texture (Karkee et al., 2008), while the InSAR technique used in SRTM may not work well in the case of shadowing, layovers or complex dielectric constant (Reuter et al., 2007). Combination of two data can take into account the advantages of each DEM source and provide complementary inputs to enhance the quality for the global

DEMs. DEM fusion workflow combines the weighted averaging and denoising algorithm (Sun et al., 2007).

4.4.1 Weighted averaging

Several authors have proposed fusion methods for digital elevation data. Karkee et al. (2008) carried out a fusion between GDEM and SRTM using fast Fourier transformation (FFT) combining with frequency domain filtering. Papasaika et al. (2011) have proposed an approach that performs DEM fusion using sparse representations. Lucca (2011) examined different DEM fusion methods, such as weighted averaging and collocation prediction, and compared the result to lidar DSM to assess the improvement of DEM fusion. Fuss (2013) has developed a DEM fusion algorithm from multiple, overlapping DEMs, using slope thresholding, K-means clustering and filtering of elevations. Tran et al. (2013a, b) have given a fusion method by selecting appropriate DEM-source-based geomorphological conditions. The most frequent DEM fusion method that has been suggested is weighted averaging. The weighted mean ($\bar{x}$) of a nonempty set of data $\{x_1, x_2, \ldots, x_n\}$ with nonnegative weights $\{\omega_1, \omega_2, \ldots, \omega_n\}$ (Papasaika, 2012) is shown:

$$\bar{x} = \frac{\sum_{i=1}^{n} \omega_i x_i}{\sum_{i=1}^{n} \omega_i} = \frac{\omega_1 x_1 + \omega_2 x_2 + \ldots + \omega_n x_n}{\omega_1 + \omega_2 + \ldots + \omega_n}, \quad (1)$$

where $x_1, x_2, \ldots, x_n$ are the input DEMs. $\omega_1, \omega_2, \ldots, \omega_n$ are the weights for DEM fusion.

However, weighted averaging applied in previous studies referred to in the earlier section considers weights based on the accuracy of the whole raster DEM source. Each raster DEM $x_1, x_2, \ldots, x_n$ is used as one input data for weighted averaging. Actually, the DEM accuracy also changes depending upon the topographic context. Therefore, in this research, a new method for DEM fusion using weighted averaging based on geomorphologic characteristics was proposed. Firstly, a landform map was extracted from SRTM. This landform classification method was done according to Dickson and Beier (2006). The algorithm is based on the topographic position index (TPI) and slope map. In general, TPI allows classifying landscape into discrete landform categories by comparison of individual cell heights with an average height of neighboring cells (Czubski et al., 2013). The TPI-based landform classification method according to Dickson and Beier (2006) can be denoted as follows:

- Valley: TPI <= −8,
- Flat: −8 < TPI <= 8, slope < 6°,
- Steep slope: −8 < TPI <= 8, slope >= 6°,
- Ridge line : TPI > 8.

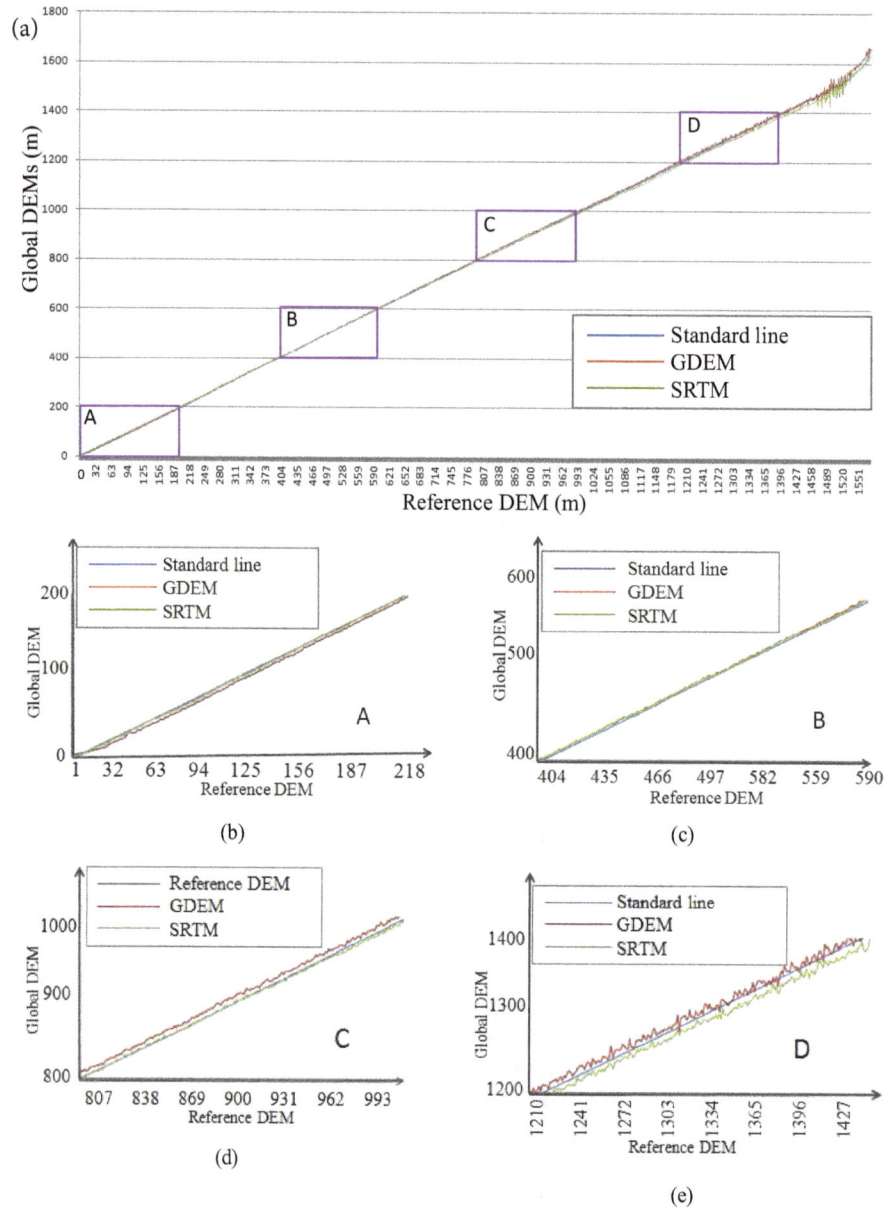

Figure 9. Behavior of GDEM and SRTM compared to the reference DEM in difference topographic contexts. (**a**) Whole area, (**b**) *A* area, (**c**) *B* area, (**d**) *C* area, (**e**) *D* area.

In this study, three categories demarcated from the landforms classification result, namely, mountain slopes (include ridge lines and steep slopes), valleys and flat areas (Fig. 10).

In order to determine the weight for the global DEMs on each landform class, the following equation (Hengl and Reuter, 2009) was applied:

$$w_i = \frac{1}{a^2}, \qquad (2)$$

where w_i is the weight for each DEM source for a given landform unit and a is the given accuracy parameter for the DEM for a given landform unit.

Terrain-related parameters were used to determine the weighting scheme for DEM fusion. Firstly, slope error (difference in slope between the global DEMs and reference DEM) was used to compare the accuracy of GDEM and SRTM in flat, valley and mountain slope areas. On each landform unit, the mean of absolute error (MAE) from the slope error map was calculated. The result is shown in Table 5.

In flat areas, GDEM has many overestimates and unstable elevation values. Therefore slope error of GDEM is larger than SRTM in this area. The weight used for GDEM can be determined according to Eq. (2): $w_1 = 1/(2.1)^2 = 0.22$; and the weight for SRTM can be shown as $w_2 = 1/(1.6)^2 = 0.39$. It can

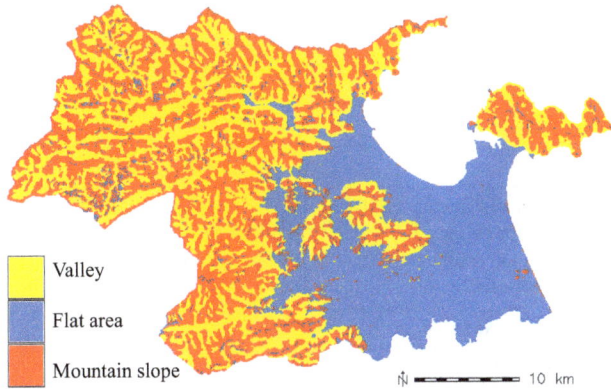

Figure 10. Landform classification map from SRTM.

Table 5. Mean of absolute error (MAE) from slope error maps of GDEM and SRTM on each landform unit.

Landform unit	GDEM (MAE)	SRTM (MAE)
Flat	2.1	1.6
Valley	5.8	5.7
Mountain slope	6.08	6.1

be seen that $w_2 \approx 2 \cdot w_1$; therefore the following formula was applied for DEM fusion in flat areas:

$$\text{Fused DEM} = (\text{GDEM} + \text{SRTM} \cdot 2)/3. \quad (3)$$

In mountain slope areas, the similar way was applied to calculate weight for DEM fusion, using MAE of slope error. In this case, GDEM and SRTM have almost same MAE (6.08 and 6.1°). Therefore, the same weights were applied for GDEM and SRTM in mountain slope areas ($w_1 = w_2$). The following equation was used in mountain slopes:

$$\text{Fused DEM} = (\text{GDEM} + \text{SRTM})/2. \quad (4)$$

In valley, GDEM and SRTM also have the similar MAE of slope error (5.8 and 5.7°). However, considering the topographic characteristic in some steep valleys, it can be seen that SRTM is ineffective in representing the valley bottom, while GDEM is still more correlative to the reference DEM (Fig. 5). In the case of valley landforms, Slope Variability (SV) (Popit and Verbovsek, 2013) was used to determine weight for DEM fusion. SV was calculated by the distance between maximum and minimum slope in the neighborhood of 3 × 3 pixels. SV errors of GDEM and SRTM with respect to the reference DEM was calculated. GDEM has a MAE of SV error of about 5.6, and SRTM has an error of about 7.3°. The weight for GDEM was calculated according to Eq. (2): $w_1 = 1/(5.6)^2 = 0.032$; and the weight for SRTM is calculated as $w_2 = 1/(7.3)^2 = 0.018$. It can be observed that $w_1 \approx 2 \cdot w_2$; therefore the following formula was used for DEM fusion in valley:

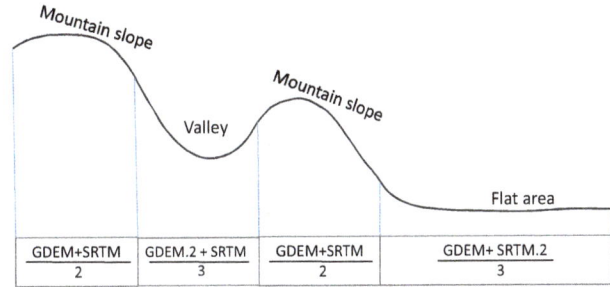

Figure 11. Weighted averaging used to fuse global DEMs.

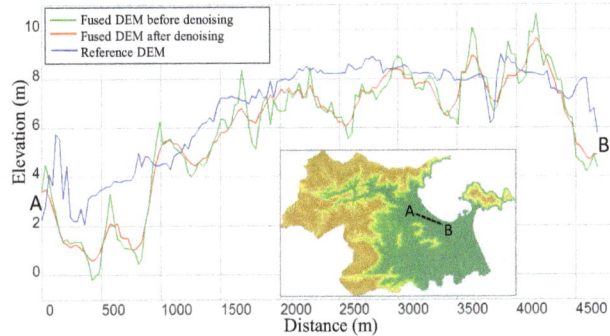

Figure 12. Result of a denoising algorithm (Sun et al., 2007) on the fused DEM.

$$\text{Fused DEM} = (\text{GDEM} \cdot 2 + \text{SRTM})/3. \quad (5)$$

The weighted averaging method based on the landform classification map is shown in Fig. 11.

4.4.2 Filtering the noises for the fused DEM

The fusion of different DEMs is problematic, since the DEMs are obtained from different sources and have different resolutions as well as accuracies (Lucca, 2011). The bias eliminations for GDEM and SRTM also use different offsets depending upon the land cover. Different weights have been used for DEM fusion in each landform type. Therefore, it is essential to filter the fusion DEM to reduce the mismatched and noisy data. In this study, the denoising algorithm (Sun et al., 2007) was used to minimize the noise effect. The level of denoising is controlled by two parameters, namely, the threshold (T) that controls the sharpness of the features to be preserved, and the number of iterations (n) that controls how much the data are smoothed. The optimum settings depend upon the nature of the topography and of the noise to be removed (Stevenson et al., 2009). Sun's algorithm (Sun et al., 2007) has been implemented in GRASS GIS as an add-on (*r.denoise*). In this denoising process, the topographic feature need to be preserved as far as possible in the fused DEM, so the parameters that were used are $T = 0.95$ and $n = 5$. As the result, the fused DEM becomes more smooth and the mismatched surfaces are minimized. The profile of the fused

Table 6. General statistics for the error of GDEM, SRTM and the fused DEM. Unit: m.

	Min error	Max error	MAE	RMSE
GDEM	−165.9	172.6	9.0	13.0
SRTM	−144.1	107	7.7	11.4
Fused DEM (before denoising)	−105.1	106.4	7.4	11.0
Fused DEM (after denoising)	−102.2	101.2	7.9	11.6

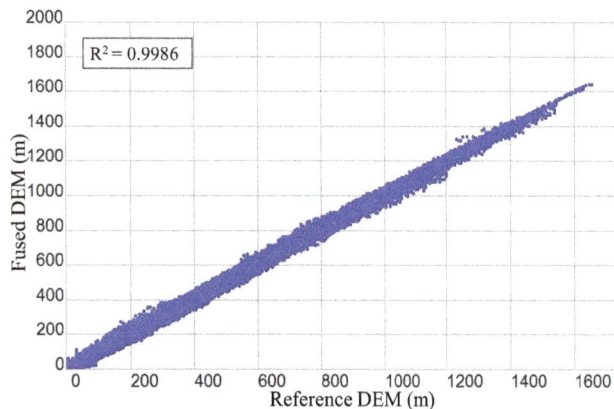

Figure 13. Correlation between the fused DEM and reference DEM.

DEM is also very much comparable to the reference DEM (Fig. 12).

5   Results and discussions

Weighted averaging based on the landform classification map has been verified as an effective method for DEM fusion. The accuracy of the fused DEM can be evaluated by statistical analysis such as RMSE, MAE and linear regression. The MAE and RMSE of the fused DEM were much improved compared to the available global DEMs. The RMSE was reduced from 75.6 m in original GDEM, 14.9 m in GDEM after removing artifacts and 13 m in GDEM after bias elimination to 11 m in the fused DEM. In SRTM, the RMSE was reduced from 14.8 m in the original SRTM and 11.4 m in the processed SRTM to 11 m in the fused DEM (Table 6).

The linear regression between the fused DEM and reference DEM also shows the significant correlation between two DEMs, with $R^2 = 0.9986$ (Fig. 13). Comparing to original data with correlation coefficient for GDEM and SRTM of 0.9976 and 0.9979, respectively, it can be, therefore, be concluded that the fused DEM shows better correlation with the reference DEM.

Statistical comparison of vertical accuracy of GDEM, SRTM and the fused DEM is shown in Table 6. The minimum error, maximum error, MAE and RMSE of the fused DEM show improvement when compared with GDEM and

Table 7. Comparison of differences in some terrain parameters of GDEM, SRTM and the fused DEM with respect to the reference DEM.

Attribute	GDEM	SRTM	Fused DEM
1. Slope			
– Mean of absolute error (MAE)	4.71	4.55	4.52
– SD of slope error	6.6	6.0	5.9
– Correlation coefficient (R) to reference DEM	0.868	0.895	0.898
2. Profile curvature			
– MAE	0.0036	0.0027	0.0026
– SD	0.0054	0.0045	0.0044
– R	0.234	0.316	0.331
3. Tangential curvature			
– MAE	0.0043	0.0036	0.0035
– SD	0.0064	0.0059	0.0059
– R	0.271	0.326	0.322
4. Topographic roughness index			
– MAE	2.79	3.02	3.01
– SD	3.9	3.7	3.6
– R	0.71	0.75	0.76

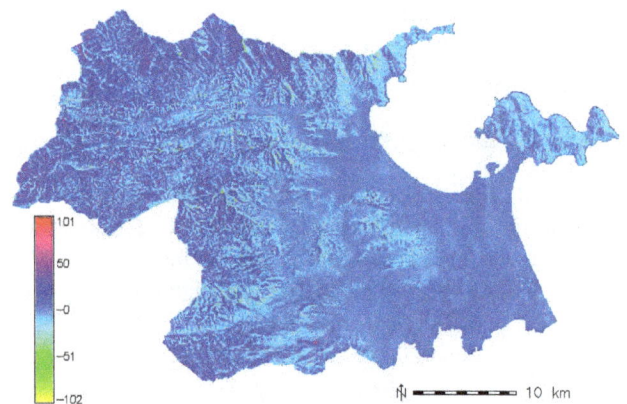

Figure 14. Difference in elevation between the fused DEM and reference DEM.

SRTM before fusion. Due to the smoothing, the final fused DEM shows a slight increase in RMSE in comparison with the fused DEM before denoising. The final fused DEM can minimize the mismatched surface and afford better extraction of topographic parameters. Based on the difference elevation map of the fused DEM (Fig. 14), it can be seen that the height error in the fused DEM is also greater in mountainous areas, especially in steep-slope areas. The minimum amount of error was observed in relatively flat areas. Figure 15 shows the histogram from the difference elevation maps of SRTM, GDEM and the fused DEM with respect to the reference DEM. In the fused DEM, the center of the histogram reaches a value of 0 m difference, and the cells that have lowest difference (0 m) are also most frequent. This result reveals that

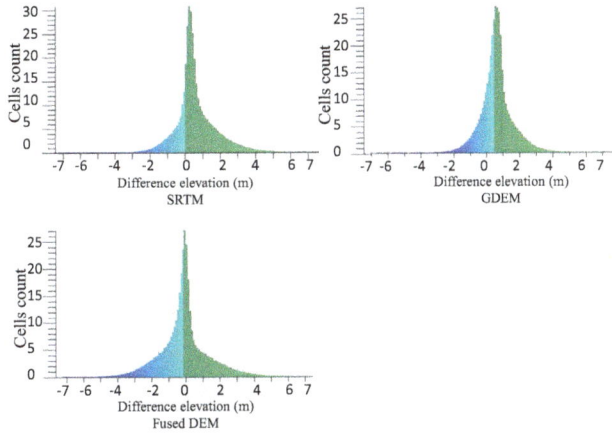

Figure 15. Histogram from the difference elevation maps of SRTM, GDEM and the fused DEM (*x* axis: cell values in tens; *y* axis: number of cells in thousands).

there is significant improvement in quality of global DEMs using the proposed DEM fusion algorithm.

The slope, profile curvature and tangential curvature maps were extracted from GDEM, SRTM and the fused DEM. Then the error maps with respect to the reference DEM were created in each terrain parameter (Table 7). Comparing to GDEM and SRTM, the fused DEM has smaller MAE and SD, and better correlation with the reference DEM. Figure 16 shows the slope, profile curvature and tangential curvature maps from the fused DEM. In these DEM derivative parameters, no major anomaly or terrace artifacts can be seen in the transition zones between landform classes.

Aspect is calculated as circular degrees clockwise from 0 to 360°, and it is therefore difficult to compare quantitatively (Deng et al., 2007). In order to assess the accuracy in aspect as well as slope, unit normal vector (NV) of topographic surface was considered. The NVs of the global DEMs and fused DEM were computed from slope and aspect values of respective DEMs. The NVs from these DEMs then were compared with the reference DEM to determine the angular difference between two NVs (Fig. 17). The NV of the terrain surface (T) can be calculated as below as suggested by Hodgson and Gaile (1999):

$$T = [x, y, z], \tag{6}$$

where $x = \sin(\text{aspect}) \cdot \sin(\text{slope})$, $y = \cos(\text{aspect}) \cdot \sin(\text{slope})$ and $z = \cos(\text{slope})$.

To derive the three-dimensional angular difference between two unit NVs (T and S) pointing away from the same origin, the following formula (Hodgson and Gaile, 1999) was applied:

$$\cos(i) = T \cdot S = t_x \cdot s_x + t_y \cdot s_y + t_z \cdot s_z. \tag{7}$$

The result of angular differences of NV is shown in Table 8. As a result, the fused DEM has a smaller mean error

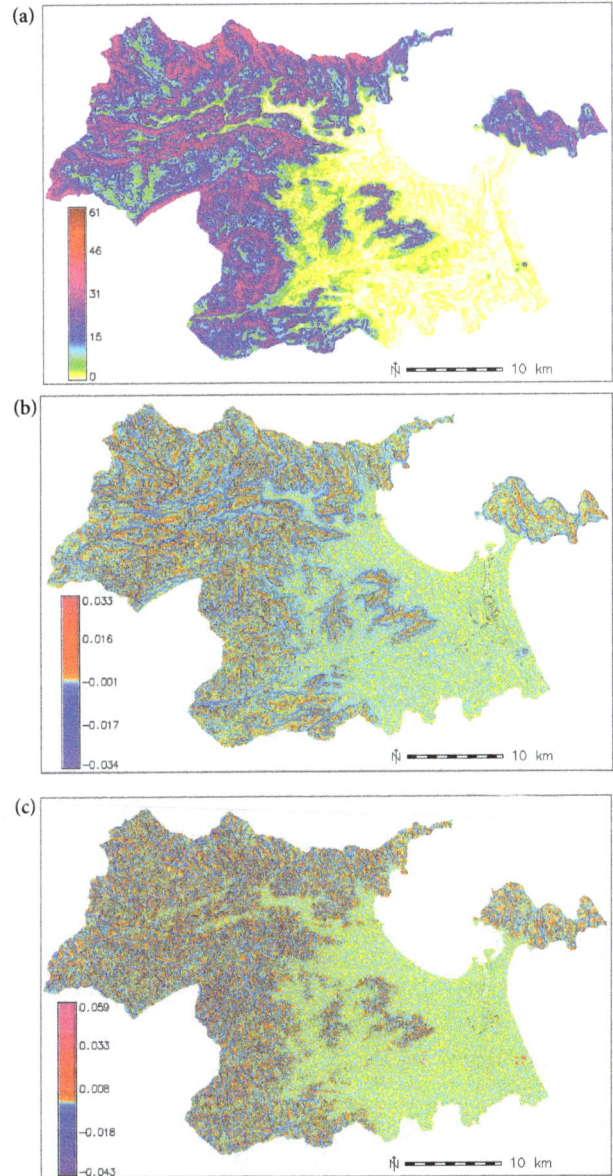

Figure 16. Slope (**a**), profile curvature (**b**) and tangential curvature (**c**) maps extracted from the fused DEM.

than GDEM and SRTM, and the SD of the fused DEM is also comparable with the global DEMs.

The topographic roughness index (TRI) was also considered to assess the quality of the fused DEM. In this study, the TRI was used as the amount of elevation difference among the adjacent cells of a DEM (Mukherjee et al., 2013b). The residuals in elevation between a grid cell and its eight neighbors were derived, and the RMSE of the elevation differences was calculated as the TRI. The TRI of the reference DEM and GDEM, SRTM as well as the fused DEM show a correlation coefficient of 0.71, 0.75 and 0.76, respectively (Table 7). The TRI derived from the fused DEM compares well with the reference DEM as compared with GDEM and SRTM.

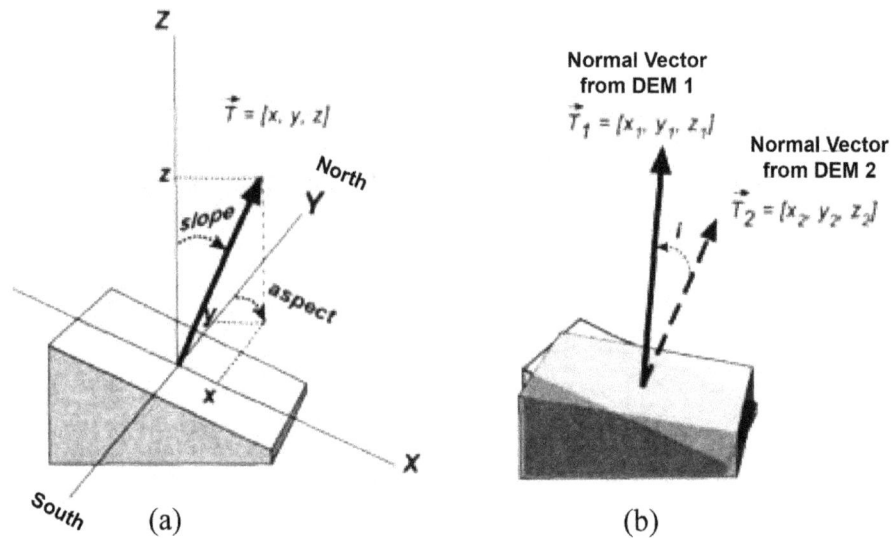

Figure 17. Normal vector of a topographic surface (**a**) and the angular difference between two normal vectors (**b**) (Hodgson and Gaile, 1999).

Table 8. Result of angular difference of unit NVs between the global DEMs, fused DEM and reference DEM.

Angular difference	GDEM	SRTM	Fused DEM
Min	0.0005	0.0015	0
Max	81.9	68.1	67.4
Mean	7.81	7.39	7.33
SD	6.85	7.03	7.06

6   Conclusions

Global, free DEMs generated from remote-sensing data always have some vertical and horizontal errors. Assessing the quality of global DEMs and validating their accuracy before use in any application is very important. In this study, the accuracies of GDEM and SRTM were determined based on height differences with the reference DEM. The artifacts with extreme high-elevation values in GDEM were eliminated by using SRTM as auxiliary data. River networks extracted from both DEMs that were used to detect and correct the horizontal errors for global DEMs can make better co-registration. The bias effect caused by tree-top canopy and building on global DEMs was also calculated by comparing these DSMs with the elevation from the reference DEM. A land cover map of Danang city in 2001 was used to calculate the height difference of GDEM and SRTM on each land cover type. Once the bias offsets were determined, effort was made to correct the elevation of these DEMs with respect to the bare-land surface.

Based on global DEM assessment in Danang city, it is observed that the accuracy of GDEM and SRTM varies depending upon the geomorphological characteristics of the target area. Fusion between two global DEMs using a geomorphological approach is an appropriate solution to enhance the quality of free DEMs for Danang city, Vietnam. The data fusion technique was applied by weighted averaging of GDEM and SRTM based on the topographic context. The weighting scheme was determined according to accuracy parameters, including MAE of slope and slope variability. The weights used for each DEM were changed locally according to the landform types. The results were compared with the reference DEM to discuss accuracy and impact of landform in variation on DEM quality. Terrain-related parameters such as slope, curvature, TRI and NV of topographic surface were considered to assess seriously the quality of the fused DEM. Results indicate that the fused DEM has improved accuracy compared to individual global DEM and most artifacts are successfully eliminated. The proposed method supports the effective utilization for the areas where the better-quality DEM is not available.

In future work, the more robust weighting scheme needs to be considered by defining a greater number of landform types. In this regard, the landform classification method may also need to be improved further. In the future, we plan to investigate landform classification using *r.geomorphon*, a new add-on that is available in GRASS 7. A "geomorphon" is a relief-invariant, orientation-invariant and size-flexible abstracted elementary unit of terrain (Stepinski et al., 2011). This landform classification map will not only be a good way to compare the height errors in micro-geomorphological classes, but will also help to compare terrain parameters extracted from fused global DEMs and the reference DEM.

The difference in elevation between DEM and DSM is useful for estimating the canopy height, especially in areas used for silviculture. Further investigation on bias effect introduced by land cover and silviculture needs to be carried out. The relationship between land cover and geomorphology also need be studied in the future, to understand the impact of topographic condition on land cover change. Several new satellite data, including ALOS-2 PRISM and PALSAR-2 (http://www.eorc.jaxa.jp/ALOS/en/index.htm), need to be incorporated to enhance the methods for multiresolution DEM fusion based on a better understanding of characteristics of DEMs derived from multiple sources.

Acknowledgements. This study was conducted under the Mobukagakusho (MEXT) scholarship supported by the Japanese government. We sincerely thank MEXT and Osaka City University for the financial and facilities support for our research. We are also grateful to the Department of Natural Resource and Environment, Danang City, Vietnam, for supporting the reference data. The authors would like to express sincere gratitude to Xianfeng Song, University of Chinese Academy of Sciences, who has given us a lot of valuable comments on geomorphological analysis. We also sincerely thank Susumu Nonogaki, National Institute of Advanced Industrial Science and Technology (AIST), Japan, for his kind help on DEM generation. The authors also thank David Hastings for his critical comments that helped in improving the quality of this paper.

Edited by: H. Mitasova

References

ASTER GDEM Validation Team: Advanced Spaceborne Thermal Emission and Reflection Radiometer (ASTER) Global digital elevation model (GDEM) Version 2, http://earthexplorer.usgs.gov/ (last access: July 2014), October 2011.

Crosetto, M. and Crippa, B.: Optical and Radar fusion for DEM generation, IAPRS, Vol. 32, Part 4, ISPRS Commission IV Symposium on GIS – Between Vision and Applications, Stuttgart, Germany, 128–134, 1998.

Czubski, K., Kozak, J., and Kolecka, N.: Accuracy of SRTM-X and ASTER elevation data and its influence on topographical and hydrological modeling: case study of the Pieniny Mts. in Poland, Int. J. Geoinform., 9, 7–14, 2013.

Deng, Y., Wilson, J. P., and Bauer, B. O.: DEM resolution dependencies of terrain attributes across a landscape, Int. J. Geogr. Inf. Sci., 21, 187–213, doi:10.1080/13658810600894364, 2007.

Dickson, B. G. and Beier, P.: Quantifying the influence of topographic position on cougar (Puma concolor) movement in southern Califonia, USA, J. Zoology, 271, 270–277, 2006.

Dowding, S., Kuuskivi, T., and Li, X.: Void fill of SRTM elevation data-Principles, processes, and performance, Images to decisions: Remote sensing foundations for GIS applications, AS-PRS 2004 Fall conference, Kansas City, MO, USA, 2004.

Fuss, C. E.: Digital elevation model generation and fusion, Master thesis in Geography, University of Guelph, Ontario, Canada, 2013.

Grohman, G., Kroenung, G., and Strebeck, J.: Filling SRTM voids: the delta surface fill method, Photogramm. Eng. Rem. S., 72, 213–216, 2006.

Hengl, T. and Reuter, H. I.: Geomorphometry, Concept, Software, Applications, Elsevier-Develop. Soil Sci., 33, 110–111, 2009.

Ho, T. K. L. and Umitsu, M.: Micro-landform classification and flood hazard assessment of the Thu Bon alluvial plain, central Vietnam via an integrated method utilizing remotely sensed data, Appl. Geogr., 31, 1082–1093, 2011.

Ho, T. K. L., Yamaguchi, Y., and Umitsu, M.: Rule-based landform classification by combining multi-spectral/temporal satellite data and the SRTM DEM, Int. J. Geoinform., 8, 27–38, 2013.

Hodgson, M. E. and Gaile, L.: A cartographic modeling approach for surface orientation-related applications, Photogramm. Eng. Rem. S., 65, 85–95, 1999.

Hofierka, J., Parajka, J., Mitasova, H., and Mitas, L.: Multivariate interpolation of precipitation using regularized spline with tension, Transact. GIS, 6, 135–150, 2002.

Kaab, A.: Combination of SRTM3 and repeat ASTER data for deriving alpine glacier flow velocities in the Bhutan Himalaya, Remote Sens. Environ., 94, 463–474, 2005.

Karkee, M., Steward, B. L., and Aziz, S. A.: Improving quality of public domain digital elevation models through data fusion, Biosyst. Eng., 101, 293–305, 2008.

Li, P., Shi, C., Li, Z., Muller, J., Drummond, J., Li., X., Li, T., Li, Y., and Liu, J.: Evaluation of ASTER GDEM using GPS bechmarks and SRTM in China, Int. J. Remote Sens., 34, 1744–1771, 2013.

Lucca, S.: Validation and fusion of digital surface models, PhD. thesis, Ph.D. Course in Environmental and Infrastructure Engineering, Department of Environmental, Hydraulic, Infracstructure and Surveying Engineering, Polytechnic University of Milan, Milan, Italy, 2011.

Mitasova, H., Mitas, L., Brown, W. M., Gerdes, D. P., Kosinovsky, I., and Baker, T.: Modeling spatially and temporally distributed phenomena: new methods and tools for GRASS GIS, Int. J. Geogr. Inf. Syst., 9, 433–446, doi:10.1080/02693799508902048, 1995.

Mukherjee, S., Joshi, P. K., Mukherjee, S., and Ghosh, A.: Evaluation of vertical accuracy of open source digital elevation model (DEM), Int. J. Appl. Earth Obs., 21, 205–217, 2013a.

Mukherjee, S., Mukherjee, S., Garg, R. D., Bhardwaj, A., and Raju, P. L. N.: Evaluation of topographic index in relation to terrain roughness and DEM grid spacing, J. Earth Syst. Sci., 122-3, 869–886, 2013b.

Nguyen, L. N. and Le, H. V.: Methods for positioning ellipsoid over the Vietnam area based on EGM96 model, Ho Chi Minh University of Technology, Ho Chi Minh, Vietnam, 1–9, 2002.

Papasaika, H., Kokiopoulou, E., Baltsavias, E., Schindler, K., and Kressner, D.: Fusion of digital elevation models using sparse representation, Photogrammetric Image Analysis, Lect. Notes Comput. Sc. 6952, Springer, Berlin, Heidelberg, 171–184, 2011.

Popit, T. and Verbovsek, T.: Analysis of surface roughness in the Sveta Magdalena paleo-landslide in the Rebrnice area, RMZ_Materials Geoenviron., 60, 197–204, 2013.

Reuter, H. I., Nelson, A., and Jarvis, A.: An evaluation of filling interpolation methods for SRTM, Int. J. Geogr. Inf. Sci., 21, 983–1008, 2007.

Ravibabu, M., Jain, K., Singh, S., and Meeniga, N.: Accuracy improvement of ASTER stereo satellite generated DEM using texture filter, Geospat. Inf. Sci., 13, 257–262, 2010.

Stepinski, T. F. and Jasiewicz, J.: Geomorphons – a new approach to classification of landforms, Geomorphometry.org/2011, 7–11 September, Redlands, CA, 109–112, 2011.

Stevenson, J. A., Sun, X., and Mitchell, N. C.: Despeckling SRTM and other topographic data with a denoising algorithm, Geomorphology, 114, 238–252, doi:10.1016/j.geomorph.2009.07.006, 2009.

Sun, X., Rosin, P. L., Martin, R. R, and Langbein, F. C.: Fast and effective feature-preserving mesh denoising, IEEE T. Vis. Comput. Gr., 13, 925–938, doi:10.1109/TVCG.2007.1045, 2007.

Suwandana, E., Kawamura, K., Sakuno, Y., Kustiyanto, E., and Raharjo, B.: Evaluation of ASTER GDEM 2 in comparison with GDEM 1, SRTM and topographic map derived DEM using inundation area analysis and RTK-d GPS data, Int. J. Remote Sens., 4, 2419–2431, doi:10.3390/rs4082419, 2012.

Tran, T. A., Raghavan, V., Masumoto, S., and Yonezawa, G.: Enhancing quality of free DEM in Danang city, Vietnam and evaluating the suitability for terrain analysis, JSGI Geoinform., 24, 54–55, 2013a.

Tran, T. A., Raghavan, V., Yonezawa, G., Nonogaki, S., and Masumoto, S.: Enhancing quality of global DEMs for geomorphological analysis, case study in Danang city, Vietnam, SC02.830-837, 34th Asian Conference on Remote Sensing, 20–24 October 2013, Bali, Indonesia, 2013b.

Yang, M. and Moon, W. M.: Decision level fusion of multi-frequency polarimetric SAR and optical data with Dempster Shafer evidence theory, in: IGRASS, 21–25 July 2003, Toulouse, France, 2003.

Zandbergen, P.: Applications of shuttle radar topography mission elevation data, Geogr. Compass, 2, 1404–1431, 2008.

Zhao, S., Cheng, W., Zhou, C., Chen, X., Zhang, S., Zhow, Z., Liu, H., and Chai, H.: Accuracy assessment of the ASTER GDEM and SRTM3 DEM, an example in the Loess Plateau and North China Plain of China, Int. J. Remote Sens., 32, 1–13, doi:10.1080/01431161.2010.532176, 2011.

3

Seismic constraints on dynamic links between geomorphic processes and routing of sediment in a steep mountain catchment

A. Burtin[1], **N. Hovius**[1], **B. W. McArdell**[2], **J. M. Turowski**[1], and **J. Vergne**[3]

[1]GeoForschungsZentrum, Helmholtz Centre Potsdam, Potsdam, Germany
[2]Swiss Federal Institute for Forest, Snow and Landscape Research WSL, Birmensdorf, Switzerland
[3]École et Observatoire des Sciences de la Terre, CNRS UMR7516, Strasbourg, France

Correspondence to: A. Burtin (burtin@gfz-potsdam.de)

Abstract. Landscape dynamics are determined by interactions amongst geomorphic processes. These interactions allow the effects of tectonic, climatic and seismic perturbations to propagate across topographic domains, and permit the impacts of geomorphic process events to radiate from their point of origin. Visual remote sensing and in situ observations do not fully resolve the spatiotemporal patterns of surface processes in a landscape. As a result, the mechanisms and scales of geomorphic connectivity are poorly understood. Because many surface processes emit seismic signals, seismology can determine their type, location and timing with a resolution that reveals the operation of integral landscapes. Using seismic records, we show how hillslopes and channels in an Alpine catchment are interconnected to produce evolving, sediment-laden flows. This is done for a convective storm, which triggered a sequence of hillslope processes and debris flows. We observe the evolution of these process events and explore the operation of two-way links between mass wasting and channel processes, which are fundamental to the dynamics of most erosional landscapes. We also track the characteristics and propagation of flows along the debris flow channel, relating changes of observed energy to the deposition/mobilization of sediments, and using the spectral content of debris flow seismic signals to qualitatively infer sediment characteristics and channel abrasion potential. This seismological approach can help to test theoretical concepts of landscape dynamics and yield understanding of the nature and efficiency of links between individual geomorphic processes, which is required to accurately model landscape dynamics under changing tectonic or climatic conditions and to anticipate the natural hazard risk associated with specific meteorological events.

1 Introduction

Geomorphic processes seldom occur in isolation. Instead, multiple processes acting on different parts of the landscape tend to occur together, in linked, two-way fashion during geomorphic events. The nature and efficiency of these interactions determines landscape response to external forcing. Hillslopes and channels in active landscapes are coupled through the effects of sediment transfer (Whipple, 2004). Hillslope processes supply sediment to streams (Hovius et al., 2000), which use it to carve their channel beds (Sklar and Dietrich,

2001; Attal and Lavé, 2006; Turowski et al., 2007; Cook et al., 2013). Channel erosion, in turn, can undercut hillslopes and cause further slope erosion (Densmore et al., 1997). This two-way link between channels and slopes permits the tectonic deformation of river long profiles (Burbank et al., 1996; Snyder et al., 2000; Attal et al., 2008) and climatic forcing to affect erosion on adjacent hillslopes (Korup et al., 2010). Similarly, the impact of climate on mass wasting can propagate downward into the fluvial system (Page et al., 1994; Wobus et al., 2010), adjusting the balance of river sediment load and transport capacity and associated channel dynamics

(Hartshorn et al., 2002; Stark et al., 2010). Even on the scale of an individual rainstorm, the transfer of sediment from hillslopes to channels and the effects of the resultant flow on the surrounding topography can propagate the impact of localized erosion to locations far outside its original footprint.

Despite their fundamental importance to the dynamics of landscapes, observational constraints on the links between geomorphic processes and the progress of eroded material are scarce (Yanites et al., 2011) because remote sensing and in situ monitoring of geomorphic activity do not have the required resolution. Remotely sensed imagery has a spatial resolution at the metre-scale (Hervas et al., 2003; Lin et al., 2004; Saba et al., 2010) but a temporal resolution that depends on the timing of overhead passage and also on cloud cover. This is not sufficient to constrain the way in which individual processes are linked in a single geomorphic event. In contrast, ground-based monitoring that includes in situ observations provides the required temporal characteristics but tends to have a spatial extent that does not cover geomorphic process systems in their entirety (Itakura et al., 2005; McArdell et al., 2007). For example, downstream, in-channel monitoring can yield frequent and localized measurements of flow properties that result from the integration of various upstream processes, but does not generally allow for this signal to be deconvolved in order to establish the pattern of geomorphic activity in the contributing catchment. To improve constraints on landscape dynamics at the catchment scale, it is, therefore, required to have observations with a sufficient spatial resolution to determine where individual geomorphic processes occur and a high temporal resolution constraining the timing of their occurrence as well as their interplay.

Seismological data have the potential to enhance high-resolution surveys of landscape dynamics. Like any environmental process, geomorphic activity displacing mass along the Earth's surface produces ground vibrations that are recorded at distant seismometers (Govi et al., 1993; Brodsky et al., 1999; Burtin et al., 2009; Lacroix et al., 2012). Seismic instruments operate with a high sampling rate, giving data coverage, potentially for years, at high temporal resolution. Moreover, with several sensors, the respective timing of seismic signals at individual stations allows the location of geomorphic sources. Finally, the amplitude and frequency characteristics of seismic signals allow the identification of individual processes (Huang et al., 2007; Burtin et al., 2013).

Thus, where background seismic noise is weak relative to the geomorphic signal, seismic records can be used to resolve erosion and sediment transport with useful spatiotemporal detail. Such an approach has been used to study incidents of landslide motion (Deparis et al., 2008; Favreau et al., 2010), rock avalanches (Dammeier et al., 2011), debris flow (Itakura et al., 2005; Arattano and Marchi, 2008; Badoux et al., 2009) and bedload transport (Burtin et al., 2010, 2011; Hsu et al., 2011). However, these studies have not typically considered the interplay of different geomorphic processes at the landscape scale. With a two-dimensional network of

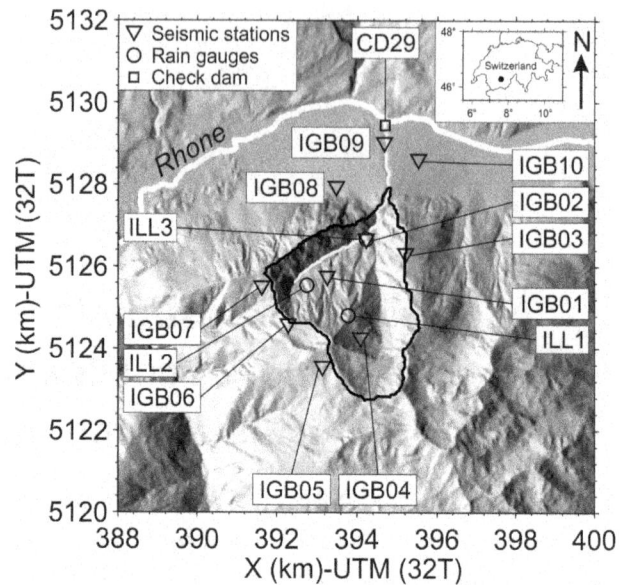

Figure 1. The Illgraben catchment. Location of the Illgraben catchment (~ 10 km², outlined in black) in Switzerland (dot in the inset map) and of the seismological stations deployed there during summer 2011 (inverse triangles, labelled IGBnn), meteorological stations from the Swiss Federal Institute for Forest, Snow and Landscape Research WSL (circles, labelled ILLn), and check dam 29 (CD29, square) where the flow depth and bedload impact rates of the study were observed.

seismometers, it is possible to scan for patterns of geomorphic activity across a landscape in continuous mode unlike any existing geomorphic technique (Burtin et al., 2013). We did this in the Illgraben, a steep mountain catchment in the Swiss Alps. With an array of ten instruments it was possible to track sediment moving from hillslopes into and along channels, obtaining constraints on the two-way link that exists between these two topographic domains. In addition, the analysis of seismic records along the main stream of the Illgraben catchment permitted observation of the downstream evolution of flow events arising from headwater erosion.

2 Experiment settings

2.1 Study area

The Illgraben catchment supplies 5–15 % of the sediment load of the Rhone River (Schlunegger et al., 2012) from a small catchment area of about 10 km² (Fig. 1). This high yield reflects the large catchment relief of > 2 km and slopes with an average gradient of 40° in fractured sedimentary rocks, making the Illgraben extremely prone to mass wasting and debris flows (Schlunegger et al., 2009). Flow events are commonly triggered in summer during convective rainstorms with measured 10 min rainfall intensities of up to 11.4 mm (Berger et al., 2011). They occur in a channel with mean slope of 16 % in bedrock that connects with the Rhone River across a debris fan with a gradient of 10 % (Badoux et al., 2009).

The channel is equipped with a debris flow monitoring system that uses geophone sensors bolted to three different check dams located inside the catchment. Sediment impacting, rolling or sliding on these check dams activates the geophones and if the recorded impulse rate exceeds a predetermined threshold, an alarm is triggered (Badoux et al., 2009). In addition, flow depth is monitored with laser sensors and sediment impact frequency with force plates at the outlet of the debris fan (CD29, Fig. 1) (McArdell et al., 2007), and flow events are registered by video cameras at this site. The setup is complemented by three automatic weather stations along an elevation transect in the catchment (ILL1-3, Fig. 1). This combination of frequent geomorphic events and existing instrumentation makes the Illgraben a suitable location for development and testing of seismic monitoring of geomorphic processes. For future deployment, seismic monitoring does not require this extreme rate of activity and can be used in locations where surface processes occur at more modest rates. Despite the anomalously high rate of erosion, the simple geomorphic structure, with a single trunk channel flanked by steep, dissected hillslopes, and the commonality of the dominant surface processes in the catchment should make our findings relevant and portable rather than unique. Notably, long-term observation of the Illgraben has given detailed insight into the meteorological preconditions for debris flow occurrence and flow mechanics (Schürch et al., 2011), but understanding of their origin and downstream evolution has remained difficult (Bennett et al., 2013).

2.2 Seismological data set

During the summer of 2011, we deployed ten seismometers in and around the Illgraben catchment (labelled IGB01 to IGB10 Fig. 1) in a 2-dimensional (2-D) geometry with an average instrument spacing of 2.88 km. Three stations were placed along the central channel to monitor flow processes, and the remaining seven were located in a ring that surrounded the catchment to record geomorphic activity on hillslopes. With this configuration, we aimed to record the activity on hillslopes and survey the spatiotemporal behaviour of flows in the Illgraben main channel. The seismic instruments were intermediate band Güralp CMG-6TD (IGB01, IGB03 to IGB07), Güralp CMG-40T (IGB02) and short period Lennartz LE-3Dlite (IGB08 to IGB10) seismometers. With the exception of the CMG-40T instrument, which has a flat response in the [0.033–50] Hz frequency band, all instruments had a flat response in at least the [1–100] Hz frequency band. This high-frequency band is well suited to the study of geomorphic processes (e.g. Helmstetter and Garambois, 2010; Burtin et al., 2011). The sampling frequency rate was set to 200 samples per second (SPS) for the intermediate band sensors and 125 SPS for the short period instruments.

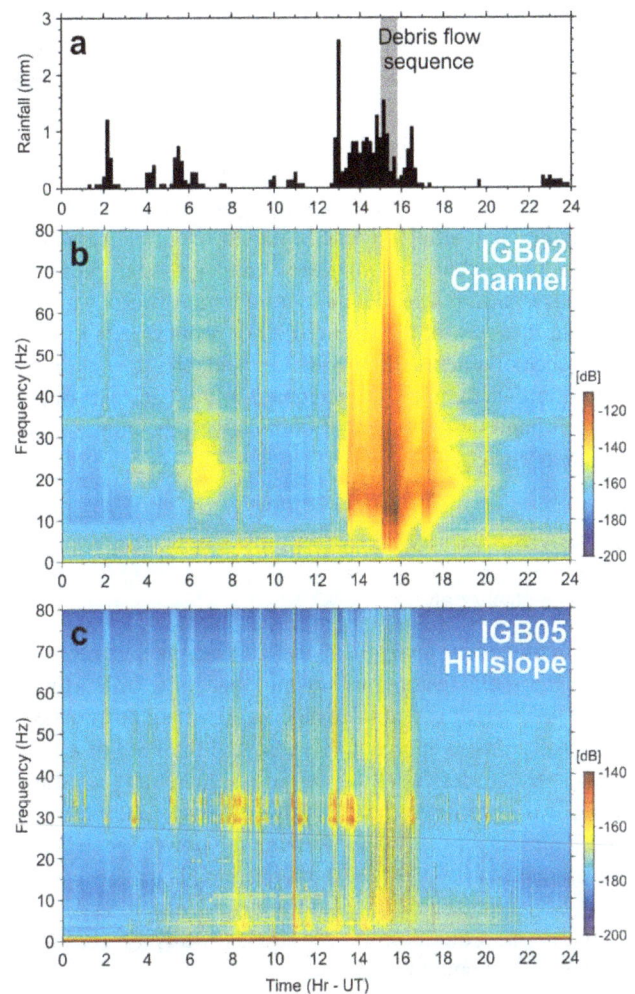

Figure 2. Spectral and temporal characterizations of surface processes. (**a**) Mean 10 min rainfall intensities recorded in the Illgraben catchment (gauges ILL1–3) on 13 July 2011. The gray shaded interval delineates the occurrence of debris flows. (**b–c**) Spectrograms for the same time period of the seismic signals recorded at station IGB02, located along the main channel, and station IGB05, outside the catchment. The amplitude is given in decibel relative to the velocity.

The seismic stations recorded for up to 100 days, but here we focus on a single rainstorm on 13 July 2011, which had the largest daily cumulative rainfall of the summer with 27 mm (Fig. 2a). This storm caused a debris flow that triggered the warning system. The debris flow propagated through the Illgraben catchment and over the debris fan and eventually entered the Rhone River.

2.3 Rainfall record

During 13 July 2011, the rainstorm did not show large temporal and spatial variability over the catchment. The correlation coefficients between the three rain gauges ranged from

0.84 to 0.91. Therefore, the precipitation rates averaged over these three locations well illustrate the meteorological conditions of this day in the Illgraben catchment. Some precipitation occurred early in the day, preceding a convective rainstorm in the afternoon with 2/3 of the daily rainfall total (Fig. 2a). Compared to historical records for the Illgraben, this convective storm with a total precipitation of 18 mm was not exceptional; similarly sized rain storms are not uncommon in the catchment during the summer season. Notably, the peak 10 min rainfall intensity of 2.6 mm, which was observed at the start of the storm, did not result directly in the occurrence of a debris flow. Instead, the debris flow warning system in the Illgraben was triggered in the later part of the storm, when 60 % (11 mm) of cumulative precipitation had already occurred.

3   Data processing

3.1   Spectral analysis

A time–frequency analysis of the continuous seismic data shows the main features of seismic signals (e.g. Burtin et al., 2008). Spectrograms were calculated with a power spectral density (PSD) approach. To compute the PSD of a time series, we first detrended the seismic signal, subtracted the mean, and deconvolved the instrument response. Then we used a multitaper method to estimate the power spectrum in one-minute time windows without overlap (Thomson, 1982). This PSD estimate offers a good frequency resolution despite the use of short duration seismic signals, which decrease the number of computed frequencies in a spectrum.

3.2   Event location method

Spectrograms of our seismic records show a number of events with durations of no more than a few tens of seconds, high-frequency content, and complex source time functions. These are the principal seismic characteristics of rockfalls (Helmstetter and Garambois, 2010). Such falls constitute well-defined sources of seismic energy, and it should be possible to determine their location. However, locating is made difficult by the high-frequency content and complexities in the source time functions of the signals of rockfalls and other erosion processes, which suppress coherence of waves between stations (Burtin et al., 2009; Lacroix and Helmstetter, 2011). Moreover, it is not possible to consistently identify specific seismic wave types such as body or surface waves in the absence of constraints on the velocity structure of the survey area, precluding the drawing of ray paths in the medium. Nevertheless, the location of hillslope processes can be determined with methods based on cross-correlation of seismic waveforms or envelops. Two types of approaches exist. One employs the maximization of coherent seismic signals (waveforms or envelops) and is called beamforming method (Almendros et al., 1999; Lacroix and Helm-

stetter, 2011). We employed another approach based on probability density function, which is computationally cheaper. It uses the cross-correlation of seismic signals between stations to determine the time delays that give optimally coherent observations across the instrument network (Burtin et al., 2009). Then the migration of these observed time delays, that is, the conversion from time to distance, can be used to retrieve the origin of an event. The cross-correlation of wave packets may include a combination of body and surface waves. For this reason, we preferred the use of a simple ballistic propagation, taking into account the topography in our migration procedure.

Specifically, for N available stations we first detrended the vertical seismic signal, removed the mean and deconvolved the instrument response. Next, we identified the frequency band with the maximum signal-to-noise ratio (SNR) for a given event. This was done by exploring frequencies ranging from 1 to 45 Hz, the dominant frequency band for hillslope processes, with a bandwidth increasing from 0.5 to 10 Hz. The seismic signals were bandpass filtered and we kept the results with the highest average SNR for all stations combined. Prior to computation of time delays in the selected frequency band, we normalized the time series to their maximum amplitude. For a pair of stations with index i_1 and i_2, we cross-correlated the seismic recordings and determined the time delay $dt_{obs}^{i_1 i_2}$ that corresponds to the maximum amplitude of the correlation function envelope. The time range of exploration should take into account the distance between stations i_1 and i_2 with respect to the topography $dl^{i_1 i_2}$ and the presumed propagation velocity V. Therefore, it corresponds to

$$dt_{obs}^{i_1 i_2} \in \left[-dl^{i_1 i_2}/V, \ +dl^{i_1 i_2}/V \right]. \tag{1}$$

With a set of $N(N-1)/2$ time delays, we implemented a migration step to convert time delays into distances for the event location, using a ballistic propagation (constant velocity) that takes into account the topography of the Illgraben catchment. The ray paths were assumed to follow the surface topography if it is the shortest path, and otherwise to cut through substrate (Fig. S1).

For each grid point (x, y) of the domain, we compared the calculated time delay dt_{calc} and the observed time delay dt_{obs} for an event source at the surface according to the probability density function

$$\rho_d(x, y, V) = \sum_{i_1=1}^{N-1} \sum_{i_2=i_1+1}^{N} e^{\left[-\frac{(dt_{calc}^{i_1 i_2} - dt_{obs}^{i_1 i_2})^2}{2\sigma_{dt}(V)^2} \right]}, \tag{2}$$

where $\sigma_{dt}(V)$ is the time error. We allowed this parameter to vary with the velocity in order to conserve a constant distance error of 0.2 km. A larger value would give event locations with a large uncertainty, whereas setting a smaller distance error might negatively affect the ability to properly locate an event. Since the propagation velocity is unknown, we explored a wide range of possible values, from 0.2–1.5 km s^{-1}, for high-frequency seismic waves travelling near the surface. To increase the accuracy of the location method, we introduced an a priori probability density function $\rho_m(x,y)$, which is centred on the location of the station that first recorded the arrival of the event, following the expression

$$\rho_m(x,y) = e^{\left[-\frac{\left(x-x_{sta}^{first}\right)^2+\left(y-y_{sta}^{first}\right)^2}{2\sigma_{prior}^2}\right]}. \quad (3)$$

where x_{sta}^{first} and y_{sta}^{first} are the coordinates of the seismic station, and σ_{prior} is the error on the assumption. This error was set at 1.60 km, the mean value of the inter-station distance of the three nearest stations of the Illgraben array. Hence, the final probability density function $\rho_{final}(x,y,V)$ is given by the relation

$$\rho_{final}(x,y,V) = \rho_d(x,y,V) \times \rho_m(x,y). \quad (4)$$

We then looked for the maximum amplitude of $\rho_{final}(x,y,V)$ to retrieve the best propagation velocity (V_{best}) and location of the event. To delimit the most likely location, we normalized $\rho_{final}(x,y,V)$ to the maximum amplitude and kept grid points that exceeded an arbitrary, conservative threshold of 0.75 (note, 0.95 is customary in seismic location methods).

To determine the time delays between seismic stations, we tested two methods based on the cross-correlation of seismic waveforms and seismic envelopes commonly used for landslides and non-volcanic tremors (e.g. Burtin et al., 2009; Zhang et al., 2010). Differences between the use of seismic waveforms and envelopes were generally but not always small. Figure S2 shows the vertical seismic signals at two paired stations (IGB03-IGB04 and IGB01-IGB05) for a rockfall event (rock 1, see Sect. 4) to which we paid particular attention, together with the cross-correlation of seismic waveforms and seismic envelopes. For the station pair IGB03-IGB04, the difference between the two methods was only 0.14 s (Fig. S2e–f). For the pair IGB01-IGB05 and in the interval dt_{obs}, which is coherent with the distance between stations and the best fit velocity (0.5 km s^{-1}, see Sect. 4), the time difference was 0.42 s (Fig. S2g–h). A difference of this magnitude has a limited impact on the accuracy of the location. However, the peak of amplitude of the cross-correlation function from envelopes was not located in the interval dt_{obs} of exploration (Eq. 1). It registered instead with a delay of 7.6 s (Fig. S2h) and was not detectable on the cross-correlation function computed from the seismic waveforms (Fig. S2g). This discrepancy is not problematic since the peak is out of the interval of exploration. However, for ballistic velocities of 0.2 and 0.3 km s^{-1}, this peak coincided with the best time delay for the pair IGB01-IGB05. This could give rise to merger or interference with peaks from other station couples and would influence the accuracy of event location. Although such a detailed analysis was not made systematically, we think that the observed behaviour may be representative. Since the use of a cross-correlation of seismic waveforms gave better constraints on the location,

we gave preference to this approach rather than the cross-correlation of seismic envelopes.

4 Rockfalls and flow pulses

4.1 Seismic signals and sources in the Illgraben

Daily spectrograms for 13 July 2011 at stations IGB02 and IGB05 illustrate the main characteristics of the seismic signal in the Illgraben catchment (Fig. 2). Along the stream, IGB02 recorded episodes of elevated high-frequency seismic energy that are consistent with the occurrence of rainfall. The episode with highest energy recorded at this station coincided with the convective afternoon storm and the ensuing flow sequence, which activated the debris flow detection and warning system of the Illgraben at 17:15 local time. This time coincidence of meteorological events, seismically recorded activity and independent flow detection are initial evidence for a seismic signal induced by channel processes. Lasting for 6 h, the seismically recorded channel activity is likely to have included bedload-transporting flows as well as debris flows, both of which can be registered by the warning system (Badoux et al., 2009).

Away from the channel, stations like IGB05 did not exhibit such long period activity (Fig. 2). During the convective rainstorm, they recorded discrete bursts of high-frequency seismic energy (> 1 Hz) lasting several tens of seconds. On seismograms, these events have short, impulsive peaks associated with multiple sources located at or near the surface. These characteristics are common for rockfalls and rock avalanches (Deparis et al., 2008; Helmstetter and Garambois, 2010; Dammeier et al., 2011). The potential increase of high-frequency seismic energy induced by the channel activity may prevent detecting the rockfall activity (Burtin et al., 2013). However, the stations deployed around the catchment, like IGB05, are not affected by such a source. Indeed, in contrast to other locations like the Himalayas (Burtin et al., 2008) or Taiwan (Hsu et al., 2011), the Illgraben channel has a limited extent, which reduces its capacity to produce an elevated background seismic noise. Therefore, the possible occurrence of a debris flow does not alter our ability to detect a significant slope activity.

High-frequency seismic signals can also have an anthropogenic origin (McNamara and Bulland, 2004). However, short-duration human disturbances in the Illgraben catchment were mainly restricted to the occasional passage of hikers, whose signals are only recorded over distances of tens of metres from a station. Noise from traffic, construction and gravel mining at the periphery of the Illgraben array was limited to a specific, narrow frequency band [2–4] Hz (Fig. 2b) and for the seismic stations deployed on the debris fan. Short-duration anthropogenic signals were not typically recorded at multiple stations in the Illgraben. In contrast, hillslope processes with larger magnitudes were observed by the entire seismic array. Such hillslope events were well expressed in

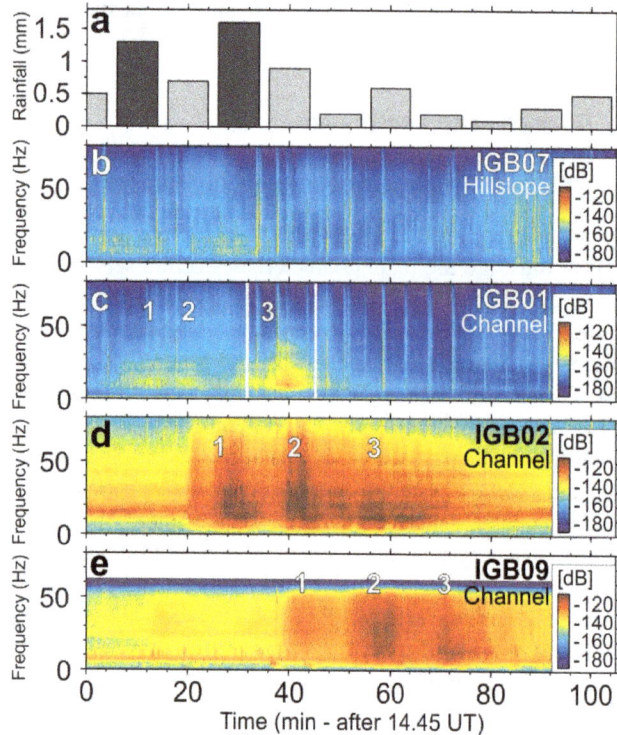

Figure 3. Seismic records of principal geomorphic activity in the Illgraben associated with rainfall on July 13, 2011. (a) Mean of 10 min rainfall intensity recorded at stations ILL2 and ILL3, inside the Illgraben (see Fig. 1). (b–e) Spectrograms in decibel of the vertical seismic signal at stations IGB07 (b), IGB01 (c), IGB02 (d) and IGB09 (e). Note the downstream propagation of seismic energy pulses 1–3. Propagation velocities ranged from 1.0–4.5 m s⁻¹. Vertical white lines on (c) delimit the time span of Fig. 4.

Figure 4. Linked mass wasting and channel flow in the upper Illgraben during flow pulse 3. (a) Spectrogram of the vertical seismic signal at station IGB01 during the flow pulse 3. The seismic energy is given in decibel relative to the velocity. Two rock avalanches (rocks 1 and 2) caused a short, sharp increase of the seismic energy at high-frequency (> 1 Hz). The gradual increase of the seismic energy over the time interval shown here reflects the increase of channel activity and the approach and passage of a flow pulse. (b–c) Vertical [1–50] Hz bandpass filtered seismograms at stations IGB01 (b) and IGB04 (c). Note the absence of channel-induced seismic signals at station IGB04 and the prominence of signals from rockfalls 1 and 2 at both stations.

the spectrograms of stations IGB01 and IGB07 (Fig. 3), located inside and at the high western periphery of the catchment, respectively. For the day of study, two local earthquakes were reported by the Swiss Seismological Service (SED). They both occurred in the morning at 04:30:59 UTC (Ml 0.7) and 05:56:24 UTC (Ml 1.2) at distances of 20 km and 28 km from the Illgraben catchment, respectively. Therefore, we can exclude local tectonic events as potential sources of the signals we recorded and as triggers of surface processes during the rainstorm.

4.2   Seismic anatomy of a debris flow sequence

Focusing on the flow sequence starting at 15:15 UTC, stations along the channel recorded seismic activity with a broad high-frequency content ([1–50] Hz) that occurred in three main episodes (pulses 1–3, Fig. 3). Lasting about 10 min each, these seismic energy pulses were timed progressively later at consecutive stations along the channel, showing the downstream propagation of their source. The coincidence between the occurrence of these propagating seismic pulses and debris flows reported by the warning system in the Illgraben

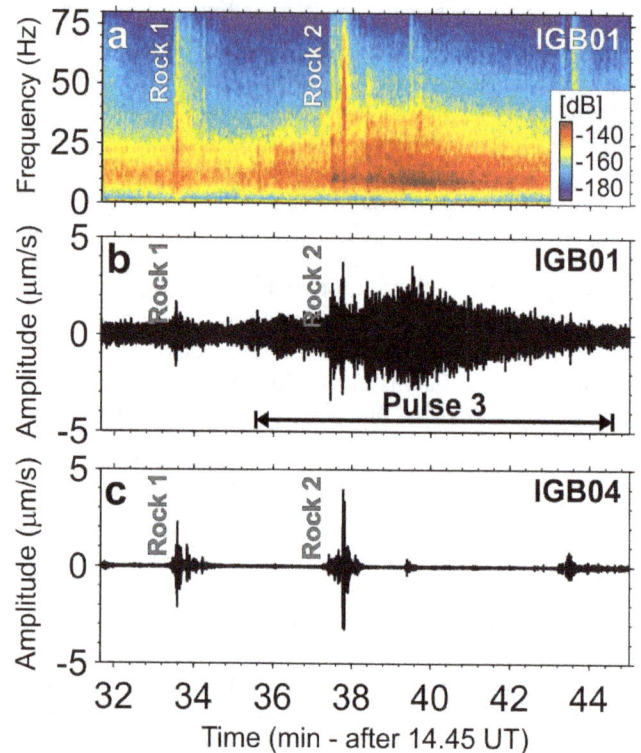

channel implies a link between them and we propose that the mobile seismic pulses represent the downstream propagation of three flow events. Furthermore, detailed analysis of the structure of these episodes reveals that at station IGB01, high on the central channel (Fig. 3c), the initial flow pulse had a gradual onset and was not clearly separated from the second pulse, even though they were so further downstream. In the upper Illgraben, the third flow pulse had a distinct and stronger seismic signal. Notably, this flow pulse was preceded by a short duration signal with rockfall characteristics, which was recorded at most stations (rock 1 at ~ 33 min, Fig. 4). The location of this event is key to understanding its possible connection to the third flow pulse.

Applying our location approach to the rock 1 event, we found that it occurred in the steep rock wall constituting the western flank of the catchment at an elevation of 1400–1900 m and within a 200 × 700 m area of uncertainty (Fig. 5).

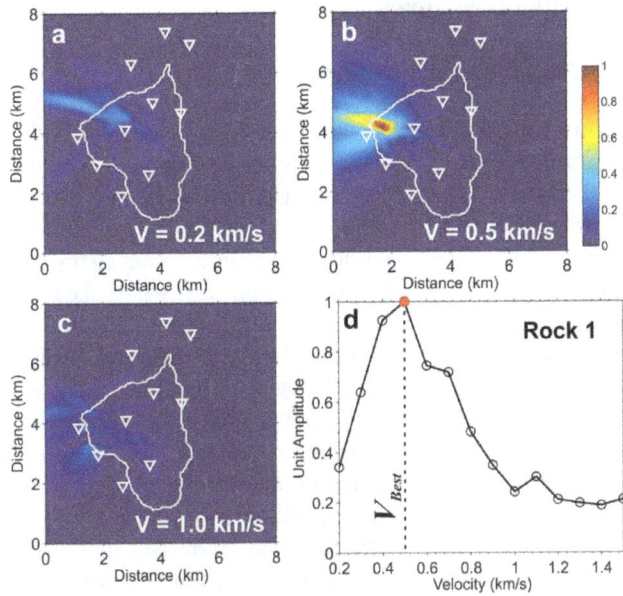

Figure 5. Location of mass wasting events in the Illgraben. (a–c) Probability density map (unit amplitude) for rock 1 location in the [29–29.5] Hz frequency band. Migration velocities are 0.2 (**a**), 0.5 (**b**) and 1.0 km s^{-1} (**c**). (**d**) Migration velocity analysis where the maximum amplitude corresponds to the best fit propagation velocity (0.5 km s^{-1} for rock 1).

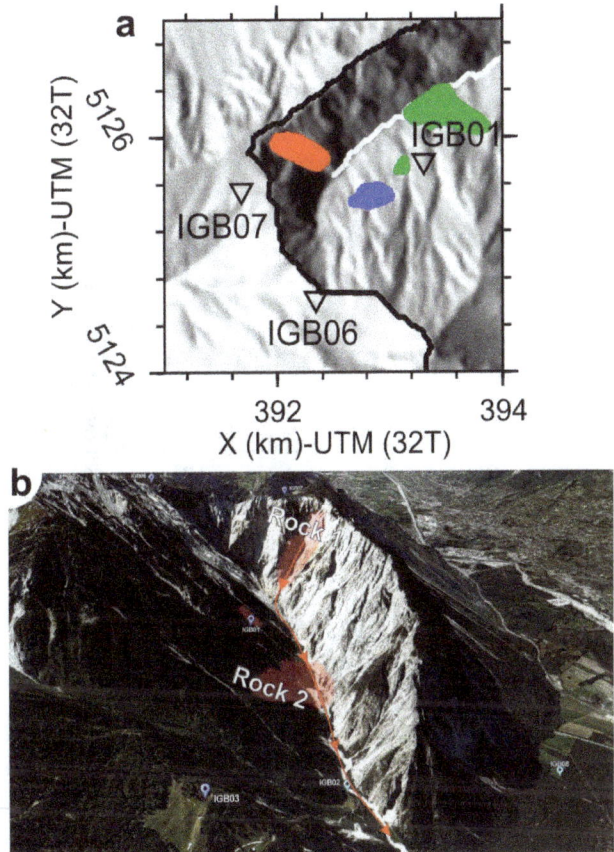

Figure 6. Location of hillslope events. (**a**) The likely location of mass wasting events related to pulse 3: rock 1 (red), rock 2 (green) and pulse 2 trigger event rock 0 (blue), all shown on a relief map of the Illgraben catchment. The colour patches indicate areas of equal probability for the event locations that correspond to the upper 75 % of the dynamic range from the event probability density maps (Fig. 4). Events located upstream stand on large, active gullies connected with the main stream. (**b**) Locations of rock avalanches (rocks 1 and 2), shown with the pathway of the flow pulse 3 with which they were associated (red curve). This series of events illustrates the two-way link between channel and slope domains.

The best-fit velocity for location of this event is 0.5 km s^{-1}, which is realistic for the propagation of shear or surface waves at shallow depths. The likely source area of rockfall rock 1 is connected to the uppermost section of the Illgraben channel, about 720 m upstream of station IGB01 (Fig. 6a). After a delay of about 160 s, an increase of seismic energy was observed at station IGB01, suggesting that rock 1 may have triggered flow pulse 3. The delay may reflect the time needed for the rockfall debris to become embedded within a channel flow and for that flow to arrive near IGB01.

During transit of flow pulse 3, a further significant, short duration event was detected at multiple stations (rock 2 at ~ 37 min, Fig. 4). This rockfall was located adjacent to the Illgraben channel, within a 400×750 m area of uncertainty, about 650 m downstream of station IGB01 (Fig. 6a). The best-fit velocity for location of this second avalanche is 0.6 km s^{-1}, which is consistent with the best-fit velocity for rock 1. This event may have been caused by ground vibrations or bank erosion during the passage of the sediment-laden flow pulse, and resulted in an immediate and sustained increase of 5 % dB in the [9–12] Hz seismic energy recorded at station IGB01. We attribute this increase to a sudden addition of sediment to the flow. Thus, our seismic data suggest that an effective, two-way link exists between the Illgraben channel and the surrounding hillslopes, whereby mass wasting during rainstorms can cause the constitution of a flow capable of transporting significant amounts of sediment, and this flow in turn can induce further mass wasting during pas-

sage (Fig. 6). Independent evidence for the occurrence and location of the seismically detected rockfalls in this sequence does not exist. However, the initiation zone coincides with active slopes in which rock avalanches have been observed previously, including a $3–5 \times 10^6$ m^3 rock avalanche in 1961 (Gabus et al., 2008; Bennett et al., 2013).

Observations on the other two flow pulses and further rockfalls that occurred during the same storm indicate that the connections between hillslope and channel processes and their role in the initiation of flow events in the Illgraben channel are diverse. Flow pulse 2 may have started in a similar way to flow pulse 3, with a rockfall (rock 0) detected in the southeast flank of the catchment (Fig. 6), in a slope known to be very active (Bennett et al., 2012). In contrast, flow pulse 1 was not directly associated with marked rockfall activity.

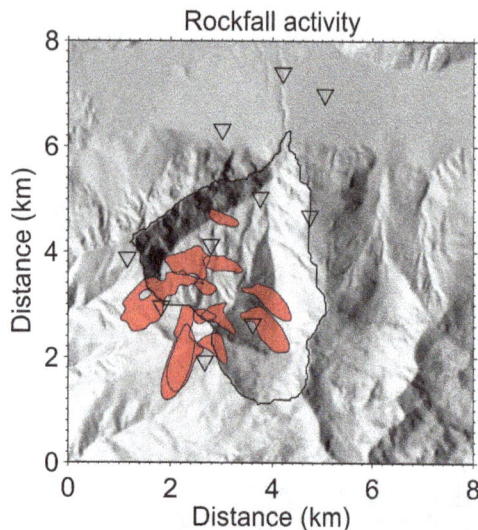

Figure 7. Location of seismic events during the convective rainstorm of 13 July 2011. The most likely area of initiation (> 0.75 of the dynamic range of the probability density function) for each event is represented in red. The station locations are indicated in inverse triangles. In contrast to rock 1 and rock 2, this slope activity cannot be associated with the initiation or downstream propagation of three flow pulses observed in the main Illgraben channel during the storm.

Instead, erosion of sediment from the headwater channel bed after sufficient runoff had accumulated may have caused this pulse. Neither flow pulse 1 nor 2 triggered any obvious secondary mass wasting during passage, but many other high-frequency short duration events were observed at most stations (Fig. 3). To locate these events, we applied the procedure also used for events rocks 0–2. In all, we managed to locate 12 further high-frequency short duration events. Within the bounds of uncertainty (> 0.75 of the dynamic range of the probability density function), most of them occurred within the Illgraben catchment (Fig. 7), many in the southeast flank, where intense erosion had been observed in previous years (Bennett et al., 2012). With temporal and frequency characteristics of rockfall events and in view of their spatial distribution in the Illgraben catchment, we have interpreted these events as rockfalls triggered by rainfall. One such event occurred in the immediate vicinity of the debris flow channel, close to the location of event rock 2 (Fig. 6), 38 min after the passage of the last flow pulse. This event may have been a bank collapse, involving colluvium exposed in the channel flank, after the passage of multiple flow pulses. All other detected rockfalls occurred without evident connection with channel processes. This illustrates the complexity of the coupling of slope and channel processes and the diversity of mechanisms by which debris flows can arise in the Illgraben.

5 Channel Dynamics

5.1 Flow propagation velocity

Having considered mass wasting in the steep flanks of the Illgraben, we now turn to activity in the central channel of the catchment. The link between pulses of seismic energy recorded at stream-side stations and the flow propagation in the Illgraben channel can be used to infer some characteristics of the channel dynamics. To do this, we tracked the downstream progress and evolution of the three flow pulses with seismic data from stations IGB01, IGB02 and IGB09, located closest to the channel (Fig. 1). The propagation velocity of flow pulses in the Illgraben channel was estimated from the envelopes of seismic energy recorded along the main stream. Prior to computing the seismic envelopes, we bandpass filtered the seismic signals of each component (vertical, north and east) between 5 and 20 Hz. We then averaged the three components of a station, and the passage of a pulse was assumed to coincide with the peak amplitude of the seismic envelopes at a station. Distances along the Illgraben channel were measured directly from ortho-rectified aerial photographs. For pulses 1 and 2, we fixed the start of propagation to match the pulse arrival at station IGB01. Flow pulse 3 is assumed to have started at the location and time of rockfall rock 1.

Seismically determined flow velocities ranged from 1.0–$4.5 \, \mathrm{m \, s^{-1}}$ and are within the range of measured debris flow velocities in the channel (0.8–$7 \, \mathrm{m \, s^{-1}}$) (Badoux et al., 2009). The propagation velocity showed some spatial variations with lower values of $\sim 1 \, \mathrm{m \, s^{-1}}$ inside the catchment (between IGB01 and IGB02) than on the debris fan ($\sim 4 \, \mathrm{m \, s^{-1}}$, Fig. 8 and Table S3). This observation may indicate that the effects of channel roughness dominated over those of channel slope in setting flow velocity. In contrast to this spatial pattern, the temporal variations were limited and the three flow pulses had similar velocity signatures.

5.2 Flow seismic energy

Despite the similarities in flow velocity, the energy level of seismic signals evolved between channel stations and differed between flow pulses (Fig. 8b). To properly compare station observations recorded at different distances from potential sources, in this case the stream, a first-order correction of the seismic energy must be applied, accounting for the geometrical spreading of seismic waves (e.g. Aki and Richards, 1980). We estimated the seismic energy along the main channel by power spectral density analysis and corrected for the travelled distance between a station and the channel as follows. For each station and each flow pulse, we computed the average seismic energy in the [5–20] Hz frequency band and in a $\pm 30 \, \mathrm{s}$ time window around the peak amplitude observed during passage of the pulse. The uncertainty on this estimate

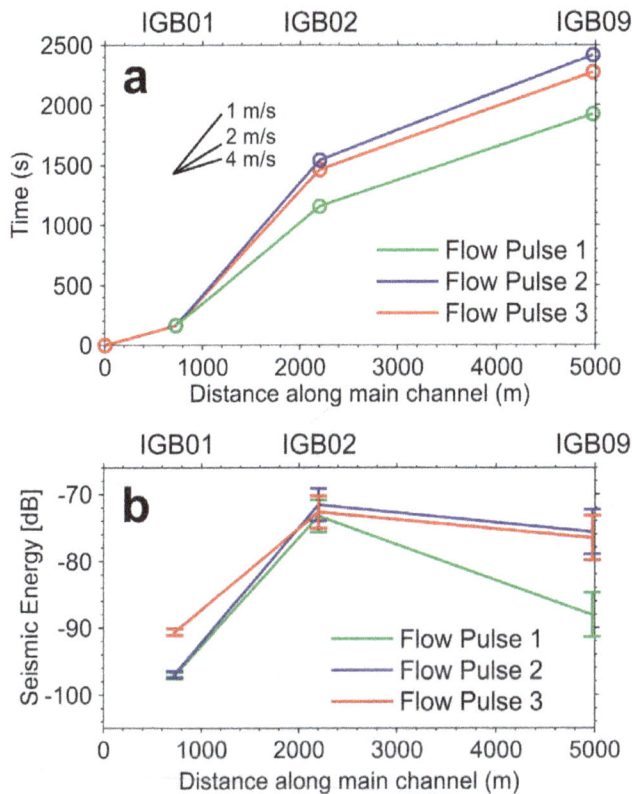

Figure 8. Velocity and energy characteristics of the debris flow sequence. (**a**) Downstream propagation of the three main flow pulses. The propagation is defined with the recordings of seismic envelopes at side-stream stations IGB01, IGB02 and IGB09. (**b**) Mean seismic energy of each flow pulse recorded at the same stations. The amplitude in decibel is corrected for the geometrical spreading of body waves.

was defined as the standard deviation from the mean seismic energy.

Taking the Illgraben channel as the principal source of energy, we calculated the average distance R from a station to the nearest 250 m stream segment. We applied this value to correct the seismic energy according to the propagation of body waves ($\sim 1/R^2$) because the channel stations are relatively close to the potential seismic sources and the travelled ray paths can be assumed to be relatively uniform in the near field. The largest distance to the channel was 432 m, at station IGB01. Stations IGB02 and IGB09 had an equal distance to stream of 121 m, making it possible to compare these two stations without correction. A geometrical spreading correction that would apply to surface waves would not lead to drastic changes in the observed trends. Naturally, a full correction should also consider the anelastic properties of the medium that account for the frequency dependence of wave attenuation. However, in the absence of an estimate of quality factors and an attenuation law, we did not attempt this correction. For a first order estimate of the energy, the correction for geometrical spreading and uniform anelastic medium

properties has to suffice at present. Hence, we do not interpret the absolute seismic energy, but instead associate the relative changes along the channel to the erosion, transfer and deposition of sediments. For the quantification of these channel processes, a careful analysis of the seismic wave content must be carried out to properly interpret the seismic energy, which is outside the scope of this study.

According to our observations, the seismic energy of all three flow pulses increased by 30–35 % dB between IGB01 and IGB02, inside the Illgraben catchment. In contrast, on the debris fan between IGB02 and IGB09, the energy decreased by 18 % dB for flow pulse 1, and only by 5 % dB for the flow pulses 2 and 3 (Fig. 8b). These variations reflect the evolution of the flows along the channel, perhaps indicating changes in the frictional characteristics of flood flows or an increase of flow discharge due to erosion or decrease in discharge due to deposition, both of which have been documented on the fan of the Illgraben (Schürch et al. 2011; Berger et al., 2011) for debris flows and debris floods.

5.3 Comparison with in situ monitoring

A comparison of the recorded seismic signals of the flow pulses with data from in situ stream monitoring yields further information about the flow properties and their evolution. For this purpose, we used data on flow depth and particle impact rate from check dam 29 (CD29, Fig. 1), located 400 m downstream of station IGB09. This distance implies a time delay between observations of flow pulses in the seismic data and flow depth and impact rate data. Moreover, a seismic sensor detects approaching flows before they reach the in-channel location nearest the station, giving rise to a progressive increase of registered seismic energy. In contrast, passage of a steep-fronted flow is registered as a sudden increase of flow depth at CD29. Acknowledging these differences, we compare the flow characteristics inferred from seismic and in-stream observations.

The spectrograms at near-channel stations showed notable shifts in the frequency content of signals and variations of seismic energy during the sequence of flow pulses (Fig. 3). At station IGB09, flow pulse 1 had relatively little seismic energy below 15 Hz, whereas pulses 2 and 3 had more energy at lower frequencies and greater seismic amplitudes (Fig. 9). In contrast, the flow depth at CD29 was similar for pulses 1 and 2 (Fig. 9c), and it peaked between pulses 2 and 3 when the seismic energy reached a temporary low (between 65 and 70 min, Fig. 9b). In addition, the flow depth of pulse 3 was relatively small, 45 % below the peak value, whereas seismic amplitude increased by 130 % for the same period. These comparisons indicate that there is no direct relation between seismic signals and the flow level, and that other flow attributes might be involved.

Bedload sediment transport is a likely source of seismic energy, which can be independently tracked from records of bedload impact rate. Flow pulse 1 had a relatively low bedload impact rate, 20 times less than flow pulse 2, even though these flows had similar depths and velocities (Figs. 9d and 8a, respectively). Meanwhile, the seismic amplitude increased by 215 % at IGB09 from flow pulse 1 to flow pulse 2. Flow pulse 3 had a moderate seismic amplitude and bedload activity. Thus, the recorded seismic amplitudes are in qualitative agreement with bedload observations rather than with flow depth. However, a clear relation between seismic amplitude and bedload impact rate is difficult to define because it is likely to be prone to grain size effects on the frequency content of the seismic signal. Moreover, the duration of a flow pulse can influence the seismic energy delivered in the streambed. A sharp, strong seismic peak can have a total energy equivalent to that of a long and smooth seismic pulse. Along the 4.25 km channel reach between IGB01 and IGB09, the three flow pulses had a similar seismic duration of 11 min (±104 s), with limited fluctuations of 5, 16 and 14 % (standard deviation) for pulses 1, 2 and 3, respectively (Fig. S4). Therefore, possible effects of stretching in time of the flow pulses can be discarded. The video camera located at CD29 recorded the passage of the debris-flow sequence. With a sampling rate of one picture per second, we noticed an elevation of flow level, as indicated by the flow height data set (Fig. 9c), but we could not extract additional information that could help to decipher the bedload fluctuation as shown by the impact rate data set (Fig. 9d).

If bedload transport has a dominant contribution to the seismic energy recorded along the Illgraben main stream, then the frequency pattern of the seismic signal should reflect an addition or loss of large sediment particles in the flow because large sediment particles produce lower frequency signals than small particles (Huang et al., 2007). At IGB09, flow pulse 1 had the lowest amplitude and little seismic energy below 15 Hz. These observations indicate a paucity of coarse bedload in the flow, which agrees with the fact that this flow pulse had low bedload impact rates at CD29 despite its relatively large discharge. Notably, pulse 1 underwent a strong reduction of seismic energy (18 % dB) across the fan, where the channel bed gradient decreases from 16 to 10 %, likely reflecting progressive deposition of sediment in the lower channel reach. This may have affected the coarsest sediment fraction in the first instance, explaining the subdued seismic activity in the channel on the distal part of the fan.

Despite a similar flow depth to flow pulse 1, flow pulse 2 had the highest seismic amplitude, with significant signal at frequencies below 15 Hz at station IGB09. Comparison of the seismic signal envelopes at IGB02 and IGB09 in the debris fan confirms the decay of amplitude for flow 1, whereas flow 2 remains the highest peak of the sequence (Fig. S4b–c). We attribute this to a greater sediment concentration and a high transport rate of coarse bedload in flow pulse 2. The increase of seismic energy, indicating higher impact energy, during pulse 2 highlights a higher potential for channel bed abrasion at the base of this denser flow. For pulse 3, the interpretation of available data is less straightforward. This pulse

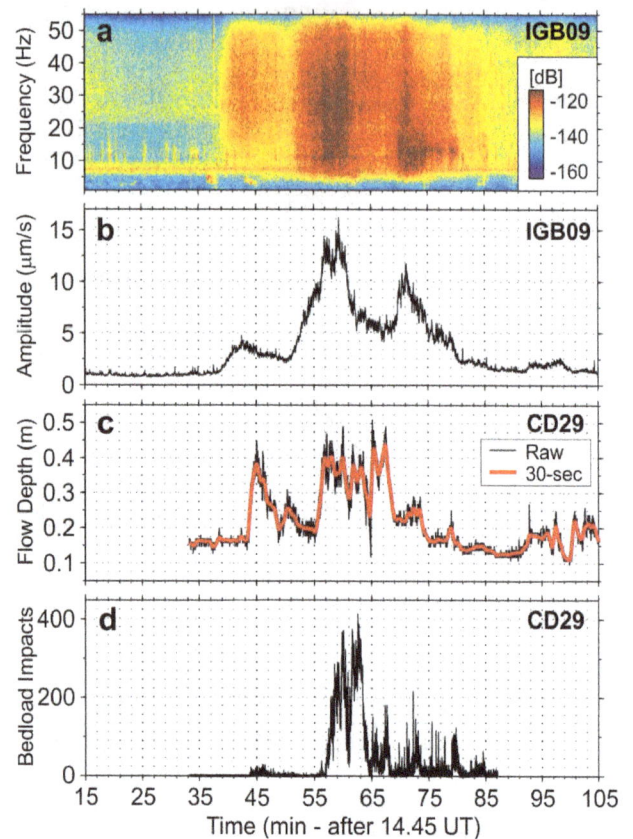

Figure 9. Flow pulse characteristics on the distal fan. **(a)** Spectrogram in decibel of the vertical seismic signal at station IGB09 during passage of flow pulses 1–3. **(b)** [5–50] Hz vertical seismic envelop at IGB09 showing three flow pulses. **(c)** Raw (black line) and 30 s smoothed (red line) flow depth data recorded at CD29, 400 m downstream of IGB09 (2 min at established flow speed). **(d)** Bedload impact rates recorded at CD29.

had the second highest seismic amplitude of the main flow sequence on this day with a substantial signal below 15 Hz, reflecting a substantial bedload transport rate. However, the pulse had only a limited flow depth and moderate bedload impact rates. Lower than expected impact rates may have resulted from a debris flow with different internal organization of the bedload material and/or from the way in which sediment particle impacts are recorded. This is done with force plates, which register an impact only when its force or acoustic amplitude (geophone) exceeds a pre-defined threshold. The absolute amplitude of such impacts is not recorded. Since the flow depth of pulse 3 was small, sediment particles may have had less energy at impact due to low drop heights and particle velocities, which could explain both reduced impact rates at CD29 and moderate seismic energy at IGB09.

6 Conclusions

Geomorphic processes generate seismic signals with distinct and different characteristics in the amplitude- and frequency-time domains, reflecting their mechanisms, granulometry, timing, location and velocity. Recording such signals with a two-dimensional seismological array, we have mapped the spatiotemporal patterns of geomorphic activity in the Swiss Illgraben, a steep mountain catchment with high erosion rates. Our array consisted of stations deployed around the catchment and along the main channel, allowing recognition, location and tracking of rockfalls on slopes as well as flow pulses in the central channel and revealing the links between individual processes. This has been done for a single convective storm.

During this storm on 13 July 2011, three separate flow pulses occurred within the Illgraben main channel, each with a significant sediment load and with the characteristics of a debris flow for at least part of the surveyed channel length. These pulses did not have common starting conditions, either in terms of precipitation, or in terms of the trigger mechanism. The first flow pulse started without detected precursor activity on catchment hillslopes, instead mobilizing sediment already present in the channel. In contrast, the other two pulses were triggered by rockfalls in the steep headwater slopes. Within the Illgraben catchment, we noticed a systematic energy increase along the bedrock channel, presumably in response to the entrainment of channel bed material (pulse 1) and/or hillslope inputs (pulses 2 and 3). During pulse 3, passage of the flow appeared to trigger a secondary mass wasting event on an adjacent hillslope. On the debris fan, pulse 1 underwent a decrease of seismic energy, whereas pulses 2 and 3 maintained their high level of energy. These trends may reflect changes in the sediment load of the propagating flows. The seismic records and independently measured particle impact rates suggest that pulse 1 had a diminished coarse sediment load, possibly causing change from a debris flow into a hyper-concentrated flow on the lower part of the debris fan. Pulses 2 and 3 maintained their energy and thus their character across the fan. Thus, our seismic observations suggest that within the time span of a single convective rainstorm, dynamic links exist between a channel and the adjacent hillslopes that can determine the onset and evolution of bedload transport in mountain catchments, and that sediment erosion and deposition during downstream propagation of these flows affect their density and rheology, and likely also their potential for erosion by particle impacts.

By recording frequency-specific amplitude information, seismic instruments register at distance many aspects of flow processes that can be confirmed with in situ observations from force plates. For many hillslope processes, such as rockfalls, quantitative in situ observation is disproportionately more difficult, and seismic records may provide insights into their mechanisms that are hard to obtain in other ways. Moreover, this seismological approach is effective on the landscape scale. The ensemble of seismic observations, made on individual, naturally occurring geomorphic process events that are tracked from inception to near termination, can reveal the ways in which separate landscape elements interact under specific meteorological conditions and how geomorphic events are constituted by multiple surface process manifestations with causal links. Thus, seismology, pursued with two-dimensional instrument networks, makes it possible for the first time to monitor distributed surface process activity with sufficient spatial as well as temporal resolution to observe and constrain the dynamics of erosional landscapes.

With telemetry and automated analysis of seismic data, this approach may give significant early warning capabilities in settings where natural hazard monitoring is now limited to localized downstream observations. Finally, combined seismological and meteorological monitoring of upland catchments over multiple annual cycles stands to yield fundamental, quantitative constraints on the role of weather as a driver of erosion and insights into the role of climate and climate change in landscape evolution. Such long-term surveys should include independent constrains on slope activity, like laser scanning, to verify locations of erosion and deposition and to calibrate the conversion from measured seismic energy to mass of rock or sediment displaced. This conversion is essential to achieving the goal of knowing the timing and location of geomorphic events in a landscape and how much material is involved.

Acknowledgements. This study was supported by the AXA Research Fund and the Isaac Newton Trust of the University of Cambridge. We thank the SEIS-UK equipment pool (NERC) and the École et Observatoire des Sciences de la Terre of Strasbourg for providing the seismic instruments, the WSL for logistic support and M. Raymond Pralong, K. Steiner, N. Federspiel, T. Glassey and F. Dufour for help in the field. Thanks are also due to A. Helmstetter, M. Attal, L. Hsu and the associate editor J. Braun for their comments and suggestions, which have improved the clarity of the paper.

Edited by: J. Braun

References

Aki, K. and Richards, P. G.: Quantitative Seismology: Theory and Methods, University Science Books, 2002, ISBN 978-1-891389-63-4, 700 pp., 2002.

Arattano, M. and Marchi, L.: Systems and sensors for debris-flow monitoring and warning, Sensors, 8, 2436–2452, doi:10.3390s8042436, 2008.

Attal, M. and Lavé, J.: Changes of bedload characteristics along the Marsyandi River (central Nepal): Implications for understanding hillslope sediment supply, sediment load evolution along fluvial networks, and denudation in active orogenic belts, edited by: Willett, S. D., Hovius, N., Brandon, M. T., and Fisher, D., Spec. Pap. Geol. Soc. Am., 398, 143–171, 2006.

Attal, M., Tucker, G. E., Whittaker, A. C., Cowie, P. A., and Roberts, G. P.: Modeling fluvial incision and transient landscape evolution: Influence of dynamic channel adjustment, J. Geophys. Res., 113, F03013, doi:10.1029/2007JF000893, 2008.

Badoux, A., Graf, C., Rhyner, J., Kuntner, R., and McArdell, B. W.: A debris-flow alarm system for the Alpine Illgraben catchment: design and performance, Nat. Hazards, 49, 517–539, 2009.

Bennett, G. L., Molnar, P., Eisenbeiss, H., and McArdell, B. W.: Erosional power in the Swiss Alps: characterization of slope failure in the Illgraben, Earth Surf. Process. Landforms, 37, 1627–1640, doi:10.1002/esp.3263, 2012.

Bennett, G. L., Molnar, P., McArdell, B. W., Schlunegger, F., and Burlando, P.: Patterns and controls of sediment production, transfer and yield in the Illgraben, Geomorphology 188, 68–82, 2013.

Berger, C., McArdell, B. W., and Schlunegger, F.: Direct measurement of channel erosion by debris flows, Illgraben, Switzerland, J. Geophys. Res., 116, F01002, doi:10.1029/2010JF001722, 2011.

Brodsky, E. E., Kanamori, H., and Sturtevant, B.: A seismically constrained mass discharge rate for the initiation of the May 18, 1980 Mount St. Helens eruption, J. Geophys. Res., 104, 29, 387–29, 1999.

Burbank, D. W., Leland, J., Fielding, E., Anderson, R. S., Brozovic, N., Reid, M. R., and Duncan, C.: Bedrock incision, rock uplift and threshold hillslopes in the northwestern Himalayas, Nature, 379, 505–510, 1996.

Burtin, A., Bollinger, L., Vergne, J., Cattin, R., and Nábělek, J. L.: Spectral analysis of seismic noise induced by rivers: A new tool to monitor spatiotemporal changes in stream hydrodynamics, J. Geophys. Res., 113, B05301, doi:10.1029/2007JB005034, 2008.

Burtin, A., Bollinger, L., Cattin, R., Vergne, J., and Nábělek, J. L.: Spatiotemporal sequence of Himalayan debris flow from analysis of high-frequency seismic noise, J. Geophys. Res., 114, F04009, doi:10.1029/2008JF001198, 2009.

Burtin, A., Vergne, J., Rivera, L., and Dubernet, P.-P.: Location of river induced seismic signal from noise correlation functions, Geophys. J. Int., 182, 1161–1173, 2010.

Burtin, A., Cattin, R., Bollinger, L., Vergne, J., Steer, P., Robert, A., Findling, N., and Tiberi, C.: Towards the hydrologic and bed load monitoring from high-frequency seismic noise in a braided river: the "torrent de St Pierre", French Alps, J. Hydrol., 408, 43–53, 2011.

Burtin, A., Hovius, N., Milodowski, D. T., Chen, Y.-G., Wu, Y.-M., Lin, C.-W., Chen, H., Emberson, R., and Leu, P.-L.: Continuous catchment-scale monitoring of geomorphic processes with a 2-D seismological array, J. Geophys. Res., 118, 1956–1974, doi:10.1002/jgrf.20137, 2013.

Cook, K. L., Turowski, J. M., and Hovius, N.: A demonstration of the importance of bedload transport for fluvial bedrock erosion

and knickpoint propagation, Earth Surf. Process. Landforms, 38, 683–695, doi:10.1002/esp.3313, 2013.

Dammeier, F., Moore, J. R., Haslinger, F., and Loew, S.: Characterization of alpine rockslides using statistical analysis of seismic signals, J. Geophys. Res., 116, F04024, doi:10.1029/2011JF002037, 2011.

Deparis, J., Jongmans, D., Cotton, F., Baillet, L., Thouvenot, F., and Hantz, D.: Analysis of rock-fall and rock-fall avalanche seismograms in the French Alps, Bull. Seismol. Soc. Am., 98, 1781–1796, 2008.

Densmore, A. L., Anderson, R. S., McAdoo, B. G., and Ellis, M. A.: Hillslope evolution by bedrock landslides, Science, 275, 369–372, 1997.

Favreau, P., Mangeney, A., Lucas, A., Crosta, G., and Bouchut, F.: Numerical modeling of landquakes, Geophys. Res. Lett., 37, L15305, doi:10.1029/2010GL043512, 2010.

Gabus J. H., Weidmann, M., Bugnon P.-C., Burri, M., Sartori, M., and Marthaler, M.: Geological Map of Sierre 1:25 000 (LK 1278, sheet 111), In Geological Atlas of Switzerland. Swiss Geological Survey: Wabern, Switzerland, 2008,

Govi, M., Maraga, F., and Moia, F.: Seismic detectors for continuous bed load monitoring in a gravel stream, Hydrol. Sci. J., 38, 123–132, 1993.

Hartshorn, K., Hovius, N., Dade, W. B., and Slingerland, R. L.: Climate-driven bedrock incision in an active mountain belt, Science 297, 2036–2038, doi:10.1126 science.1075078, 2002.

Helmstetter, A. and Garambois, S.: Seismic monitoring of Séchilienne rockslide (French Alps): analysis of seismic signals and their correlation with rainfalls, J. Geophys. Res., 115, F03016, doi:10.1029/2009JF001532, 2010.

Hervas, J, Barredo, J. I., Rosin, P. L., Pasuto, A., Mantovani, F., and Silvano, S.: Monitoring landslides from optical remotely sensed imagery: the case history of Tessina landslide, Italy, Geomorphology, 54, 63–75, 2003.

Hovius, N., Stark, C. P., Chu, H.-T., and Lin, J.-C.: Supply and Removal of Sediment in a Landslide-Dominated Mountain Belt: Central Range, Taiwan, The Journal of Geology, 118, 73–89, doi:10.1086/314387, 2000.

Hsu, L., Finnegan, N. J., and Brodsky, E. E.: A seismic signature of river bedload transport during storm events, Geophys. Res. Lett., 38, L13407, doi:10.1029/2011GL047759, 2011.

Huang, C.-J., Yin, H.-Y., Chen, C.-Y., Yeh, C.-H., and Wang, C.-L.: Ground vibrations produced by rock motions and debris flows, J. Geophys. Res., 112, F02014, doi:10.1029/2005JF000437, 2007.

Itakura, Y., Inaba, H., and Sawada, T.: A debris-flow monitoring devices and methods bibliography, Nat. Hazards Earth Syst. Sci., 5, 971–977, 2005.

Korup, O., Densmore, A. L., and Schlunegger, F.: The role of landslides in mountain range evolution, Geomorphology, 120, 77–90, 2010.

Lacroix, P. and Helmstetter, A.: Location of seismic signals associated with microearthquakes and rockfalls on the Séchilienne landslide, French Alps, Bull. Seismol. Soc. Am., 101, 341–353, 2011.

Lacroix, P., Grasso, J.-R., Roulle, J., Giraud, G., Goetz, D., Morin, S., and Helmstetter, A.: Monitoring of snow avalanches using a seismic array: Location, speed estimation, and relationships to meteorological variables, J. Geophys. Res., 117, F01034, doi:10.1029/2011JF002106, 2012.

Lin, C.-W., Shieh, C.-L., Yuan, B.-D., Shieh, Y.-C., Liu, S.-H., and Lee, S. Y.: Impact of Chi-Chi earthquake on the occurrence of landslides and debris flows: example from the Chenyulan River watershed, Nantou, Taiwan, Engineering Geology, 71, 49–61, 2004.

McArdell, B. W., Bartelt, P., and Kowalski, J.: Field observations of basal forces and fluid pore pressure in a debris flow, Geophys. Res. Lett., 34, L07406, doi:10.1029/2006GL029183, 2007.

McNamara, D. E. and R. P. Buland: Ambient noise levels in the continental United States, Bull. Seismol. Soc. Am., 94, 1517–1527, 2004.

Page, M. J., Trustrum, N. A., and DeRose, R. C.: A high-resolution record of storm-induced erosion from lake sediments, New Zealand, J. Paleolimnol., 11, 333–348, 1994.

Saba, S. B., van der Meijde, M., and van der Werff, H.: Spatiotemporal landslide detection for the 2005 Kashmir earthquake region, Geomorphology, 124, 17–25, 2010.

Schlunegger F., Badoux A., McArdell B. W., Gwerder C., Schnydrig D., Rieke-Zapp D., and Molnar P.: Limits of sediment transfer in an alpine debris-flow catchment, Illgraben, Switzerland, Quat. Sci. Rev., 28, 1097–1105, 2009.

Schlunegger, F., Norton, K., and Caduff, R.: Hillslope processes in temperate environments, in: Treatise in Geomorphology 3, edited by: Marston, R. and Stoffel, M., Mountain and Hillslope Geomorphology, Elsevier, London, 2012.

Schürch, P., Densmore, A. L., Rosser, N. J., and McArdell, B. W.: Dynamic controls on erosion and deposition on debris-flow fan, Geology, 39, 827–830, 2011.

Sklar, L. and Dietrich, W. E.: Sediment and rock strength controls on river incision into bedrock, Geology, 29, 1087–1090, 2001.

Snyder, N. P., Whipple, K. X., Tucker, G. E., and Merritts, D. J.: Landscape response to tectonic forcing: Digital elevation model analysis of stream profiles in the Mendocino triple junction region, northern California, Geol. Soc. Am. Bull., 112, 1250–1263, 2000.

Stark, C. P., Barbour, J. R., Hayakawa, Y. S., Hattanji, T., Hovius, N., Chen, H., Lin, C.-W., Horng, M.-J., Xu, K.-Q., and Fukahata, Y.: The climatic signature of incised river meanders, Science, 327, 1497–1501, 2010.

Thomson, D. J.: Spectrum estimation and harmonic analysis, Proc. IEEE, 70, 1055–1096, 1982.

Turowski, J. M., Lague, D., and Hovius, N.: Cover effect in bedrock abrasion: a new derivation and its implications for the modeling of bedrock channel morphology, J. Geophys. Res., 112, F04006, doi:10.1029/2006JF000697, 2007.

Whipple, K. X.: Bedrock Rivers and the Geomorphology of Active Orogens, Ann. Rev. Earth Planet. Sci., 32, 151–185, 2004.

Wobus, C. W., Tucker, G. E., and Anderson, R. S.: Does climate change create distinctive patterns of landscape incision?, J. Geophys. Res., 115, F04008, doi:10.1029/2009JF001562, 2010.

Yanites, B. J., Tucker, G. E., Hsu, H.-L., Chen, C., Chen, Y.-G., and Mueller, K. J.: The influence of sediment cover variability on long-term river incision rates: An example from the Peikang River, central Taiwan, J. Geophys. Res., 116, F03016, doi:10.1029/2010JF001933, 2011.

Zhang, H., Nadeau, R. M., and Toksoz, M. N.: Locating nonvolcanic tremors beneath the San Andreas Fault using a station-pair double-difference location method, Geophys. Res. Lett., 37, L13304, doi:10.1029/2010GL043577, 2010.

4

Transitional relation exploration for typical loess geomorphologic types based on slope spectrum characteristics

S. Zhao[1] and W. Cheng[2]

[1]Department of Surveying and Mapping, College of Mining Technology, Taiyuan University of Technology, Taiyuan 030024, China
[2]State Key Laboratory of Resources and Environmental Information System, Institute of Geographic Sciences and Natural Resources Research, CAS, Beijing 100101, China

Correspondence to: S. Zhao (zhaoshangmin@tyut.edu.cn)

Abstract. Based on the Chinese Geomorphologic Database at 1 : 1 000 000 scales, the distribution of the typical loess geomorphologic types (such as the loess tableland, loess ridge and loess knoll) is acquired in the Loess Plateau of China. Then, based on the SRTM (Shuttle Radar Topography Mission) digital elevation model (DEM) data and topographic analysis methods, the slope spectrums are computed for the typical loess geomorphologic types and their subtypes. Through achieving the tendency line of the slope spectrum and analysing the slope spectrum characteristics of the loess typical geomorphologic types, the transitional relationships are explored: (1) the general rule is that loess tableland transitions to loess ridge, and then to loess knoll. (2) The specific relationships for the subtypes are as follows: in loess tableland, the transition is from loess terrace to complete tableland, then to residual tableland, and finally to beam tableland. In the loess ridge, the transition is from oblique ridge to knoll ridge, and the final stage is the loess knoll.

1 Introduction

Topography is formed through relief-generation processes (such as uplift and stream incision) and slope-driven denudation processes, so the resulting topographic slopes can gives clues for these processes, and many topographic studies are carried out using slope distributions to explore these processes (Wolinsky and Pratson, 2005). For example, early studies focused on the hill slope profiles and modal characteristics slopes (Strahler, 1956; Carson, 1971; Anderson et al., 1980). Subsequent studies mainly attempted to use slope statistics from natural and simulated digital elevation models (DEMs) to relate the topography to geomorphologic processes (Burbank, 1992; Roering et al., 1999; Iwahashi et al., 2001). Finally, recent studies have mainly taken slope as an important topographic factor for characterizing and automatically classifying the landform types (Crevenna et al., 2005; Bue and Stepinski, 2006; Tang and Li, 2008).

This study takes the Loess Plateau of China as an experimental area, which attracts world attention in geographic research because of its unique geomorphologic features. In the Loess Plateau, the slope spectrum index is applied to study the loess geomorphology (Tang et al., 2008). The slope spectrum, as a microscopic terrain index, can reveal macro-geomorphologic features, which makes it a valuable topographic index in geomorphologic research (Tang et al., 2005).

In previous studies, loess geomorphology has been classified and characterized by using the slope spectrum index (Li et al., 2007; Wang et al., 2008; Zhou et al., 2010). The achievement of 1 : 1 000 000 scales in the Chinese Geomorphologic Database (CGD) provides the distribution of the loess geomorphologic types (Cheng et al., 2011a). As to the slope spectrum index, it can be computed based on digital elevation model (DEM) data, which, in this study, are the Shuttle Radar Topography Mission (SRTM) data.

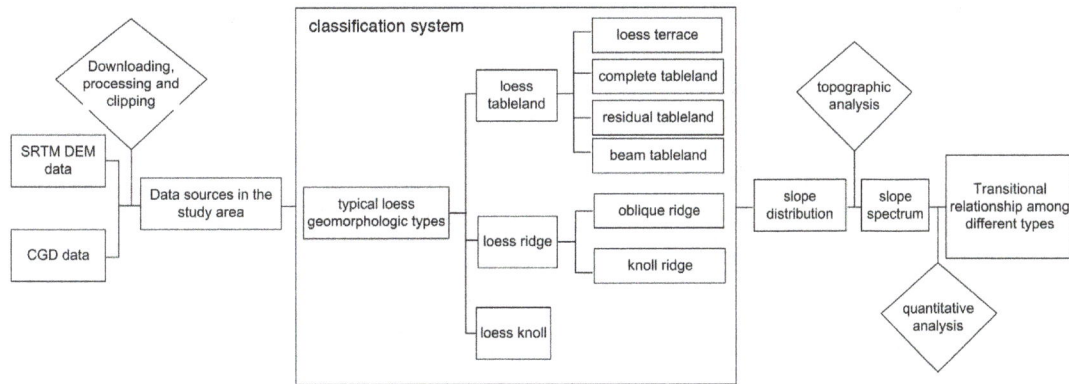

Figure 1. The workflows in this research.

Hence, this study aims to analyse the characteristics of the slope spectrums of the typical loess geomorphologic types using CGD and SRTM DEM data, and then explore the transitional relationships among the types and their subtypes based on the changing rules of slope characteristics.

2 Study area and data sources

2.1 Study area

This study selects the Loess Plateau of China as the study area. Located in the upper and middle reaches of the Yellow River, the Loess Plateau is on the west side of Taihang Mountain, the east side of the Qinghai–Tibet Plateau, the north side of Qinling Mountain and the south side of the Mongolian Plateau. The Loess Plateau is one of the four largest plateaus in China, and the widest and deepest loess in the world is distributed within it, forming as a result the most typical loess geomorphology. Loess geomorphology has a close relation with soil erosion, which is the one of the most serious problems for ecological and environmental safety in the Loess Plateau. Thus research on loess geomorphology has both scientific significance and economic importance.

2.2 Data sources

The main data sources used in this study are from the CGD at 1 : 1 000 000 scales and the SRTM DEM data.

The CGD data at 1 : 1 000 000 scales were recently achieved by using remote sensing visual interpretation and geographic information system methods from multi-source data, such as remote sensing images, SRTM3 DEM data, published geomorphologic maps, geologic data, geographic base data, etc. (Cheng et al., 2011a). These data are the source data to compile the 1 : 1 000 000 set of geomorphologic atlases of China (Li et al., 2009) and can be divided into seven layers: relief and altitude, genesis, sub-genesis, morphology, micro-morphology, slope and aspect, and material and lithology (Cheng et al., 2011b). In this study, the fourth

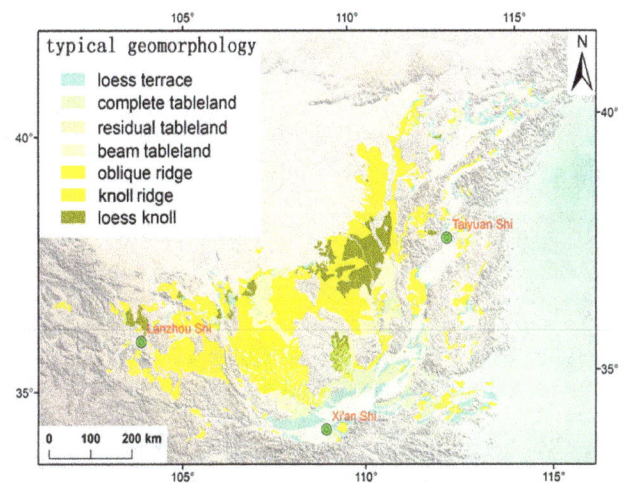

Figure 2. Typical loess geomorphology distribution in the Loess Plateau.

layer (morphology) and the fifth layer (micro-morphology) are used to acquire the distribution of the typical loess geomorphologic types and their subtypes.

The SRTM DEM data used here are the SRTM3 DEM data, which have 3° spatial resolution (SRTM3) and are processed by the Consortium for Spatial Information of the Consultative Group for International Agricultural Research. It can provide continuous elevation surface information and wide coverage (60° N–56° S), and as a result it has been used in many research fields. In this study, the SRTM3 DEM data are the version 4 data, and they are used to compute the slope spectrums for the typical loess geomorphologic types and their subtypes.

3 Methodology

The workflow of this study is shown in the Fig. 1.

This workflow shows the following: the main data sources (CGD data and SRTM3 DEM data) are firstly downloaded,

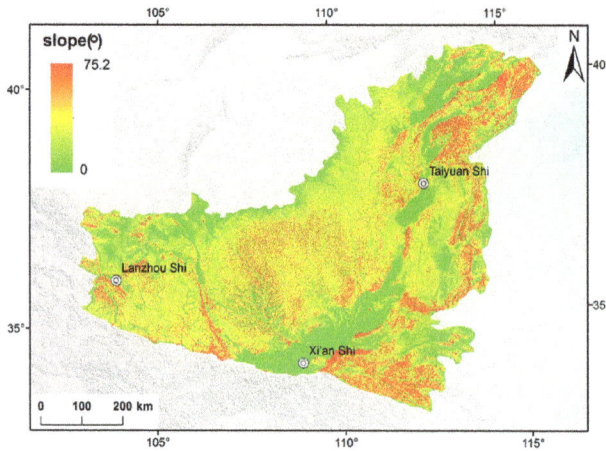

Figure 3. Slope distribution status in the Loess Plateau of China.

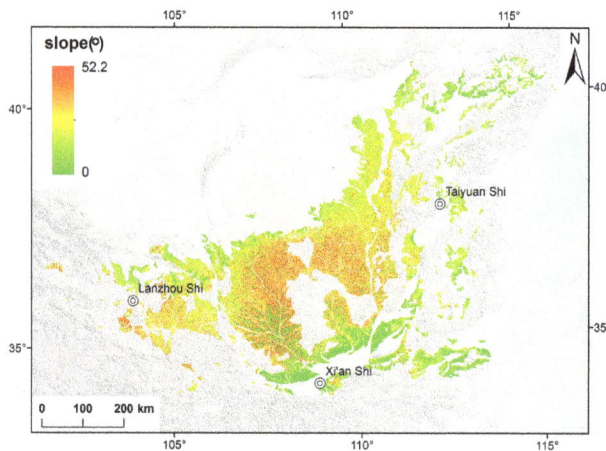

Figure 4. Slope distribution status in the typical loess geomorphologic area of the Loess Plateau.

Table 1. Numerical statistics for the slope of different geomorphologic units (°).

Geomorphologic unit	Min	Max	Mean	Standard deviation
Loess Plateau	0	75.2	10.7	8.0
Typical geomorphologic area	0	52.2	9.8	6.7
Loess tableland	0	51.0	7.1	6.7
Loess ridge	0	52.2	11.5	6.2
Loess knoll	0	46.8	10.6	5.8

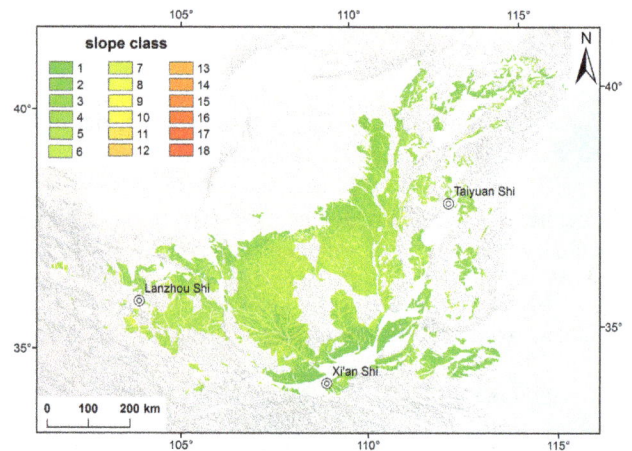

Figure 5. Slope class distribution status in the typical loess geomorphologic area of the Loess Plateau.

processed and clipped; then, based on the CGD data, the distribution of the typical loess geomorphologic types and their subtypes is acquired, and the slope spectrums are computed for these geomorphologic types using SRTM3 DEM data and the topographic analysis method; finally, after the quantitative analysis to the characteristics of the slope spectrums (such as the tendency line, the regression equation and its corresponding R^2), the transitional relationship is achieved.

3.1 Acquisition of the distribution of the typical loess geomorphologic types and their subtypes

The typical loess geomorphology types are loess tableland, loess ridge and loess knoll. The distribution of these typical loess geomorphologic types is determined using the CGD data. The areas covered by the three types are 6.52×10^4, 9.12×10^4 and 1.91×10^4 km^2, respectively.

The transitional rule among typical loess geomorphologic types is widely acknowledged as from loess tableland to loess

ridge and finally to loess knoll (Sang et al., 2007). Hence, in this study, more attention is paid to the transitional rules of the subtypes of typical loess geomorphology.

Considering the distribution areas and situations of the typical loess geomorphology types, the loess tableland is divided into four subtypes: loess terrace, complete tableland, residual tableland and beam tableland. The loess ridge is divided into oblique ridge and knoll ridge, and the loess knoll is not divided further. From the CGD data, the distribution of the subtypes of the typical loess geomorphology can be acquired and is shown in Fig. 2.

Figure 2 shows the distribution of the typical loess geomorphology: loess tableland is mainly distributed in the northern part of the city of Xi'an and eastern part of the city of Lanzhou; in addition, sparse distribution can be seen around the city of Taiyuan. The loess ridge has the greatest area, and is mainly distributed in the eastern part of Lanzhou and western part of Taiyuan. Loess knoll has the least area, and is distributed in the western part of Taiyuan, the northern part of Xi'an, and around Lanzhou.

The acquisition of the distribution of the typical loess geomorphology provides the chance to compute the slope spectrum index for each type.

Table 2. Distribution status of the slope intervals for the typical loess geomorphologic types.

Slope interval (°)	Loess tableland		Loess ridge		Loess knoll	
	Percentage	Grid number	Percentage	Grid number	Percentage	Grid number
0–3	39	3 481 883	6	728 129	7	195 155
3–6	18	1 608 957	15	1 832 410	16	422 065
6–9	12	1 054 711	19	2 333 408	20	511 864
9–12	9	805 219	18	2 253 837	19	490 007
12–15	7	647 609	15	1 896 942	15	401 575
15–18	6	509 414	11	1 408 602	11	281 653
18–21	4	365 281	8	952 910	6	168 136
21–24	3	229 533	5	571 302	3	84 814
24–27	1	121 041	2	287 704	1	34 178
27–30	1	52 343	1	120 561	0	10 974
30–33	0	18 310	0	41 779	0	2870
33–36	0	5269	0	11 749	0	579
36–39	0	1415	0	2543	0	103
39–42	0	324	0	577	0	28
42–45	0	98	0	160	0	6
45–48	0	15	0	52	0	1
48–51	0	0	0	11		
51–54	0	1	0	1		
Total	100	8 901 423	100	12 442 677	100	2 604 008

Table 3. Distribution status of the slope intervals for the subtypes of the loess tableland.

Slope interval (°)	Loess terrace		Complete tableland		Residual tableland		Beam tableland	
	Percentage	Grid number	Percentage	Grid number	Percentage	Grid number	Percentage	Grid number
0–3	65	1 713 075	41	1 568 793	10	36 401	8	163 851
3–6	20	516 158	19	737 272	15	56 426	15	297 644
6–9	8	211 427	12	467 274	15	58 501	16	316 833
9–12	4	100 710	9	342 276	15	55 785	15	306 259
12–15	2	51 825	7	258 675	14	51 875	14	285 322
15–18	1	27 116	5	194 177	12	44 530	12	243 966
18–21	1	13 296	4	135 348	9	34 529	9	182 496
21–24	0	6507	2	83 790	6	22 236	6	117 388
24–27	0	2992	1	43 770	3	11 469	3	63 067
27–30	0	1433	0	18 460	1	4503	1	28 140
30–33	0	598	0	5728	0	1466	1	10 578
33–36	0	226	0	1315	0	291	0	3470
36–39	0	76	0	240	0	31	0	1084
39–42	0	23	0	20	0	1	0	285
42–45	0	11	0	1			0	85
45–48	0	2					0	13
48–51	0	0						
51–54	0	1						
Total	100	2 645 476	100	3 857 139	100	378 044	100	202 0481

3.2 Slope spectrum computation for the typical loess geomorphologic types and their subtypes

Based on SRTM3 DEM data and the topographic analysis method, the slope distribution status is computed for the whole Loess Plateau, and the result is shown in Fig. 3.

Through clipping the slope distribution map for the whole Loess Plateau by the distribution area of the typical loess ge-omorphologic types, the slope distribution status in the typical loess geomorphologic area of the Loess Plateau is acquired, which is shown in Fig. 4.

Based on Figs. 3 and 4, the numerical statistics can be applied to the slope distribution status of the whole Loess Plateau, the typical loess geomorphology area, the loess tableland area, the loess ridge and the loess knoll; the results of this are shown in Table 1.

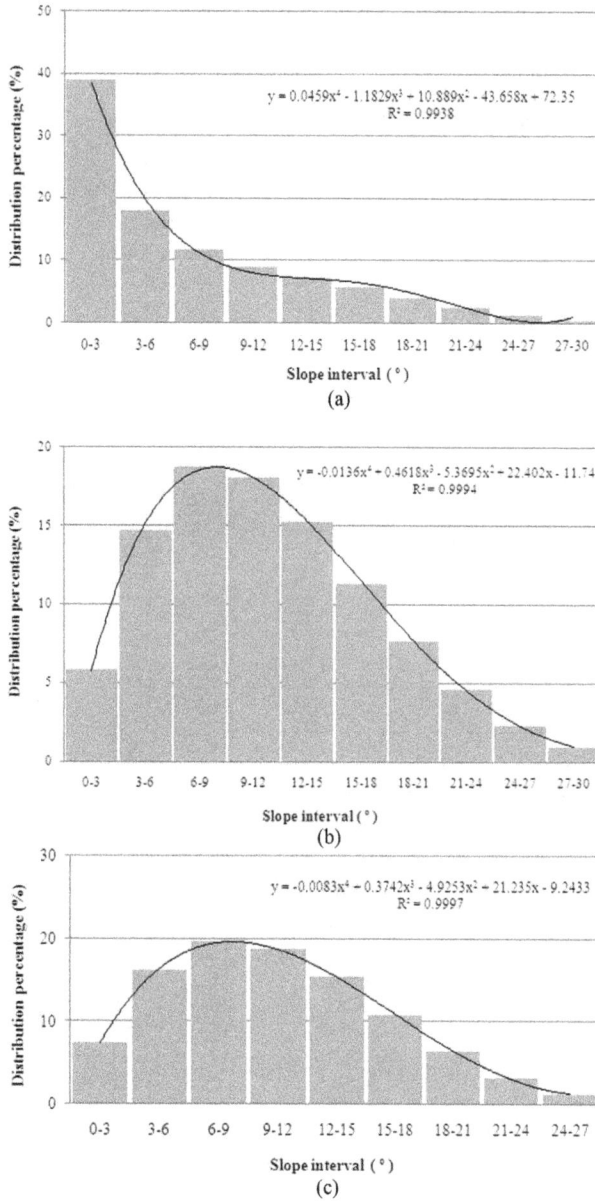

Figure 6. Slope spectrum and its quantitative analysis for the typical loess geomorphologic types. (**a**) Loess tableland, (**b**) loess ridge and (**c**) loess knoll.

Table 1 gives the maximum, minimum, mean and standard deviation values for all geomorphologic units, and shows the general slope distribution discrepancy among the geomorphologic units.

In order to quantitatively analyse the slope distribution status among all typical loess geomorphologic types, the slope is classified with an interval of 3°, which is applicable in slope spectrum studies (Zhu and Li, 2009). The slope class distribution status in the typical loess geomorphologic area can be computed, and the results are shown in Fig. 5.

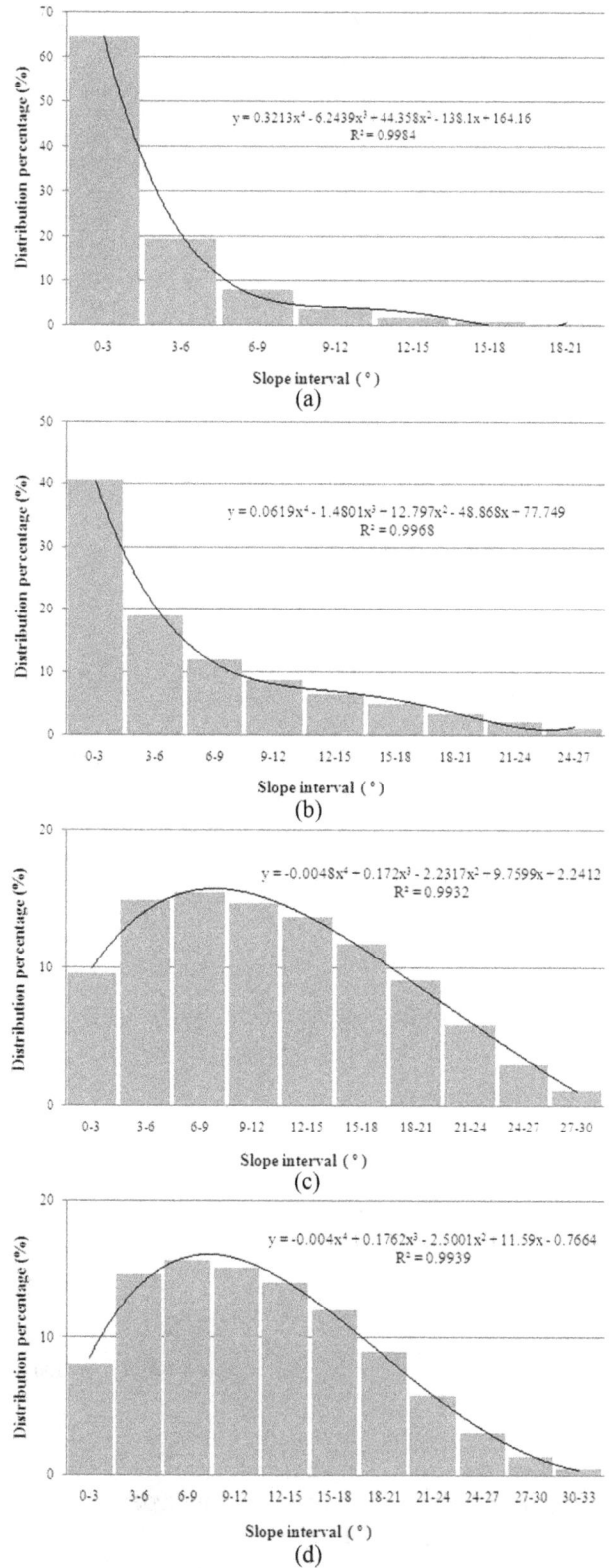

Figure 7. Slope spectrum and its quantitative analysis for the sub-types of the loess tableland. (**a**) Loess terrace, (**b**) complete tableland, (**c**) residual tableland and (**d**) beam tableland.

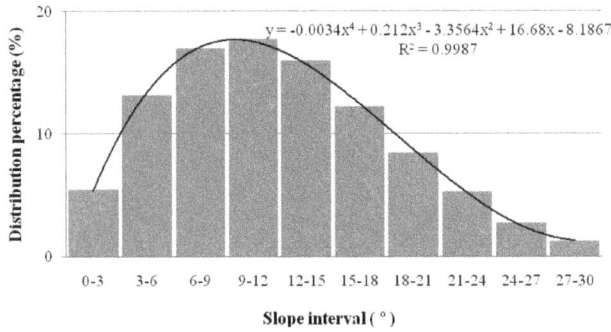

(a)

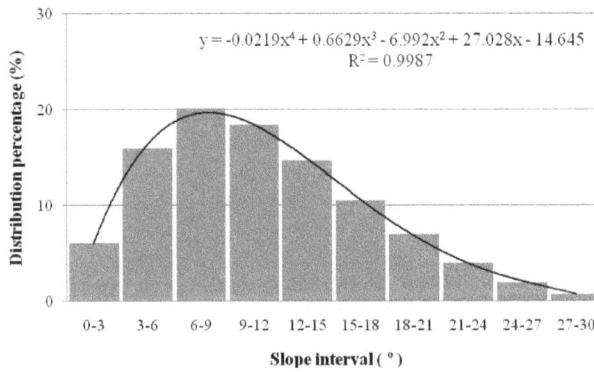

(b)

Figure 8. Slope spectrum and its quantitative analysis for the loess ridge. **(a)** Oblique ridge and **(b)** knoll ridge.

Based on Fig. 5, overlapped with the classified slope data and the distribution data of typical loess geomorphologic types, the slope distribution status is acquired for all the types; through numerical statistics, the slope spectrum index is achieved for every type.

3.3 Quantitative analysis for the slope spectrum

Through use of Microsoft Excel, the slope spectrums of the typical loess geomorphologic types are quantitatively analysed.

Through several experiments, the tendency line for the slope spectrum is firstly acquired. In order to maintain high similarity, the function of the fourth-degree polynomial regression is chosen for the tendency prediction; under this condition, the R^2 is above 0.99.

3.4 Exploration to the transitional relationships among the typical loess geomorphologic types

Based on the quantitative analysis results of the slope spectrums of all typical loess geomorphologic types, the transitional relationships can be explored among these types by

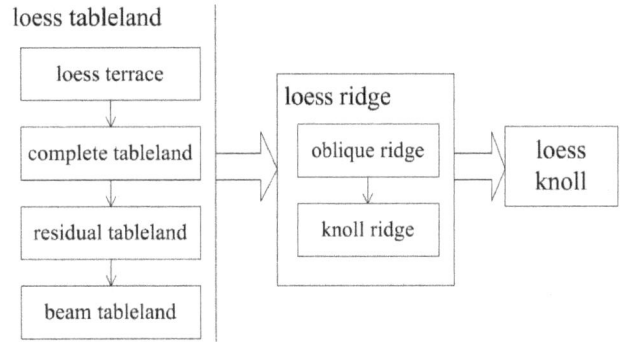

Figure 9. Transitional relationships among typical loess geomorphological types.

carefully analysing the coefficients of the regression equations and referencing the general transitional status.

4 Results

4.1 The slope spectrum analysis for the typical loess geomorphologic types

The slope class distribution status for the typical loess geomorphologic types is computed as shown in Table 2.

In Table 2, according to the 3° slope interval, the areal distribution percentages of the three loess geomorphologic types – loess tableland, loess ridge and loess knoll – are computed and listed.

Based on the computed results of Table 2, the slope spectrum indexes for the typical loess geomorphologic types are acquired as shown in Fig. 6.

In Fig. 6, the slope spectrums are mainly acquired by using slope intervals whose percentage is above 1 %; intervals below 1 % are ignored because they are too low. Furthermore, the quantitative analysis of the slope spectrums – including the tendency lines, the regression equations and the R^2 – is conducted and given. The illustration in Fig. 6 is the same as the following slope spectrums.

Table 2 and Fig. 6 show that loess tableland is mainly distributed in the 0–9° slope area, especially in the 0–3° interval, and loess ridge and loess knoll are both mainly distributed in the 3–15° slope area, especially in the 6–12° interval. Hence, the slope spectrum index for the loess ridge is similar to that for the loess knoll, with the difference being that the loess ridge is a little steeper than loess knoll. As to the loess table, it is much flatter than the other two types; thus it has distinct slope spectrum characteristics compared to loess ridge and loess knoll.

4.2 The slope spectrum analysis for the subtypes of the loess tableland

The subtypes of the loess tableland are loess terrace, complete tableland, residual tableland and beam tableland. Based

Table 4. Distribution status of the slope intervals for the subtypes of the loess ridge.

Slope interval (°)	Oblique ridge		Knoll ridge	
	Percentage	Grid number	Percentage	Grid number
0–3	5	302 457	6	423 662
3–6	13	729 610	16	1 102 216
6–9	17	942 728	20	1 390 490
9–12	18	985 107	18	1 268 763
12–15	16	885 598	15	1 011 807
15–18	12	678 737	11	730 640
18–21	8	469 594	7	483 916
21–24	5	293 507	4	278 244
24–27	3	153 879	2	134 075
27–30	1	66 960	1	53 610
30–33	0	24 294	0	17 478
33–36	0	7270	0	4478
36–39	0	1653	0	884
39–42	0	428	0	149
42–45	0	136	0	29
45–48	0	45	0	7
48–51	0	10	0	1
51–54	0	1		
Total	100	5 542 014	100	6 900 449

on the distribution status of these subtypes, SRTM DEM data and the slope spectrum computation method, the slope distribution with a 3° interval is computed for these subtypes and shown in Table 3.

Table 3 gives the slope percentage distribution of these subtypes of the loess tableland. Based on Table 3, the slope spectrum is computed and results are shown in Fig. 7.

Table 3 and Fig. 7 show that (1) loess terrace is mainly distributed in the 0–6° slope area, especially in the 0–3° slope area, which is the flattest subtype of the loess tableland; (2) complete tableland is mainly distributed in the 0–9° slope area, especially in the 0–3° slope area, which is similar to the whole loess tableland; and (3) residual tableland and beam tableland are steep and similar, they are both almost evenly distributed in the 0–21° slope area, and the slope spectrums for the two subtypes are similar to that for the loess ridge.

4.3 The slope spectrum analysis for the subtypes of the loess ridge

The subtypes of loess ridge are oblique ridge and knoll ridge. The slope distribution status for the two subtypes is shown in Table 4.

Based on Table 4, the slope spectrum for the oblique ridge and knoll ridge is computed, and results of which are shown in Fig. 8.

From Table 4 and Fig. 8 we can see that (1) the oblique ridge and knoll ridge are both mainly distributed in the 3–18° slope area, and the slope spectrum characteristics for the two subtypes are similar; (2) the oblique ridge is a little steeper

than the knoll ridge; and (3) compared to oblique ridge, the slope for the knoll ridge is more similar to that for the loess knoll.

4.4 Transitional relationship exploration for the typical loess geomorphologic types

Based on the tendency lines and regression equations of the slope spectrums, the transitional order of the loess geomorphologic types is as follows: in the loess tableland, with the slope from flat to steep, the subtypes transition from loess terrace to complete tableland, then to residual tableland and finally to beam tableland; in the loess ridge, the subtypes transition from oblique ridge to knoll ridge. Hence, the transitional relationships among the typical loess geomorphologic types are achieved as shown in Fig. 9.

Figure 9 shows the general transitional rule of the typical geomorphologic types: from loess tableland to loess ridge, and a final transition to loess knoll. As to the subtypes, the slope change rule is adopted; thus every subtype transitions according to the elevating slope.

5 Discussions

5.1 Innovations

Based on DEM data and the topographic analysis methods, slope spectrum in previous studies has mostly been used in landform analysis, landform classification and geomorphologic classification (Iwahashi et al., 2001; Tang and Li, 2008; Tang et al., 2008).

In this study, the geomorphologic boundary is determined in advance using visual interpretation and multi-source data, such as remote sensing images, DEM data, etc. (Cheng et al., 2011a, b). The accurate boundary of the loess geomorphologic types provides much more precise results for slope spectrum characteristic analysis; moreover, the transitional relationships among the typical loess geomorphologic types are explored based on the slope spectrum characteristic analysis, especially for the subtypes. Hence, this study gives a way for the relationships among the loess geomorphologic types to be studied in depth.

5.2 Prospects

The following aspects can be improved in future studies:

1. In this study, the boundary of the geomorphologic types is determined by multi-source data, whereas the slope spectrum characteristics are analysed based on the DEM data. If the boundary can be revised using DEM data, the slope spectrum analysis could have better results.

2. Topographic analysis in this study mainly uses the slope spectrum. In the future, more topographic indexes will be used to acquire more reasonable results.

3. The transitional relationships among geomorphologic types are explored in this study. In the future, the relation could be explored for every geomorphologic type. Through analysing the difference between the same geomorphologic type in different regions, the threshold values of the topographic indexes among the geomorphologic types can be found, and thus the fundamental difference among the geomorphologic types can be acquired.

6 Conclusions

The results can be summarized as the following:

1. The slope spectrum of the typical loess geomorphologic types and their subtypes are acquired by using the CGD and DEM data; then, based on the tendency line, regression equation and R^2, the slope spectrum characteristics are quantitatively analysed, which provides a way to obtain the relationships among the loess geomorphologic types.

2. The transitional relationships among the typical loess geomorphologic types and their subtypes are as follows: in general, for the typical loess geomorphologic types, the transition is from loess tableland to loess ridge and finally to loess hill. Specifically, for the subtypes of the loess tableland, there is a transition from loess terrace to complete terrace, and then to the residual tableland

and finally to beam tableland. As to the loess ridge, the transition is from oblique ridge to knoll ridge.

Acknowledgements. This study was supported by the National Natural Science Foundation of China (41301469 and 41171332), the National Science Technology Support Plan Project (2012BAH28B01-03), the Qualified Personnel Foundation of Taiyuan University of Technology (tyutrc-201221a) and the Opening Foundation of LREIS. We would like to express our gratitude to the editors and the anonymous reviewers for suggestions that improved our paper.

Edited by: H. Mitasova

References

Anderson, M. G., Richards, K. S., and Kneale, P. E.: The role of stability analysis in the interpretation of the evolution of threshold hillslopes, Institute of British Geographers Transactions, 5, 100–112, 1980.

Bue, B. D. and Stepinski, T. F.: Automated classification of landforms on Mars, Comput. Geosci., 32, 604–614, 2006.

Burbank, D. W.: Characteristic size of relief, Nature, 359, 483–484, 1992.

Carson, M. A.: An application of the concept of threshold slopes to the Laramie Mountains, Wyoming, Institute of British Geographers Special Publication, 3, 31–47, 1971.

Cheng, W. M., Zhou, C. H., Chai, H. X., Zhao, S. M., and Zhou, Z. P.: Research and compilation of the Geomorphologic Atlas of the People's Republic of China (1:1,000,000), J. Geogr. Sci., 21, 89–100, 2011a.

Cheng, W. M., Zhou, C. H., and Li, B. Y.: Structure and contents of layered classification system of digital geomorphology for China, J. Geogr. Sci., 21, 771–790, 2011b.

Crevenna, A. B., Rodríguez, V. T., Sorani, V., Frame, D., and Ortiz, M. A.: Geomorphometric analysis for characterizing landforms in Morelos State, Mexico, Geomorphology, 67, 407–422, 2005.

Iwahashi, J., Watanabe, S., and Furuya, T.: Landform analysis of slope movements using DEM in Higashikubiki area, Japan, Comput. Geosci., 27, 851–865, 2001.

Li, F. Y., Tang, G. A., Jia, Y. N, and Cao, Z. D.: Scale Effect and Spatial Distribution of Slope Spectrum's Information Entropy, Geo Information Sci., 9, 13–18, 2007.

Li, J. J., Zhou, C. H., and Cheng, W. M.: Geomorphologic Atlas of the People's Republic of China, Science Press, Beijing, China, 2009.

Roering, J. J., Kirchner, J. W., and Dietrich, W. E.: Evidence for nonlinear, diffusive sediment transport on hillslopes and implications for landscape morphology, Water Resour. Res., 35, 853–870, 1999.

Sang, G. S., Chen, X., Chen, X. N., and Che, Z. L.: Formation model and geomorphic evolution of loess hilly landforms, Arid Land Geogr., 30, 375–380, 2007.

Strahler, A. N.: Quantitative slope analysis, Geol. Soc. Am. Bull., 67, 571–596, 1956.

Tang, G. A. and Li, F. Y.: Landform Classification of the Loess Plateau Based on Slope Spectrum from Grid DEMs, Advances in Digital Terrain Analysis, edited by: Zhou, Q. M., Lees, B., and Tang, G. A., Springer Berlin Heidelberg, 107–124, 2008.

Tang, G. A., Ge, S. S., Li, F. Y., and Zhou, J. Y.: Review of Digital Elevation Model (DEM) Based Research on China Loess Plateau, J. Mountain Sci., 2, 265–270, 2005.

Tang, G. A., Li, F. Y., Liu, X. J., Long, Y., and Yang, X.: Research on the slope spectrum of the Loess Plateau, Sci. China Ser. E: Technol. Sci., 51, 175–185, 2008.

Wang, C., Tang, G. A., Li, F. Y., Zhu, X. J., and Jia, Y. N.: The Uncertainty of Slope Spectrum Derived from Grid Digital Elevation Model, Geo. Information Sci., 10, 539–544, 2008.

Wolinsky, M. A. and Pratson, L. F.: Constraints on landscape evolution from slope histograms, Geology, 3, 477–480, 2005.

Zhou, Y., Tang, G. A., Yang, X., Xiao, C. C., Zhang, Y., and Luo, M. L.: Positive and negative terrains on northern Shaanxi Loess Plateau, J. Geogr. Sci., 20, 64–76, 2010.

Zhu, M. and Li, F. Y.: Influence of slope classification on slope spectrum, Sci. Surv. Mapping, 34, 165–167, 2009.

5

Observations of the effect of emergent vegetation on sediment resuspension under unidirectional currents and waves

R. O. Tinoco and G. Coco

Environmental Hydraulics Institute IH Cantabria, University of Cantabria, Santander, Spain

Correspondence to: R. O. Tinoco (tinocor@unican.es)

Abstract. We present results from a series of laboratory experiments on a wave and current flume, where synchronous velocity and concentration measurements were acquired within arrays of rigid cylinders, representative of emergent vegetation and benthic communities, under different flow conditions. The density of an array of rigid cylinders protruding through a sandy bed affects the velocity field, sediment motion and resuspension thresholds when subjected to both unidirectional currents and regular waves. We compare the measured resuspension thresholds against predictions of sediment motion on non-obstructed flows over sandy beds. The results show that even if flow speeds are significantly reduced within the array, the coherent flow structures and turbulence generated within the array can enhance sediment resuspension depending on the population density.

1 Introduction

Aquatic vegetation and benthic populations alter their habitat in many different ways. Some species act as ecosystem engineers, modifying their habitats to guarantee their survival. We focus our study on the case of sparse to dense, rigid, randomly distributed arrays of emergent (protruding through the water free surface) elements, with scales representative of populations of tube worms and rigid vegetation, such as mangrove roots.

Two species would be directly represented by the designed arrays (Fig. 1): (1) European fan worm (*Sabella spallanzanii*), a filter feeding tube worm that creates canopies of feeding fans over the sediment, generally found in shallow subtidal areas, and (2) Black mangrove (*Avicennia germinans*), a subtropical woody shrub found in salt marshes.

The European fan worm is considered as an ecosystem engineer, building large platforms on the sea floor, creating sheltering areas for other organisms to grow and live, trapping particles, blocking light, improving oxygen supply in the sediment, altering current velocities, and stabilizing/destabilizing the sediment within and around the patch (Wallentinus and Nyberg, 2007; van Hoey et al., 2008). Black

mangrove, living in intertidal areas, forms a network of pneumatophores, protruding through the water surface to allow for root respiration (Houck and Neill, 2009), forming an array that shelters the sediment and organisms around it.

The effects of such populations on sediment resuspension and deposition as a function of the properties of the array (i.e. density, porosity, solid volume fraction, individual spacing) are still not well understood (Eckman et al., 1981; Carey, 1983; Luckenbach, 1986; Graf and Rosenberg, 1997; Friedrichs et al., 2000; Coco et al., 2006), and one even finds contradictory results between field and experimental studies (Madsen et al., 2001), which emphasizes the need to distinguish between purely physical effects, as the ones we address, and purely biological effects, as the ones generated by macrophytes, microphytobenthos or macrofauna (LeHir et al., 2007). For example, an increase in turbulence by the physical presence of benthic organisms may enhance resuspension, but the mucus produced by those same organisms can form a protective film that suppresses it.

As Nikora (2010) points out, the growing new field of hydrodynamics of aquatic ecosystems needs to eliminate the multiple knowledge gaps between fluid mechanics, ecology and biomechanics. The work presented herein adds another

Figure 1. Species represented in the study. Black mangrove roots (left) and European fan worm (right). Pictures obtained from the USDA-NRCS and MESA websites, respectively.

step by incorporating biomorphodynamics as a critical component of the system.

1.1   Thresholds of sediment motion for a flat, sandy bed

There are many studies on critical stresses and velocities for sediment motion and resuspension on a flat, sandy bed under currents and waves (e.g. van Rijn, 2007 and references therein). However, literature on thresholds of motion for sediment within patches of vegetation or benthic populations is very scarce (see for example the laboratory work of Eckman et al., 1981 for unidirectional flows). We present herein a few of the existing criteria describing critical velocities and critical bed shear stress for sediment motion, as well as critical velocities for resuspension applicable to our investigated case. While those refer to non-populated beds, they will offer a starting point for comparison against the results from populated cases, allowing us to investigate the impact of the added physical processes induced by the presence of a vegetation patch.

We consider hereafter the formulations of van Rijn (1984) and Soulsby (1997) for critical velocities (a), the formulation of Soulsby (1997) for critical bed shear stress (b), and the resuspension threshold formulations of Soulsby (1997), van Rijn (1984), and Komar and Miller (1973) (c).

a. Critical velocities.

From van Rijn (1984):

$$U_{cr} = 0.19 d_{50}^{0.1} \log_{10}(4D/d_{90}) \; ; \; 100 < d_{50} < 500 \, \mu m$$
$$U_{cr} = 8.5 d_{50}^{0.6} \log_{10}(4D/d_{90}) \; ; \; 500 < d_{50} < 2000 \, \mu m,$$

where D is the water depth, and d_{50} and d_{90} are the correspondent percentiles on the granulometric curve.

From Soulsby (1997):

$$U_{cr} = 7 \left(\frac{D}{d_{50}} \right)^{1/7} \left[g(s_\rho - 1)d_{50}f(D_*) \right]^{1/2},$$

where

$$s_\rho = \frac{\rho_s}{\rho} = \frac{\text{density of the sediment}}{\text{density of the fluid}}$$

$$D_* = \left(\frac{g(s_\rho - 1)}{v^2} \right)^{1/3} d_{50}$$

$$f(D_*) = \frac{0.30}{1 + 1.2D_*} + 0.055 \left(1 - e^{-0.020D_*} \right)$$

and g is gravity, for values of $D_* > 0.1$.

b. Critical bed shear stress.

From Soulsby and Whitehouse (Soulsby, 1997), we get values for critical bed shear stress from the Shields curve, and Soulsby fit to the Shields curve as

$$\theta_{cr} = \frac{0.24}{D_*} + 0.055 \left(1 - e^{-0.020D_*} \right)$$

$$\theta_{cr} = \frac{0.30}{1 + 1.2D_*} + 0.055 \left(1 - e^{-0.020D_*} \right),$$

where one can calculate

$$\tau_{cr} = \theta_{cr} g(\rho_s - \rho)d_{50}$$

$$u_* = \left(\frac{\tau_{cr}}{\rho} \right)^{1/2}$$

such that a critical velocity can be estimated as

$$U_{cr} = \frac{u_*}{\kappa} \ln \left(\frac{z}{z_0} \right)$$

using von Karman's $\kappa = 0.41$, and estimating z_0 from Nikuradse's roughness:

$$z_0 = \frac{k_s}{30} \left(1 - \exp \left[\frac{-u_* k_s}{27v} \right] \right) + \frac{v}{9u_*}$$

with $k_s = 2.5d_{50}$.

c. Resuspension thresholds.

We use two criteria based on the settling velocity of the sand particles, w_s, given empirically for natural sands as (Soulsby, 1997):

$$w_s = \frac{\nu}{d_{50}}\left[\left(10.36^2 + 1.049D_*^3\right)^{1/2} - 10.36\right]$$

First, considering local turbulent bursts of sediment particles are lifted when (van Rijn, 1984):

$$u_{*cr} = \frac{4w_s}{D_*} , D_* < 10$$
$$u_{*cr} = 0.4w_s , D_* > 10.$$

Second, assuming no material can be suspended unless the friction velocity exceeds the settling velocity (Soulsby, 1997):

$$u_* \geq w_s.$$

In the case of waves, the wave-orbital velocity theory from Komar and Miller (1973) has been shown to give accurate predictions for sand resuspension (Green, 1999). A critical orbital velocity, U_{wc} can be calculated from

$$\frac{\rho U_{wc}^2}{(\rho_s - \rho)gd_{50}} = 0.21\left(\frac{d_w}{d_{50}}\right)^{1/2}$$

where $d_w = \frac{U_{wc}T}{2\pi}$ is the near-bottom orbital excursion.

1.2   Studies on flow through aquatic vegetation

Several research groups have studied flow through aquatic vegetation. Nepf (2012a) summarizes some of the experimental work done on rigid and flexible, emergent and submerged model vegetation subject to unidirectional currents on a flat, rigid bed. Nepf (2012b) highlights the challenges on the study of vegetated flows, specifically mentioning the need for a better understanding of sediment motion processes within vegetated regions, pointing out that (a) there is no reliable method to estimate bed stresses within the vegetation, (b) it is not known if bed shear stress is the only relevant parameter, (c) the role of the turbulence generated by the vegetation must be studied, (d) a new parameterisation is needed to account for spatial variability of the vegetation, and (e) the feedbacks between flow, vegetation and sediment are not yet understood.

A similar case occurs with studies on waves through vegetation. We find studies focusing on wave dissipation both through laboratory experiments (e.g. Fonseca and Cahalan, 1992), and field studies (e.g. Infantes et al., 2012), as well as analytical and numerical models to predict wave damping and velocities within the vegetated field (e.g. Kobayashi

Table 1. Relevant parameters for the studied arrays.

n (m^{-2})	a (m^{-1})	ϕ	$\langle s \rangle/d$	aD	ad
0	–	–	–	–	–
25	0.5	0.008	4.45	0.08	0.01
150	3.0	0.047	1.70	0.48	0.06
250	5.0	0.079	1.40	0.80	0.10

et al., 1993; Maza et al., 2013), but none accounting for a mobile bed.

In terms of sediment studies, recent field and laboratory works on currents through submerged (Bouma et al., 2007; Borsje et al., 2011) and emergent vegetation (Follett and Nepf, 2012) share similar findings. Specifically, they found an increase in velocities and turbulent kinetic energy at the edges of the patch of vegetation, and increased scour with increased densities of the patch. The added effects of the mixing layer created at the plant–open-water interface for the submerged case might create differences in the observed deposition patterns. Follett and Nepf (2012) observed the maximum accumulation of sediment downstream of an emergent patch, while Bouma et al. (2007) reported maximum accumulations within the submerged patch. In the case of waves, to the authors knowledge, there are no experimental works on determination of sediment resuspension thresholds in the presence of arrays of rigid cylinders.

Our goal is to investigate how the presence of emergent arrays of sparse and dense rigid cylinders affects the onset of sediment resuspension under currents and under wave conditions. We record synchronous, collocated measurements of velocity and suspended sediment concentration to get mean and instantaneous values, to provide novel information to identify the role in sediment transport of turbulent quantities and high frequency fluctuations generated by the arrays.

2   Laboratory experiments

The experiments were conducted at the wave and current flume at the Environmental Hydraulics Institute of the University of Cantabria (IH Cantabria), a 54 m long, 2 m wide flume capable of running currents along (0°) and opposing (180°) the waves generated by a 2 m by 2 m piston-type wave maker with a maximum stroke of 2.0 m. Two pumps run in parallel to generate flow in the flume, and can be set independently to run at frequencies, f_p, between 20 and 50 Hz.

An 18 m long, 0.20 m deep, sand bed was built, where a 6 m-long array of randomly placed, rigid cylinders was located, as shown in Fig. 2. Water depth was set at $D = 0.16$ m, to ensure emergent conditions at all times during the waves series. The sediment is well-sorted, $d_{50} = 0.31$ mm, silica sand.

PVC cylinders, with a diameter $d = 0.02$ m, were used to represent on a 1 : 1 scale either populations of tube worms

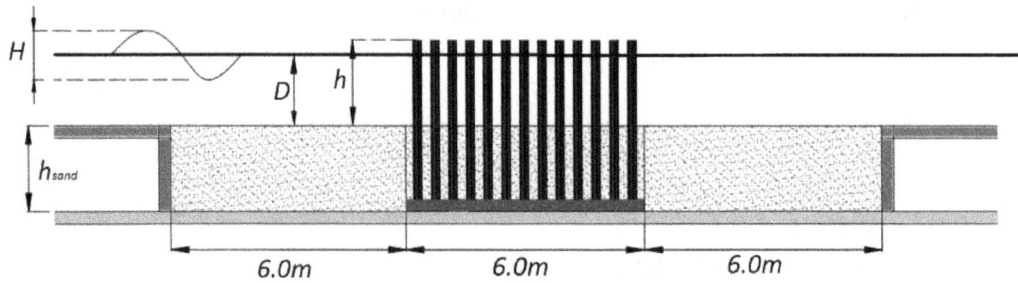

Figure 2. Random array of emergent cylinders, $d = 2.0$ cm, protruding $h = 21$ cm above the sand bed. Water depth was set at $D = 0.16$ m for all cases.

(e.g. *Sabella spallanzanii*) and mangrove roots (e.g. *Avicennia germinans*). The arrays were designed by defining the number of elements per square meter, n, and a minimum distance, $s_{min} = d$, allowed between the edges of adjacent cylinders. The 0.40 m long cylinders were individually attached with screws to six 1 m long, 2 m wide PVC plates fitted to the bottom of the flume and later covered by the 0.20 m-deep sand bed, creating the array of cylinders protruding 0.21 m above the bed. Three densities were considered to represent a sparse, intermediate and dense population: $n = \{25, 150, 250\}$ m^{-2}. The study began with the densest case, and selected cylinders were removed to achieve progressively lower densities. Once a cylinder was removed, it was not possible to install it again, such that a single random configuration was used for each density. Three relevant parameters of the array are calculated for each density: (1) volumetric frontal area of the array, a (m^{-1}), given as the frontal area of the cylinders, $A_f = N_{cylinders}dD$, per volume, $V = A_{planform}D$, which reduces to $a = nd$; (2) solid volume fraction ϕ ( ), given as the ratio between the volume of solids and fluid volume, or $\phi = n\pi d^2/4$ for emergent cylinders; and (3) the mean separation between cylinders, $\langle s \rangle$ (m), given in non-dimensional form as $\langle s \rangle/d$ (Table 1). To ensure fully covering the range from sparse to dense conditions, the selected densities satisfy criteria proposed by Friedrichs et al. (2000), defining dense arrays as those with values of $\phi > 0.045$, and by Nepf (2012a), considering sparse arrays with $aD \ll 0.1$, and dense if $aD \gg 0.1$. We chose the intermediate density, $n = 150$ m^{-2} to be in the limit between sparse and dense arrays.

Acoustic Doppler velocimeters (ADV – Nortek Vectrino) and optical backscatter sensors (OBS – Seapoint turbidity meter, SeaPoint Sensors Inc.), recording at a sampling frequency $f_s = 50$ Hz, were used to measure synchronous records of velocities and concentrations, respectively. For the emergent case presented herein, ADVs and OBSs were located 5 cm above the sand bed, at the beginning ($x_c = 0$ m) and at the middle ($x_c = 3$ m) of the array. The distance of 5 cm above the bed was determined by the physical dimensions of the instruments, such that the sensors were not in contact with the bottom and would not be buried in the sand

during the experiments. A total of 19 capacitive wave gauges (Akamina technologies) were used for free surface measurements along the length of the flume.

For the study on unidirectional currents, different flow rates were ran for each array density, increasing the frequency of one of the pumps from $f_p = 20$–50 Hz. The sand was flattened manually before each series, using a floor squeegee for the non-populated areas and a cold water high pressure cleaner within the cylinders array, flattening again after the maximum velocity was studied. The range of velocities considered falls in a range of Reynolds numbers $Re_D = UD/\nu < 4 \times 10^4$, and $Re_d = Ud/\nu < 5 \times 10^3$ for all cases. Flow remained subcritical, with $Fr < 0.2$ for all cases.

For studies on waves, five cases with regular waves were conducted for each density, with a constant period $T = 2.5$ s and nominal wave height $H = 0.01$–0.05 m. For a water depth $D = 0.16$ m, the dispersion relationship yields a wavelength $\lambda = 3.08$ m, fitting two wavelengths within the 6 m array. Series of 50 waves were generated for each H. The maximum values presented are for $H = 0.05$ m, given that breaking occurred at $H = 0.06$ m.

3 Results

3.1 Currents

For each density, the series was stopped once scour became noticeable (see Figs. 3 and 4). Once sand starts accumulating behind the cylinders and bedforms start to appear within the array, vortices are generated in the now perturbed bed, enhancing resuspension as compared to a flat bed case. Once such a state is reached and the threshold values have been exceeded, the series has been stopped. As seen in Fig. 5, as the density of the array increased, scour appeared for lower velocities, thus reducing the range of velocities studied for the denser cases (fewer data points for $n = 250$ m^{-2} in Fig. 5).

Time averaged longitudinal, x velocities are shown in Fig. 5 (left) for the same flow conditions for each array density. The measured response by the OBS for all densities studied are also shown in Fig. 5 (right). We notice an abrupt jump in the concentration values for the two denser cases

Table 2. Wave parameters for all densities at locations $x_c = 0$ m and $x_c = 3$ m.

n (m^{-2})	$x_c = 0$ m				$x_c = 3$ m			
	H_m (m)	U_{w+} (m s^{-1})	U_{w-} (m s^{-1})	d_w (m)	H_m (m)	U_{w+} (m s^{-1})	U_{w-} (m s^{-1})	d_w (m)
0	0.013	0.076	−0.093	0.030	0.015	0.083	−0.098	0.033
0	0.029	0.143	−0.067	0.057	0.031	0.120	−0.162	0.048
0	0.047	0.228	−0.076	0.091	0.047	0.188	−0.127	0.075
0	0.062	0.289	−0.092	0.115	0.060	0.221	−0.198	0.088
0	0.079	0.304	−0.112	0.121	0.079	0.245	−0.245	0.098
25	0.014	–	–	–	–	–	–	–
25	0.031	0.128	−0.068	0.051	0.029	0.131	−0.047	0.052
25	0.050	0.224	−0.062	0.089	0.044	0.200	−0.084	0.079
25	0.068	0.293	−0.077	0.116	0.055	0.222	−0.163	0.088
25	0.078	0.313	−0.081	0.125	0.063	0.257	−0.171	0.102
150	0.012	0.068	−0.078	0.027	0.013	0.071	−0.086	0.028
150	0.029	0.141	−0.134	0.056	0.024	0.086	−0.076	0.034
150	0.047	0.233	−0.129	0.093	0.032	0.113	−0.098	0.045
150	0.064	0.288	−0.168	0.114	0.038	0.127	−0.126	0.050
150	0.070	0.289	−0.157	0.115	0.042	0.148	−0.137	0.059
250	0.015	0.047	−0.042	0.019	0.014	0.053	−0.040	0.021
250	0.031	0.106	−0.078	0.042	0.022	0.102	−0.038	0.041
250	0.047	0.214	−0.085	0.085	0.027	0.130	−0.053	0.052
250	0.062	0.259	−0.122	0.103	0.031	0.137	−0.082	0.055
250	0.078	0.291	−0.128	0.116	0.036	0.144	−0.101	0.057

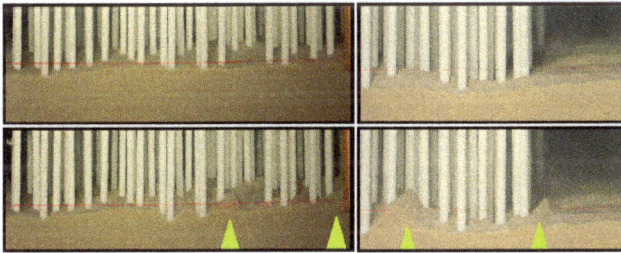

Figure 3. Side view: $n = 150$ m^{-2}, upstream (left) and downstream (right) edge of the array, before (up) and after (down) 5 min running at $f_p = 30$ Hz. Notice the scour and bedforms forming within the array at the up- and downstream sections (as pointed by the green arrows). Red line is the initial reference for the bed level. Notice the factor of 2 increase in the vertical scale.

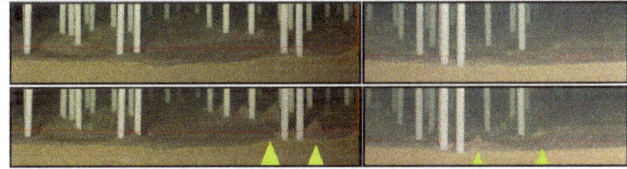

Figure 4. Side view: $n = 25$ m^{-2}, upstream (left) and downstream (right) edge of the array, before (up) and after (down) 5 min running at $f_p = 30$ Hz. Notice the scour and bedforms forming within the array at the up- and downstream sections (as pointed by the green arrows). Red line is the initial reference for the bed level. Notice the factor of 2 increase in the vertical scale.

once a critical value is reached, in contrast with the trend seen for the non-populated case.

Figure 6 (left) shows the values of turbulent kinetic energy, $k = \frac{1}{2}\left(u'^2 + v'^2 + w'^2\right)$, for all densities and flow rates studied, with higher values being reached for the denser cases at very low speeds compared to the non-populated case. We also notice in Fig. 6 (right) that even when the highest speeds for the non-populated case reach similar values of k as the densest array at lower speeds, the recorded concentration of suspended sediment is much lower.

As noticed in Fig. 7, large fluctuations were observed in the transverse, y velocities, corresponding with transverse standing waves generated by the array of cylinders. As the shredding frequency of the cylinders approaches a natural frequency of the flume, the corresponding standing wave appears (see Tinoco and Cowen, 2013).

The natural frequencies, f, of a flume of width B, can be calculated for each wave mode m (see Tinoco and Cowen, 2013 for details), as $f = \frac{1}{2\pi}\left(\frac{\pi m}{B}g\tanh\left(\frac{\pi m B}{D}\right)\right)^{1/2}$. In the case of our flume, for the first four modes, $m = \{1, 2, 3, 4\}$, we have $f(m) = \{0.62, 0.88, 1.08, 1.25\}$ Hz. By obtaining the frequency spectra of all concentration and velocity records, we notice clear peaks at the frequencies corresponding to modes 2 and 3 at lower velocities, and a clear peak corresponding to mode 3 as we increase the flow. Figure 7 shows spectra for

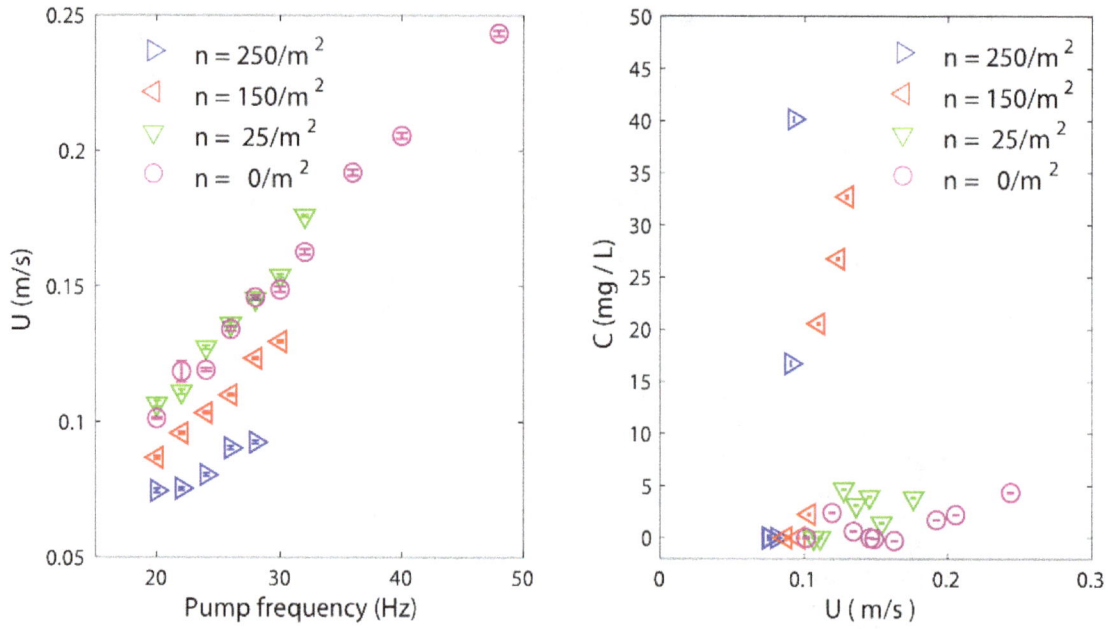

Figure 5. Left: mean longitudinal x velocities as a function of the pump frequency. Right: mean suspended sediment concentrations recorded at each velocity for all densities. Velocities measured at the center of the array $x_c = 3$ m, at $z = 0.05$ m above the bed. Vertical error bars for the 95 % CI. Notice the range of velocities studied decreased for the denser cases to avoid disruptions in the bed.

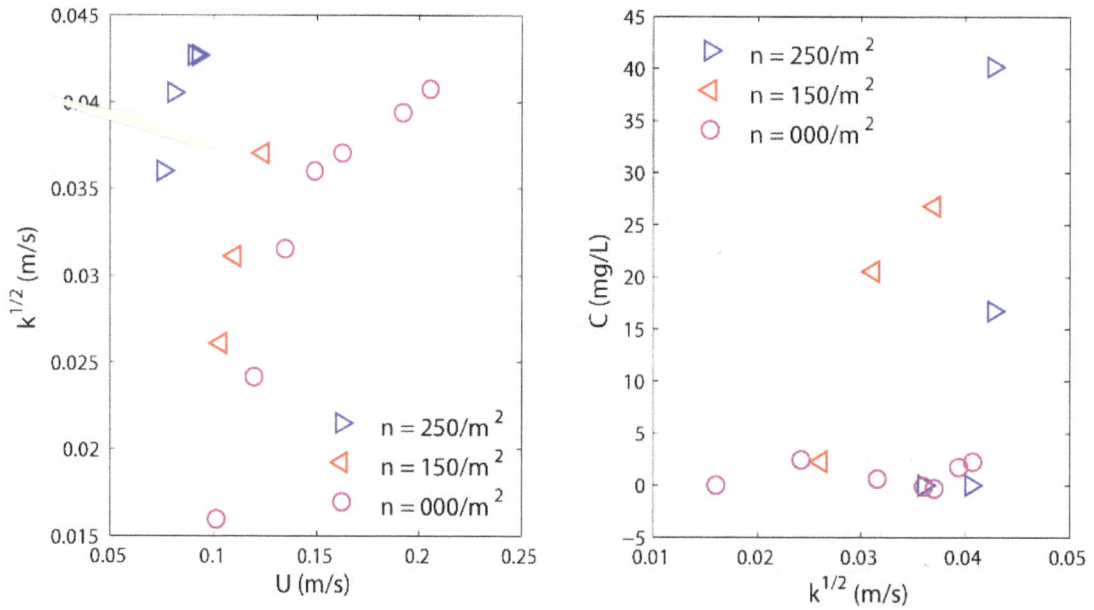

Figure 6. Turbulent kinetic energy, k, as a function of mean velocity U (left), and concentrations against k for all densities (right). Measurements at the center of the array $x_c = 3$ m, at $z = 0.05$ m above the bed.

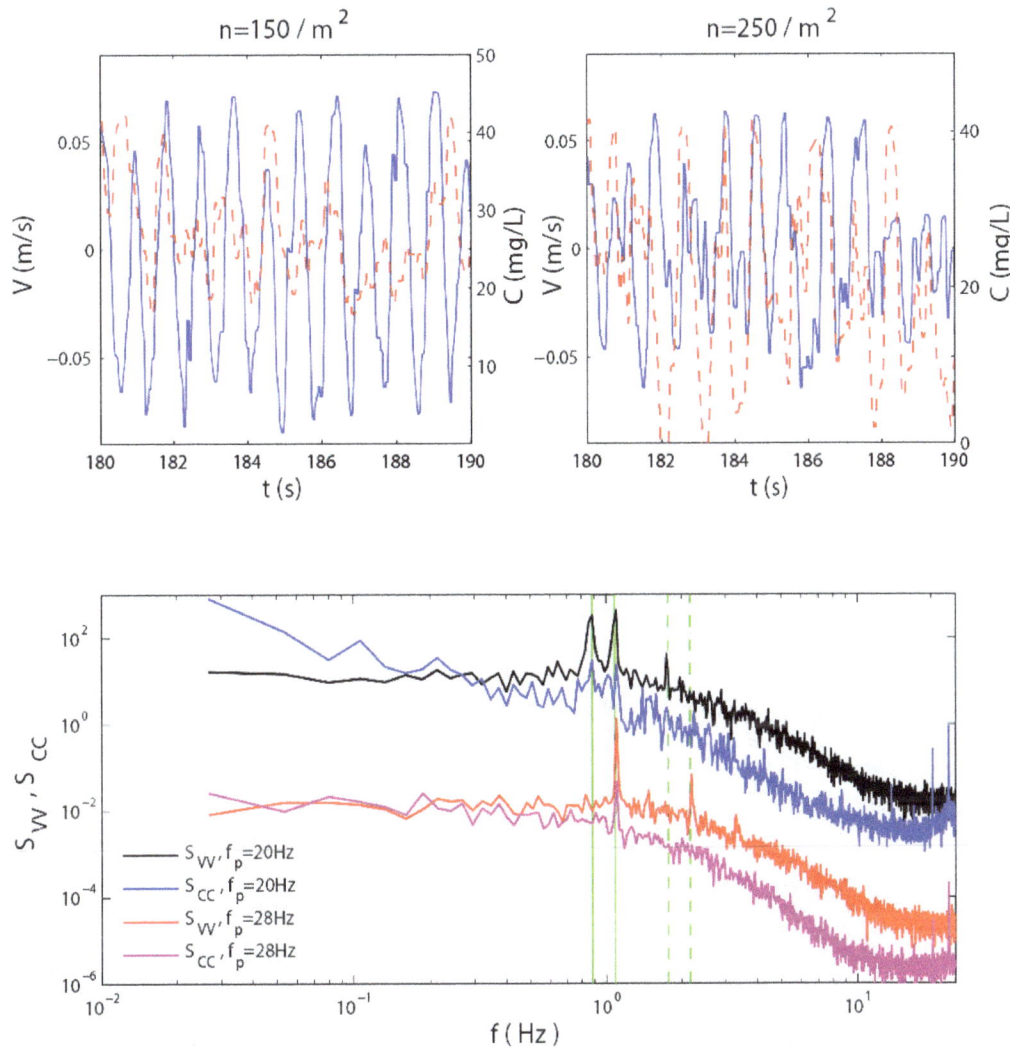

Figure 7. Top: transverse velocity (blue solid line) and concentration (red dashed line) fluctuations due to transverse standing waves generated with increasing density. Bottom: velocity and concentration spectra for $n = 250\,\text{m}^{-2}$, at the minimum and maximum flow rates considered. Vertical solid green lines show the expected frequencies for modes 2 (0.88 Hz) and 3 (1.08 Hz) of the flume, and dashed green lines show twice such frequencies . Actual magnitudes of S_{vv} and S_{CC} have been altered to improve visualisation.

the densest case, $n = 250\,\text{m}^{-2}$, highlighting the match of frequencies for the slowest and fastest flow conditions studied.

Figure 7 shows that the OBS not only captured the bulk increase in concentration of suspended sediment due to the oscillations in V, but also the concentration oscillations they caused.

3.2 Waves

The measured mean wave heights H_m at relevant x locations: near the paddle ($x_c = -16\,\text{m}$), at the beginning of the sand section ($x_c = -6\,\text{m}$), and within the array of cylinders ($x_c = \{0, 1, 2, 3, 5, 6\}\,\text{m}$), are shown in Fig. 8. To build the sand bed, ramps were built on each side of the flume, changing its initial configuration and resulting in the wave shoaling ob-

served in Fig. 8. Reflection effects from the downstream end of the flume were more evident in the non-populated case (noticed by the fluctuations in H_m), since the cylinders were very efficient also at damping reflected waves.

While wave dissipation is not the main focus of our work, the effect of the array at dissipating the incoming waves is evident, with reductions of more than 50 % of the approaching wave height for the densest cases (Fig. 8). With energy being dissipated as the wave advances through the array, the bottom shear stresses also decrease, reducing resuspension in favour of sediment deposition. Figures 9 and 10 show the up- and downstream edges of the array before and after a series of 50 waves. Scour and bedforms are clearly present at the upstream section, whereas the downstream area shows little, if any, bed disruptions. This behaviour is contrary to the one

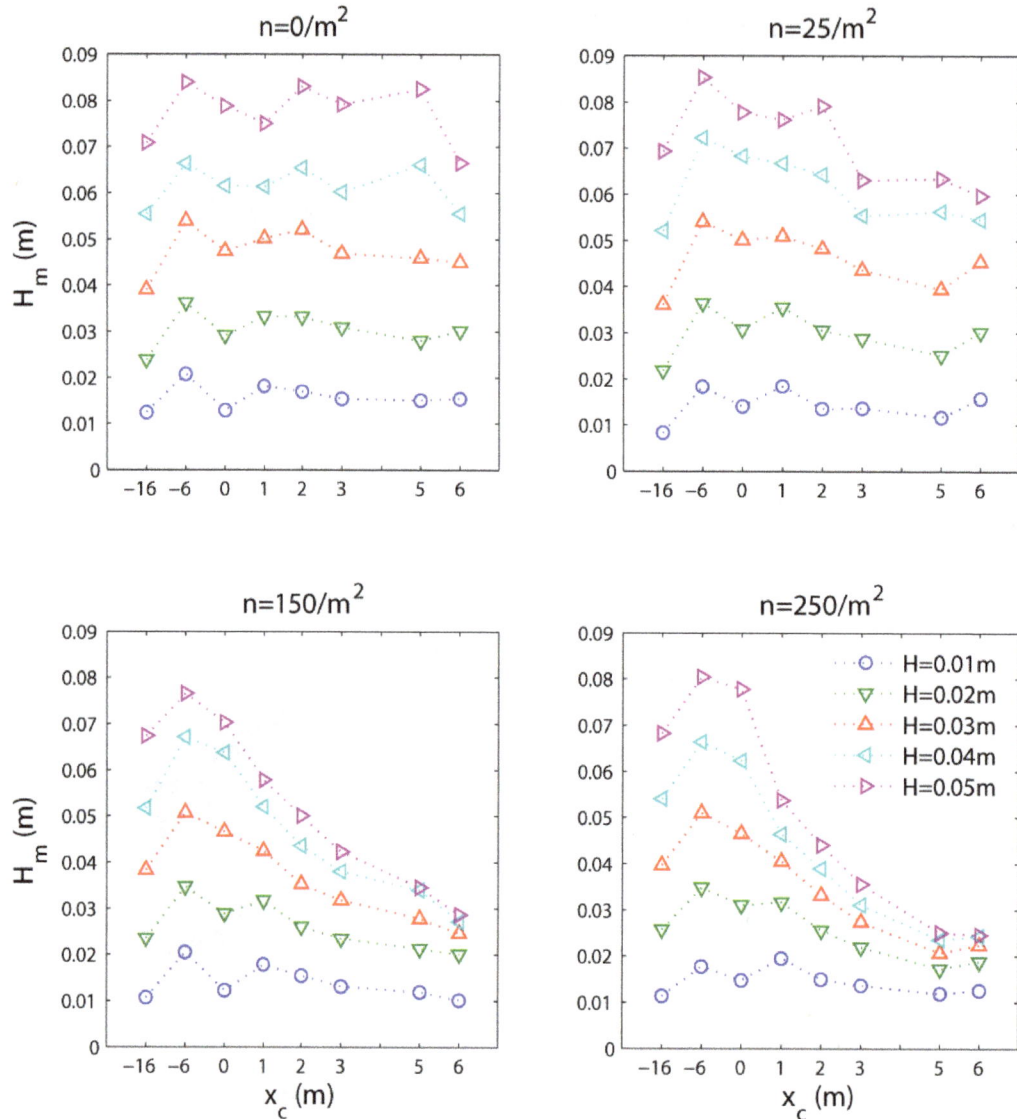

Figure 8. Mean wave height, H_m, recorded along the flume, from the nearest sensor to the wave maker ($x_c = -16$ m), to the beginning of the sand bed ($x_c = -6$ m) to the end of the array ($x_c = 6$ m. Notice negative x_c coordinates are not in scale.

observed for currents, where the array effectively slows down the flow, yielding smaller velocities at the upstream section than at the downstream exit.

To account for the relationship between orbital velocities within the array and the measured concentrations, the maximum velocities in both directions (U_{w+} and U_{w-}) were obtained from the records. The near-bottom orbital excursion, $d_w = \frac{U_{w+}T}{2\pi}$, was calculated using U_{w+} as the near-bottom orbital speed (Table 2). The relationship between the near bottom orbital speed and recorded values of suspended sediment concentration appears in Figs. 11 and 12. From the results from the station at $x_c = 3$ m it is clear the difference between sparse ($n = 0, 25\,\mathrm{m}^{-2}$) and dense ($n = 150, 250\,\mathrm{m}^2$) arrays,

with the data almost collapsing into two distinctive curves in Fig. 12.

We also compare the behaviour of free surface elevation and concentration for the non-populated and the densest case, at the two instrumented locations (Figs. 13 and 14). The correlation between the time series is evident, and the amount of suspended sediment increases as the wave travels through the sand bed for $n = 0$, while the opposite occurs for the densest case. The effect was clearly visible during the experiments, with a cloud of sediment being lifted as the wave encountered the array. Even when monochromatic, sinusoidal waves were generated at the paddle, shoaling due to the ramps built to contain the sand, and the changes in the bottom material, concrete–steel–sand, result in significant wave transforma-

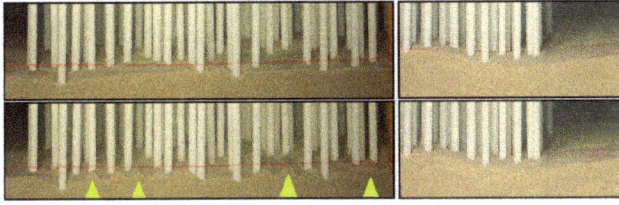

Figure 9. Side view: upstream (left) and downstream (right) edge of the array, before (up) and after (down) 50 waves, $H = 0.05\,\mathrm{m}$, $T = 2.5\,\mathrm{s}$, $n = 150\,\mathrm{m}^{-2}$. Notice the scour and bedforms forming within the array at the upstream section (as pointed by the green arrows) while the downstream section remains unaltered. Notice the factor of 2 increase in the vertical scale.

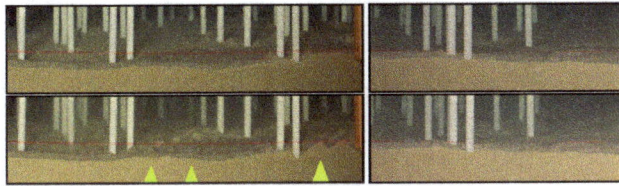

Figure 10. Side view: upstream (left) and downstream (right) edge of the array, before (up) and after (down) 50 waves, $H = 0.05\,\mathrm{m}$, $T = 2.5\,\mathrm{s}$, $n = 25\,\mathrm{m}^{-2}$. Notice the scour and bedforms forming within the array at the upstream section (as pointed by the green arrows) while the downstream section remains unaltered. Notice the factor of 2 increase in the vertical scale.

tion. In Fig. 13 it is noticed the skewed, saw-tooth shape nature of the waves as they reach the middle of the array ($x_c = 3\,\mathrm{m}$). The effects of velocity and acceleration skewness on sediment transport are an active area of research and, as pointed out by Hoefel and Elgar (2003) and van der A et al. (2010), are not well understood. However, it is known that as the waves enter shallow water (representative of the habitat of our considered species), sinusoidal waves can transform into skewed, saw-tooth shaped waves with a sharp front as the ones observed in Fig. 13 leading to higher onshore velocities that may result in onshore net transport. Since we do not have enough data to compare different skewness and asymmetry conditions, we limit the study to the measured velocities under the achieved wave conditions and their respective suspended sediment concentrations.

To look for similar trends in the recorded velocities and concentrations, the frequency spectra are calculated for the $n = 0$ and $n = 250\,\mathrm{m}^{-2}$ cases, as shown in Fig. 15. By comparing the densest against the non-populated case, we notice the passing of energy towards higher frequencies, as eddies break down to smaller scales through the array and energy passes from wave scales to individual cylinder scales.

Table 3. Measured critical velocities for currents, U_c, and critical orbital velocities, U_{wc}, according to the measurements of suspended sediment concentration at $z = 0.05\,\mathrm{m}$ for all densities at $x_c = 3\,\mathrm{m}$.

n (m^{-2})	U_c (m s^{-1})	U_{wc} (m s^{-1})
250	0.080	0.063
150	0.096	0.071
25	0.111	0.131
0	0.163	0.120

Table 4. Theoretical critical values for sediment motion (SM) and resuspension (R).

Criteria	U_c (m s^{-1})
van Rijn 1984 (SM)	0.297
Soulsby 1997 (SM)	0.259
Soulsby and Whitehouse 1997 (SM)	0.312
van Rijn 1984 (R)	0.532
Soulsby 1997 (R)	0.917

4 Discussion

4.1 Currents

The results present some clear trends relevant to the impact of benthos on sediment motion.

Even when longitudinal, x velocities are being damped by the presence of the array (Fig. 5), turbulent kinetic energy, $k = \frac{1}{2}\left(u'^2 + v'^2 + w'^2\right)$, is generated within the cylinder array (Fig. 6), enhancing the resuspension process and accelerating the onset of sediment transport, consistent with the findings of Sumer et al. (2003), who observed significant increases in sediment transport if they increase turbulence levels while keeping the same flow speeds.

For the denser cases, even at low speeds, scour begins to occur, the wake behind each cylinder contributes to lifting the sediment, interacting with the adjacent wakes and the reduced current, allowing for the sediment to stay in suspension for the instruments to capture. Nepf and Koch (2013) observed a similar behaviour, with vertical secondary flows generated by vertical pressure gradients behind obstructions protruding from the bottom of a rigid bed. Comparing Figs. 5 and 6, it is apparent that neither the mean flow speed, nor the turbulent kinetic energy, are solely responsible for the resuspension: looking at Fig. 5 (right), lower values of k are required to reach similar values of C at $n = 150\,\mathrm{m}^{-2}$ than at $n = 250\,\mathrm{m}^{-2}$, consistent with velocities within the $n = 150\,\mathrm{m}^{-2}$ array higher than those within the densest one. Follett and Nepf (2012) also noted the competing effect between the reduced velocities and increase of turbulent production as the density increases, resulting in increased scour at higher densities, concluding that given the small size of their patch ($< 0.22\,\mathrm{m}$), it behaved as the leading edge of a

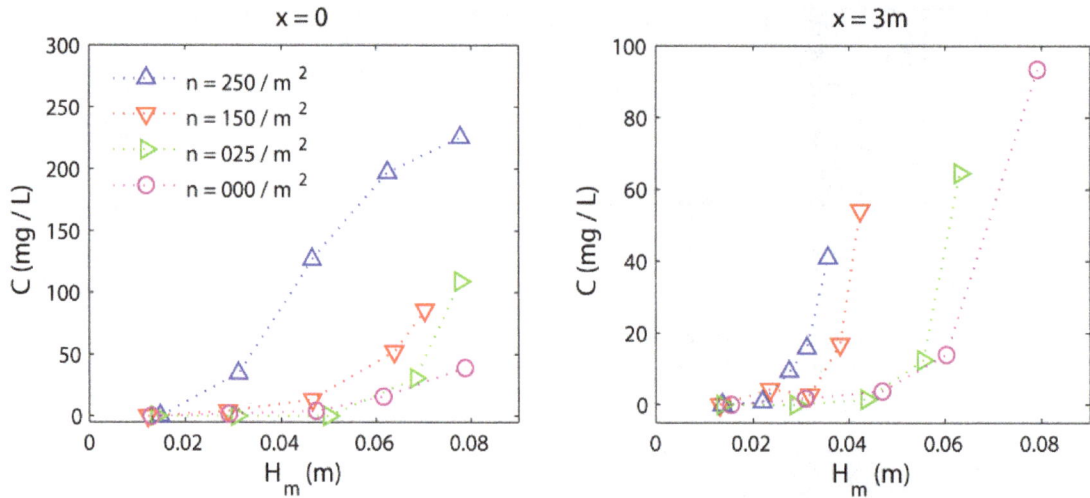

Figure 11. Mean concentrations against measured mean wave heights for all densities at two instrumented locations. Notice the different vertical scales.

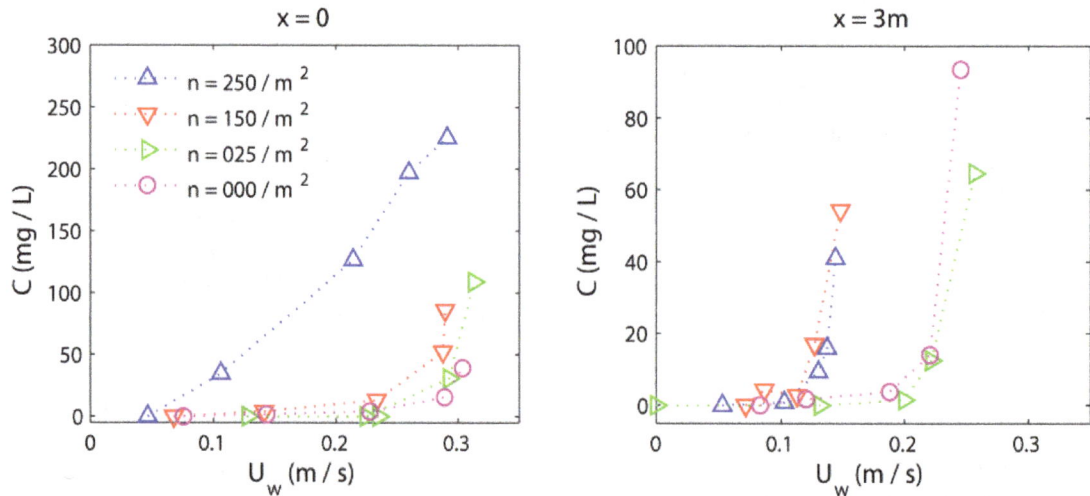

Figure 12. Mean concentrations against measured near-bottom orbital speed for all densities at two instrumented locations. Notice the different vertical scales.

longer patch. However, our measurements at the middle of our array, where fully developed flow is expected, continue to show the same increase in scour.

As a reference point, we can compare the threshold values obtained from the OBS readings (Table 3) against theoretical values from non-populated mobile beds (see Sect. 1.1). The determination of thresholds from experimental data can be a topic of discussion by itself. In our case, the thresholds were determined by finding the velocity at which the instrument starts recording concentrations higher than the background levels. The predicted values (Table 4) are considerably higher than the ones observed during the experiments, not only in the populated cases, where the cylinders clearly enhance sediment resuspension, but also in the smooth sand bed case, which can be explained either by irregularities in

the flat initial conditions of the sand bed, or by the effect of the intrusive instrumentation deployed (steel rods protruding through the sand to hold ADVs and OBSs). The height of the measurements (5 cm above the bed) must also be considered, since the physical dimensions of the instruments prevent us from getting closer to the bed, where material could be already in suspension at lower elevations before the OBSs are able to record it.

4.2 Waves

Data presented in Figs. 11 and 12 hint at a density independence within certain density ranges. Madsen et al. (2001) found contradictory results in studies arguing whether or not density is a relevant parameter for velocity reduction. Our

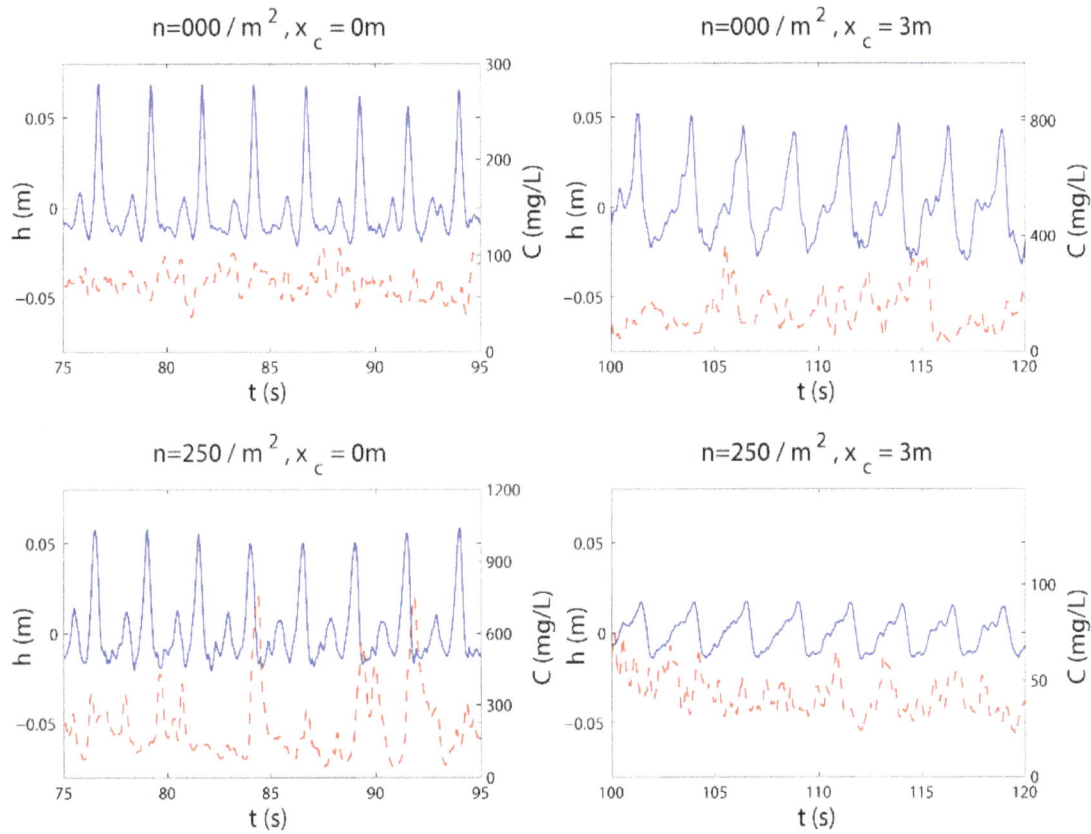

Figure 13. Time series of free surface elevation (blue solid line) and concentration (red dashed line) for the non-populated (up) and densest (down) cases, at locations $x_c = 0$ m (left) and $x_c = 3$ m (right), for waves $H = 0.05$ m. Notice the scale for η is conserved in all figures, while the scale of C changes between figures.

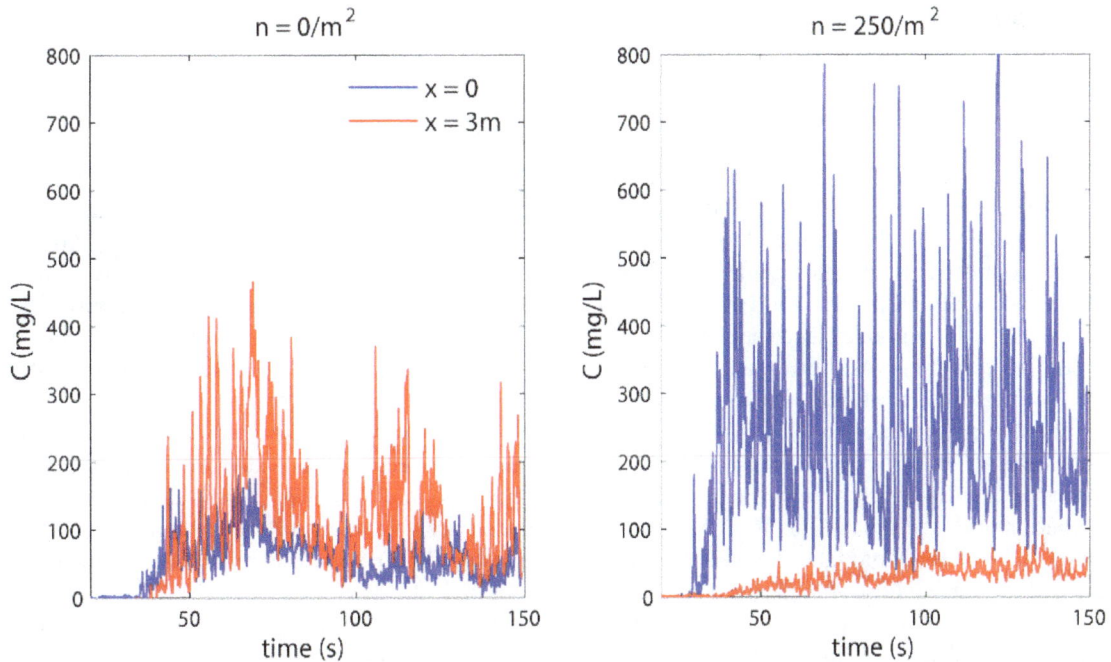

Figure 14. Time series of suspended sediment concentration for the non-populated (left) and densest (right) cases, at locations $x_c = 0$ m and $x_c = 3$ m, for waves $H = 0.05$ m.

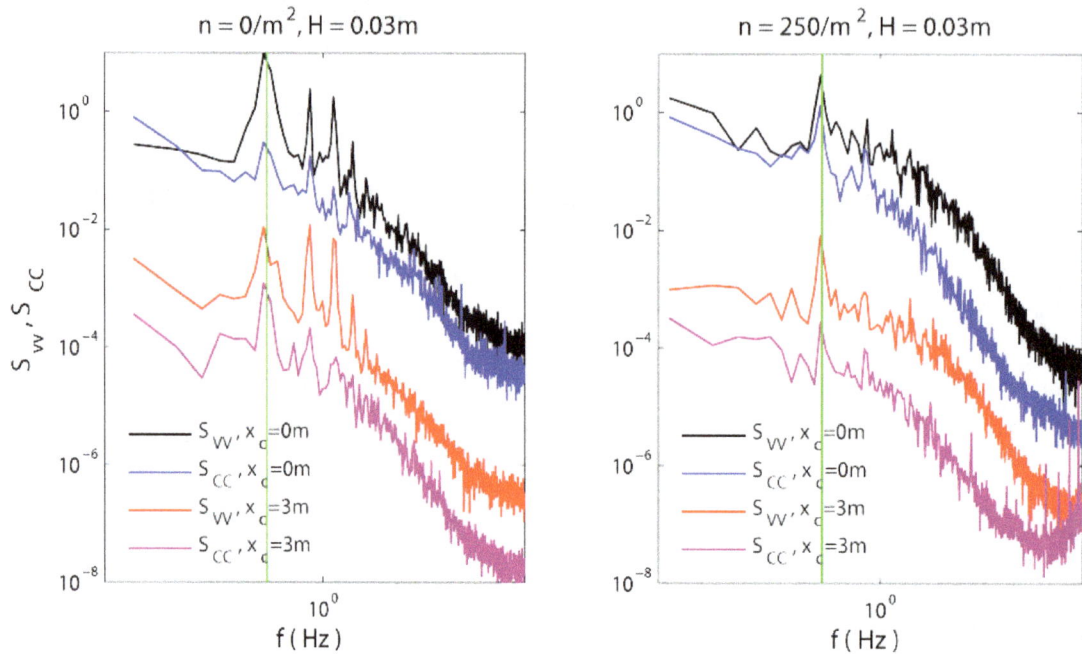

Figure 15. Frequency spectra of transverse velocity and concentration for the cases $n = 0\,\mathrm{m}^{-2}$ (left) and $n = 250\,\mathrm{m}^{-2}$ (right), at longitudinal locations $x_c = 0$ and $x_c = 3\,\mathrm{m}$. Vertical green line shows the considered wave period $T = 2.5\,\mathrm{s}$. Actual amplitudes altered to improve visualisation.

data shows that (a) results depend significantly on the location of the measurement, and (b) density values within the same regime (sparse, intermediate or dense) can share similar behaviours, but the relevance of the density can not be overlooked (e.g. from Figs. 11 and 12, looking only at the results for $n = 25$ and $150\,\mathrm{m}^{-2}$, the effect of density is clear, while looking only at the $n = 250$ and $150\,\mathrm{m}^{-2}$ can mislead into conclusions of density independence).

The data presented in Fig. 12 allow us to estimate a critical orbital velocity for each density. We notice that an abrupt change in concentration is apparent for all cases at specific velocities, but the determined thresholds (velocities at which the concentration exceeds the background levels) occur at lower values. We compare those values, presented in Table 3, against the theoretical threshold, $U_{wc} = 0.112\,\mathrm{m\,s}^{-1}$ from Komar and Miller (1973) for non-populated mobile beds. The measured values for the sparse cases, $n = 0$ and $n = 25\,\mathrm{m}^{-2}$, agree with the predicted value, while the densest cases show a near 40 % reduction of the critical velocity.

Additional effects of the array density were observed by looking at the spectra to investigate similarities between velocities and concentration records. Figure 15 shows two relevant features: (1) the presence of higher frequency harmonics in all cases, (2) the smoother spectra for the densest case, with none of the large peaks observed in the non-populated case, showing a redistribution of energy within the array, with more energy contained at smaller scales and higher frequencies, and (3) the same redistribution appears as we move

from $x_c = 0$ to $x_c = 3\,\mathrm{m}$ through the array, with an evident shift in the energy cascade.

Madsen et al. (2001) found differences in the interactions between flow, vegetation and sediment, among regions dominated by either waves or unidirectional currents, finding contradictory results in (a) whether or not the density is relevant for velocity reduction, (b) if turbulence decreased or increased within vegetation, and (c) whether sediment resuspension is higher in wave-dominated areas than in tide-dominated ones. As pointed out through the present manuscript, one must be extremely cautious when comparing: (1) emergent and submerged conditions: the added mixing layer and the appearance of skimming flow in the submerged case produce contradictory results within the vegetation; (2) rigid and flexible model vegetation: plants bending under currents create a different and increased sheltering effect damping possible coherent structures that would appear on rigid structures, while the flapping of flexible vegetation under waves increases the bed exposition to the flow and changes the energy dissipation with respect to rigid element; or (3) laboratory and field studies: although density effects on resuspension have already been reported (e.g. Widdows and Brinsley, 2002), it is worth reiterating that biological interactions in the field between the organisms and the sediment can produce an effect contrary to the expected from their physical presence (e.g. plants roots or biofilms holding the sediment together).

5 Conclusions

Arrays of randomly placed emergent cylinders, simulating benthic populations, change the flow–sediment interactions around and within the array.

The presence of arrays of rigid cylinders enhances sediment resuspension under waves and currents. The measured critical velocities for the densest cases are considerably smaller than those for the sparsest cases, and smaller than predicted velocities for non-populated beds.

With unidirectional currents, by slowing down the flow, emergent arrays might be expected to suppress bottom shear stress, reduce resuspension and favour deposition. However, the cylinder wakes enhance scour, the bed shear stresses change, and the increased turbulence levels within the array effectively enhance resuspension.

For the sparse cases under currents, the values of suspended sediment concentration begin to increase slowly when the threshold is reached, as opposed to the dense cases where once the threshold is reached there is an abrupt jump in such values.

For both waves and currents, eddies break down to smaller scales as they move through the array, with energy passing from wave scales to individual cylinder scales.

The density of the array is a determining parameter on the early onset of sediment resuspension. There is a clear distinction between the resuspension behaviour of sparse and dense arrays.

More studies are needed to further improve understanding and predictive capability of resuspension in these fragile environments.

Acknowledgements. The authors gratefully acknowledge the support of the Augusto Gonzalez Linares program at the University of Cantabria, and the comments of two anonymous reviewers to improve the manuscript.

Edited by: F. Metivier

References

Borsje, B., Kruijt, M., Werf, J., Hulscher, S., and Herman, P.: Modeling biogeomorphological interactions in underwater nourishments, Coast. Engineer. Proc., 1, 1–11, 2011.

Bouma, T., van Duren, L., Temmerman, S., Claverie, T., Blanco-Garcia, A., Ysebaert, T., and Herman, P.: Spatial flow and sedimentation patterns within patches of epibenthic structures: Combining field, flume and modelling experiments, Cont. Shelf Res., 27, 1020–1045, 2007.

Carey, D.: Particle resuspension in the benthic boundary layer induced by flow around polychaete tubes, Can. J. Fish. Aquat. Sci., 40, 301–308, 1983.

Coco, G., Thrush, S., Green, M., and Hewitt, J.: Feedbacks between bivalve density, flow, and suspended sediment concentration on patch stable states, Ecology, 87, 2862–2870, 2006.

Eckman, J., Nowell, A., and Jumars, P.: Sediment destabilization by animal tubes, J. Mar. Res., 39, 361–374, 1981.

Follett, E. M. and Nepf, H. M.: Sediment patterns near a model patch of reedy emergent vegetation, Geomorphology, 179, 141–151, 2012.

Fonseca, M. and Cahalan, J.: A preliminary evaluation of wave attenuation by four species of seagrasses, Estuar. Coast. Shelf Sci., 35, 565–576, 1992.

Friedrichs, M., Graf, G., and Springer, B.: Skimming flow induced over a simulated polychaete tube lawn at low population densities, Mar. Ecol.-Prog. Ser., 192, 219–228, 2000.

Graf, G. and Rosenberg, R.: Bioresuspension and biodeposition: a review, J. Marine Syst., 11, 269–278, 1997.

Green, M. O.: Test of sediment initial-motion theories using irregular-wave field data, Sedimentology, 46, 427–441, 1999.

Green, M. O.: Very small waves and associated sediment resuspension on an estuarine intertidal flat, Estuarine, coastal and shelf science, 93, 449–459, 2011.

Hoefel, F. and Elgar, S.: Wave-induced sediment transport and sandbar migration, Science, 299, 1885–1887, 2003.

Houck, M. and Neill, R.: Plant fact sheet for black mangrove (Avicennia germinans (L.) L.), Tech. rep., USDA-Natural Resources Conservation Service, Louisiana Plant Materials Center, 2009.

Infantes, E., Orfila, A., Simarro, G., Terrados, J., Luhar, M., and Nepf, H.: Effect of a seagrass (Posidonia oceanica) meadow on wave propagation, Mar. Ecol.-Prog. Ser., 456, 63–72, 2012.

Kobayashi, N., Raichle, A., and Asano, T.: Wave attenuation by vegetation, J. Waterway, Port, Coastal, Ocean Eng., 119, 30–48, 1993.

Komar, P. D. and Miller, M. C.: The threshold of sediment movement under oscillatory water waves, J. Sediment. Petrol., 43, 1101–1110, 1973.

LeHir, P., Monbet, Y., and Orvain, F.: Sedimenr erodability in sediment transport modelling: Can we account for biota effects?, Cont. Shelf Res., 27, 1116–1142, 2007.

Luckenbach, M.: Sediment stability around animal tubes: The roles of hydrodynamic processes and biotic activity, Limnol. Oceanogr., 31, 779–787, 1986.

Madsen, J., Chambers, P., James, W., Koch, E., and Westlake, D.: The interaction between water movement, sediment dynamics and submersed macrophytes, Hydrobiologia, 444, 71–84, 2001.

Maza, M., Lara, J., and Losada, I.: A coupler model of submerged vegetation under oscillatory flow using Navier-Stokes equations, Coast. Eng., 80, 16–34, 2013.

Nepf, H.: Flow and transport in regions with aquatic vegetation, Annu. Rev. Fluid Mech., 44, 123–142, 2012a.

Nepf, H.: Hydrodynamics of vegetated channels, J. Hydraul. Res., 50, 262–279, 2012b.

Nepf, H. M. and Koch, E. W.: Vertical secondary flows in submersed plant-like arrays, Limnol. Oceanogr., 44, 1072–1080, 1999.

Nikora, V.: ydrodynamics of aquatic ecosystems: an interface between ecology, biomechanics and environmental fluid mechanics, River Res. Appl., 26, 367–384, 2010.

Soulsby, R.: Dynamics of marine sands, Thomas Telford Publications, 1997.

Sumer, B., Chua, L., Cheng, N., and Fredsoe, J.: Influence of turbulence on bed load sediment transport, J. Hydraul. Eng., 129, 585–596, 2003.

Tinoco, R. and Cowen, E.: The direct and indirect measurement of boundary stress and drag on individual and complex arrays of elements, Exp. Fluids, 54, 1–16, 2013.

van der A,D.A., O'Donoghue, T., and Ribberink, J.S.: Measurements of sheet flow transpor t in acceleration-skewed oscillatory flow and comparison with practical formulations, Coast. Eng., 57, 331–342, 2010.

VanHoey, G., Guilini, K., Rabaut, M., Vincx, M., and Degraer, S.: Ecological implications of the presence of the tube-building polychaete Lanice conchilega on soft-bottom benthic ecosystems, Mar. Biol., 154, 1009–1019, 2008.

VanRijn, L.: Sediment transport, Part II: Suspended load transport, J. Hydraul. Eng., 110, 1613–1641, 1984.

VanRijn, L.: Unified view of sediment transport by currents and waves. I: Initiation of motion, bed roughness, and bed-load transport, J. Hydraul. Eng., 133, 649–667, 2007.

Wallentinus, I. and Nyberg, C.: Introduced marine organisms as habitat modifiers, Mar. Pollut. Bull., 55, 323–332, 2007.

Widdows, J. and Brinsley, M.: Impact of biotic and abiotic processes on sediment dynamics and the consequences to the structure and functioning of the intertidal zone, J. Sea Res., 48, 143–156, 2002.

6

Extracting topographic swath profiles across curved geomorphic features

S. Hergarten[1]**, J. Robl**[2]**, and K. Stüwe**[3]

[1]Universität Freiburg i. Br., Institut für Geo- und Umweltnaturwissenschaften, Freiburg, Germany
[2]Universität Salzburg, Institut für Geographie und Geologie, Salzburg, Austria
[3]Universität Graz, Institut für Erdwissenschaften, Graz, Austria

Correspondence to: S. Hergarten (stefan.hergarten@geologie.uni-freiburg.de)

Abstract. We present a new method to extend the widely used geomorphic technique of swath profiles towards curved geomorphic structures such as river valleys. In contrast to the established method that hinges on stacking parallel cross sections, our approach does not refer to any individual profile lines, but uses the signed distance from a given baseline (for example, a valley floor) as the profile coordinate. The method can be implemented easily for arbitrary polygonal baselines and for rastered digital elevation models as well as for irregular point clouds such as laser scanner data. Furthermore it does not require any smoothness of the baseline and avoids over- and undersampling due to the curvature of the baseline. The versatility of the new method is illustrated by its application to topographic profiles across valleys, a large subduction zone, and the rim of an impact crater. Similarly to the ordinary swath profile method, the new method is not restricted to analyzing surface elevations themselves, but can aid the quantitative description of topography by analyzing other geomorphic features such as slope or local relief. It is even not constrained to geomorphic data, but can be applied to any two-dimensional data set such as temperature, precipitation or ages of rocks.

1 Introduction

Traditional cross-section analysis of topography always involves some randomness in the choice of the profile line. Swath profiles are a widespread tool to reduce this randomness and to focus on the morphologic features of interest. This technique has been used in numerous geomorphic studies. Some of them use swath profiles mainly for illustrating the topography and focus on profiles of surface elevation itself (e.g., Rehak et al., 2008; Robl et al., 2008), while several others also analyze further geomorphic properties such as slope and local relief (e.g., Fielding et al., 1994; Montgomery, 2001; Mitchell and Montgomery, 2006b). In some studies swath profiles were even used to examine the relationship between topography and other data that are potentially related to topography, for example precipitation (Fielding et al., 1994), ages of rocks (Reiners et al., 2003), and exhumation rates (Mitchell and Montgomery, 2006a). Several further examples of applying swath profiles were reviewed

by Telbisz et al. (2013). As mentioned by Grohmann (2004), the basic idea even dates back to the 1920s. The implementation in geographic information systems, which was the main focus of the paper by Grohmann (2004), has also been addressed in several online tutorials and discussions.

In principle, creating a swath profile is nothing else but stacking several parallel profiles. As illustrated in Fig. 1a, computing a swath profile involves extending the original profile line (green) to a rectangle of a given width (the swath; black frame), and the elevation data are stacked along red lines normal to the profile line. In this context, stacking comprises computing at least a mean value of the elevations along the red lines, but in many cases, minimum and maximum values or the standard deviation are determined and analyzed, too.

Considering wider swaths reduces the effect of variations in topography normal to the profile line. This leads to smoother profiles and probably helps to recognize the characteristics of the along-profile topography. But in return, swath

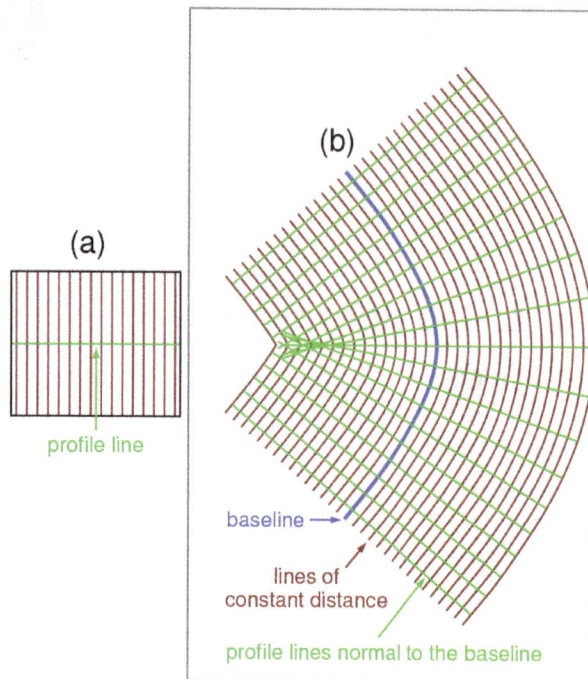

Figure 1. (a) Illustration of the idea of ordinary swath profiles. Green: profile line. Red: lines along which the elevation data are stacked. **(b)** Two ways of generalizing swath profiles towards curved baselines (blue). Green: profile lines normal to the baseline which might be stacked. Red: lines of constant signed distance from the baseline (the method proposed here). The gray frame defines the region of all points that potentially contribute to the profile as discussed in Sect. 3.

per it was also shown that swath profiles are in principle not restricted to rectangular swaths, but also allow the consideration of curved structures. As an example, the application of "central swath profiles" to the morphology of a volcano was presented where the profile coordinate is the distance from a given point. Beyond this, even an extension towards curvilinear profile lines was suggested, and two different ways of stacking the data of the surface points not situated on the profile line were discussed. As an example, the application to a longitudinal profile over a mountain range was given.

In this paper we present an alternative idea of generalizing swath profiles towards curved geomorphic structures which differs from the idea of Telbisz et al. (2013) not only technically, but also in its spirit. While their method is designed to take a profile along (and in some sense parallel to) a given curved profile line, our approach generates swath profiles across a given (curved) baseline (for example, a river). The examples presented in Sect. 4 will illustrate that this concept has even a wider field of application in geomorphology than taking profiles along curved lines.

2 The generalized swath profile method

Our new approach hinges on the finding that the profile line of a topographic cross section is not a distinct morphological object in most cases. This particularly applies to swath profiles and is most obvious in the example of river valleys where the river is the distinct morphological feature. For a longitudinal river profile (i.e., a profile along the river line), swath profiles are not useful as they may be strongly biased by the cross-sectional shape of the valley even if the curvature of the river line is taken into account. In return, taking a swath profile makes sense if we are interested in the shape of the valley and choose a profile line across the valley.

In contrast to Telbisz et al. (2013) we therefore do not start from a distinct (potentially curved) profile line, but define a (curved) baseline (the river line for a profile across a river valley) that fixes the origin of the profile coordinate. The two more or less obvious ways to define a swath profile across (i.e., in some sense normal to) this baseline are discussed in the following.

The first way consists in defining several nonparallel profiles where each of them is normal to the baseline at different points on the baseline. The data of these profiles are stacked in such a way that the origin of each individual profile line is located at its intersection with the baseline. The green lines in Fig. 1b illustrate this concept. Although somewhat straightforward, this method still suffers from some problems. First, the density of the individual profiles varies within the considered domain, so that points close to the center of the baseline's curvature contribute more to the resulting swath profile than regions at the convex side. This results in an over- or undersampling depending on the curvature of the baseline. It is also easily recognized in Fig. 1b that the same surface

profiles are subject to a systematic bias as soon as the topography contains curved structures such as river valleys. Profiles across valleys are among the most widely used types of topographic profiles. Even if the original profile line is normal to the the valley floor, stacking along the red lines in Fig. 1a loses track of a curved valley floor depending on its curvature. As a consequence, even a perfectly V-shaped valley will look more like a U-shaped valley where the width of the artificially flattened valley floor increases with both the curvature of the valley and the width of the swath. This phenomenon is exemplified in Fig. 2 using a part of the Taugl Gorge in the Austrian Alps. This example will be discussed in more detail when demonstrating the capabilities of our new method in Sect. 4. We will find there (Fig. 4d) that the considered part of the gorge is very narrow, and that the walls become particularly steep towards the river. In Fig. 2 a rather wide (1.5 km) swath normal to a 1 km long profile across the valley is considered. Due to the nonstraight course of the river, the valley floor which is in fact very narrow appears to be about 100 m wide, and the steepness of the walls is underestimated.

In a recent paper, Telbisz et al. (2013) considered theoretical aspects of the swath profile method in detail. In this pa-

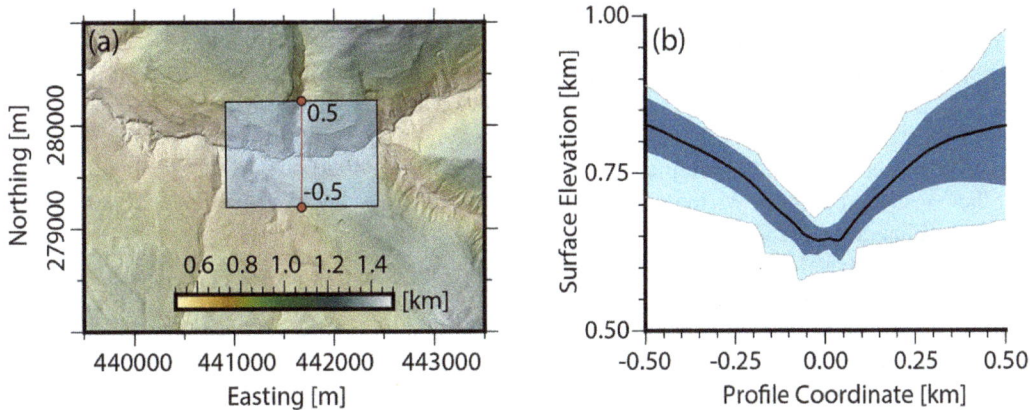

Figure 2. (a) Ordinary swath profile across the central part of the Taugl Gorge (Austria). The curves in **(b)** comprise the mean elevations (black line), extreme values (hull of the light blue area) and mean values ±1 standard deviation (hull of the dark area). The data were derived from lidar based DEM data (spatial resolution 1 m) in the MGI (Military Geographical Institute) Austria M31 coordinate system provided by the federal government of Salzburg.

point may even contribute to several of the stacked profiles, but at different profile coordinates. As a second problem, this method requires some smoothing of the baseline (in particular if it was automatically derived from a digital elevation model) since the orientation of the individual profile lines is very sensitive to the small-scale roughness of the baseline.

The alternative concept refrains from defining individual profiles, but directly uses the signed distance of any surface point from the baseline instead. Signed distance (sometimes called oriented distance) shall mean that all points to the right-hand side of the baseline have a positive distance, while those to the left-hand side have a negative distance. This signed distance defines the position of the considered surface point on the resulting generalized swath profile. In principle, this method stacks the elevation data along the red lines of constant signed distance in Fig. 1b.

Computing these lines of constant distance would, of course, be rather demanding. However, we do not need these lines explicitly, but simply travel through the entire domain instead, i.e., along all points of the DEM (digital elevation model) in the discrete case. We therefore only need to compute the signed distance of each DEM point towards the baseline of the profile, which is discussed in the next section in more detail.

Apart from the advantage of its simplicity, this procedure guarantees that each DEM point contributes exactly once to the profile, immediately avoiding over- or undersampling. Furthermore it is robust against the roughness of the baseline. It is easily recognized that the lines of small distances follow this roughness, while those of larger distances become smooth (see, for example, the red lines in Fig. 4a).

In view of these advantages over the method of stacking individual profile lines, we only consider the approach based on the signed distance in the following.

3 Implementation

The implementation of our method based on a DEM and a baseline defined by a polygonal line is in principle straightforward. In a first step, a bounding box, i.e., a tight rectangle with edges parallel to the coordinate axes around the baseline, is computed. This bounding box is extended by half the length of the desired profile in each direction, so that it includes all points of the surface that may contribute to the profile. The resulting rectangle is illustrated in Fig. 1b by the grey frame.

Then, all DEM points in this rectangle are iterated, and the signed distance of each point from the polygonal baseline is computed. As long as the number of line segments of the baseline is not too high, this can be done individually for each DEM point p without using further information. Let b_1, ..., b_n be the points defining the baseline. We then determine the distance of p towards each segment, which is assumed to be a straight line between two adjacent baseline points b_i and b_{i+1}. This means that we compute λ_i for each segment $i = 1, \ldots, n-1$ in such a way that the distance between the point p and the point

$$q_i = (1 - \lambda_i)b_i + \lambda_i b_{i+1} \qquad (1)$$

is minimal under the side condition $0 \leq \lambda_i \leq 1$. The minimum of all obtained distances yields the distance of the point p to the baseline.

Practically, this algorithm is sufficient in all cases where the baseline is generated manually, so that its number of segments does not exceed some hundreds. A more efficient algorithm is only necessary for automatically generated baselines with a high resolution, as when flow channels derived from the DEM are considered. If the length of the baseline segments is similar to the mesh width of the DEM, the numerical effort increases like the baseline length raised to the power of three (or alternatively, to the DEM resolution raised to the power of three for a given baseline length), so that the procedure becomes expensive for long baselines or high DEM resolutions.

In this case, we suggest an iterative algorithm using available information on the nearest point on the baseline from the neighborhood of the considered point p. We then keep track of the index of the baseline segment that contains the presumably nearest point on the baseline for each DEM point p. When searching for the baseline point nearest to p, we consider the information on the nearest baseline segment in the neighborhood of p. Let $s_1, \ldots, s_9$ be the indices of the segments nearest to p among p and its eight direct and diagonal neighbors. We then search for the nearest point only in the baseline interval from segment $\min(s_1, \ldots, s_9) - 1$ to $\max(s_1, \ldots, s_9) + 1$. As a consequence, only a small part of the baseline has to be searched in each step. This iteration is repeated until the results do not change any more. In order to achieve a high efficiency and to avoid getting stuck at local minima, e.g., for meander-like baselines, the points of the DEM should be iterated in varying directions. A straightforward scheme would be row-wise forward, row-wise backward, column-wise forward, and then column-wise backward. Practically, only the variation between forward and backward iteration is important and provided a convergence after a few iterations in all tests that were performed.

The two end points of the baseline require a specific treatment anyway. As our approach is based on the distances towards the baseline, it extends the swath by semicircles around the end points of the baselines. These semicircles may introduce a strong bias, for example, for profiles across valleys. In order to avoid this artefact we include only these points of the DEM in the analysis where the nearest point on the baseline is none of the two end points of the baseline. The red lines in Fig. 1b already take this constraint into account.

As a last step of the analysis, the obtained pairs of elevation versus profile coordinate data must be aligned in bins. The bin width (on the profile-coordinate axis) defines the spatial resolution of the resulting profile.

Although the examples provided in Sect. 4 were derived from DEMs on regular lattices (either Cartesian coordinates or in longitude and latitude), the method can easily be applied to arbitrary point clouds without any loss of efficiency. Furthermore, it is not restricted to topographic data, but applicable to any two-dimensional data set, for example, hydrological data or the distribution of elements in a mineral.

An implementation of the method is available as an online tool at http://hergarten.at/geomorphology/swathprofile.php using data from several available DEMs with almost global coverage. Furthermore, a command line tool (C++ source code without requirement of any specific libraries) is included as a supplement, and the latest version can be directly obtained from the corresponding author.

4   Applications

In order to illustrate the new method for the construction of swath profiles, the applications presented in the following focus on topographic features that are inherently curved in plan view. Beyond the obvious application to valleys we present examples of a large subduction zone and an impact structure.

4.1   Fluvial and glacial valleys

The morphological analysis of valleys is probably the best example to illustrate the advantages of the method presented here. The cross-sectional shape of river valleys is often used to infer the incision process of the valley. In particular, two end members of valley shapes are usually discerned: V-shaped valley cross sections are considered to be characteristic for actively incising streams and a fluvial erosional history (e.g., Bull, 1979), while U-shaped cross sections are interpreted in terms of glacial erosion (e.g., Bonney, 1874). However, identifying these shapes quantitatively is often nontrivial: locations appropriate for representative individual cross sections are often difficult to find, for example, because of tributaries. And as discussed in the introduction, river valleys are typically curved, so that ordinary swath profiles suffer from artefacts that may even make a V-shaped valley look like a U-shaped valley.

In Fig. 3 we use the generalized swath profile method to characterize two valleys that are famous for their fluvial and glacial erosional history, respectively: a segment of the Colorado River upstream of the Grand Canyon, which is a typical actively incising bedrock channel (e.g., Pederson et al., 2006; Wernicke, 2011) and the glacially carved Yosemite Valley of California (Gutenberg et al., 1956).

For both generalized swath profiles the baseline was chosen along the actual course of the rivers draining the valleys. Our analysis reveals that both cross sections are qualitatively similar with a flat valley floor, steep faces, and well-marked ledges forming distinct shoulders on average. However, the apparently flat valley floor is just the water level for the Colorado River, while the Yosemite Valley is indeed much wider than the river itself.

Our next example (Fig. 4) refers to a V-shaped valley with a very narrow river and illustrates the power of morphological analysis of different river segments. The Taugl River is a tributary of the Salzach River that drains a major region of the Eastern Alps. It is characterized by several segments that flow through an ancient uplifted landscape, over a knickpoint in the channel profile, and in a young incised landscape. In the headwater region (Fig. 4b, c), the average cross section of the Taugl River is perfectly V-shaped with a narrow valley floor and an average topographic gradient of about 0.6 for the corresponding hillslopes. The central segment (Fig. 4d, e) starts downstream of a distinct knickpoint (waterfall) and describes a narrow gorge with about 50 m high, almost vertical faces bordering the river. Above this very steep slope

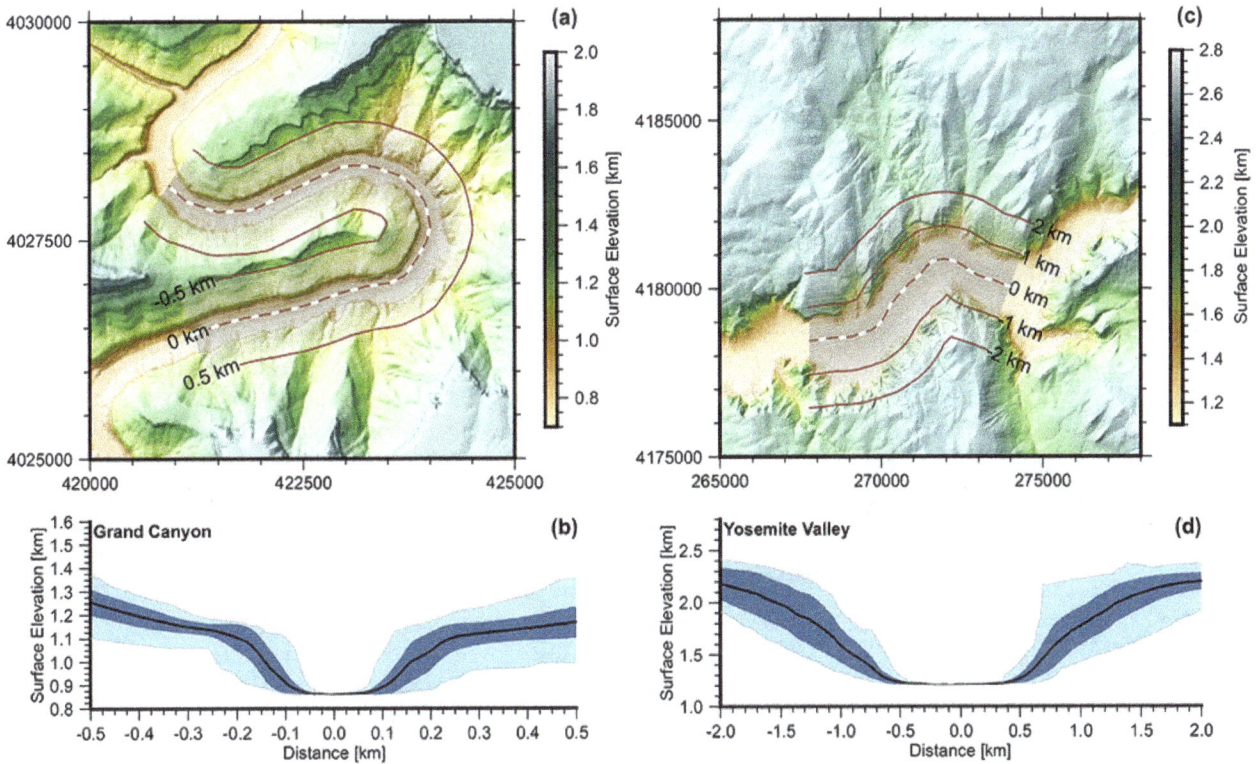

Figure 3. Fluvially versus glacially eroded valleys. The maps show sections of (**a**) the Colorado River formed by fluvial incision and (**c**) the Yosemite Valley shaped by glacial scouring. Constant distance lines are red, only the baseline is dashed red and white. The pale region covers all values included in the swath profiles shown in (**b**) and (**d**). Mean elevation (black line), extreme values (hull of the blue area) and mean elevation ±1 standard deviation (hull of the dark blue area) are plotted. Bin width is 10 m. Maps and swath profiles were derived from the National Elevation Dataset in the UTM (Universal Transverse Mercator) coordinate system with a spatial resolution of 10 m (Gesch et al., 2002).

segment the topographic gradient declines with increasing distance to the stream forming convex hillslopes. When comparing this profile with the ordinary swath profile shown in Fig. 2, the advantage of the new method is obvious. The third segment (Fig. 4f, g) is similar to the middle segment except for an apparently flat region on the orographic left-hand side of the river. However, the standard deviations amounting to some tens of meters and the high local topographic gradients reveal that it is not a simple horizontal planation surface. Such a result may arise from the roughness of the surface, but may also occur if the overall slope of this surface parallel to the profile baseline differs from the slope of the valley itself. This uncertainty is, of course, an inevitable limitation when two-dimensional information is condensed into a one-dimensional cross section.

In these profiles it is also clearly visible that the variability (both standard deviation and difference between minimum and maximum elevation) does not vanish at the baseline itself due to the downstream gradient of the river. In order to avoid this, our software optionally allows us to consider the elevations of all DEM points relative to their nearest point on the baseline and to transform the obtained profile (consisting of elevations relative to the baseline) back to absolute values afterwards.

4.2 Subduction zones and related oroclines

Subduction zones are a key to understanding plate tectonics (Stern, 2002). Their topographic and bathymetric representation often shows similar characteristic features over several hundred to some thousand kilometers along strike, but strong variations perpendicular to these zones. Abyssal plains, forebulge and the deep sea trench are located on the subducting lower plate while the related mountain ranges are located on the upper plate. In general, these geomorphic features are located at a certain distance to the subduction zone and profiles normal to the strike of these zones are useful to characterize the morphological features. However, small-amplitude features like the elastic forebulge on the subducting plate may easily be missed because they are swamped by irregularities along strike. The construction of swath profiles is therefore an essential aid in the topographic analysis, but may be hampered by the curvature of the subduction zone.

In order to illustrate the value of our new method, we present a topographic characterization of the curved plate

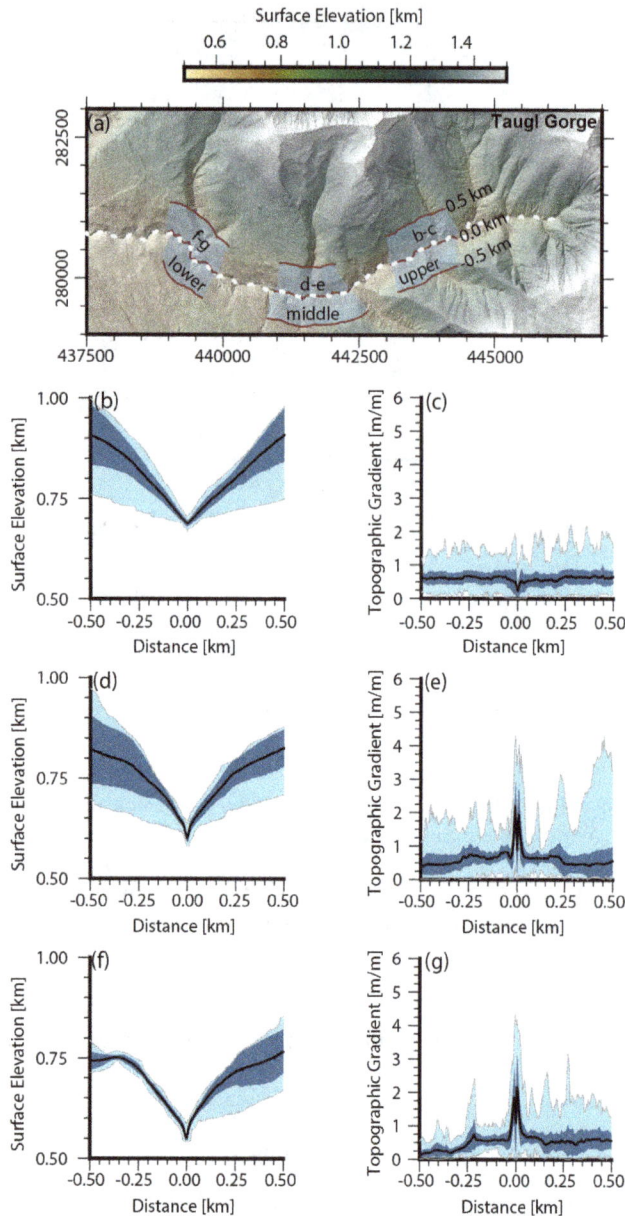

Figure 4. (a) Average geomorphic parameters of three morphologically different sections of the Taugl Gorge (Austria). The baseline (white and red dashed line) follows the course of the Taugl River. The blue transparent regions cover all values included in the profiles as annotated. Topography (**b, d, f**) and topographic gradient (**c, e, g**) are shown, comprising mean values (black line), extreme values (hull of the light blue area) and mean values ±1 standard deviation (hull of the dark area). The bin width of all profiles is 5 m. Maps and profiles were derived from lidar based DEM data (spatial resolution 1 m) in the MGI Austria M31 coordinate system provided by the federal government of Salzburg.

boundary between the Nazca Plate and the South American Plate (e.g., Isacks, 1988) in Fig. 5. As the subduction of the Nazca Plate is clearly related to the topographic development of the Andes (e.g., Gephardt, 1994), the bathymetry and to-

pography perpendicular to the plate boundary should exhibit characteristic features at similar distances to the plate boundary although the age of the Nazca Plate and its mechanical properties are spatially diverse (e.g., Capitanio et al., 2011). For our analysis we have chosen the deep sea trench between the two plates as the best feature to align the baseline of the general swath profile.

Even for the long baseline of more than 2000 km along strike used here, the standard deviations of the averaged elevations are quite small, especially for the submarine part of the swath. The analysis shows that the average surface elevation of the abyssal plains of the Nazca plate is −4.3 km with a standard deviation of less than 0.5 km. Local extrema in the curves of minimum and maximum elevation (light blue area) are predominantly seamounts (local maxima of the maximum elevation) and graben structures (local minima of the minimum elevation). The distribution of the seafloor topography is positively skewed due to the existence of some high seamounts that do not necessarily have a strong influence on the mean elevation and its standard deviation. In return, the distribution of the land surface elevations is negatively skewed because even the entire line of minimum elevation may arise from a single deep valley.

The forebulge at a distance of −50 to −100 km to the trench accounts for an average increase of about 0.2 km of surface elevation relative to the abyssal plains. The deep sea trench is on average 6.5 km deep with a standard deviation of about 0.7 km, indicating a very uniform depth distribution on the subsiding plate. On the side of the continental South American Plate, the surface elevation reaches sea level at an average distance of about 100 km from the trench. The standard deviation is small (approximately 0.7 km), indicating a uniform active continental margin over more than 2000 km in strike. The highest average surface elevation reaches about 4 km (with peaks above 6 km) and appears at a distance of 250 km from the trench. Here the standard deviation amounts to 0.8 km going along with the variable width of the Andes with a maximum west–east extent approximately at the maximum curvature of the trench. At a distance of 250 km and beyond the average surface elevation declines, while the standard deviation increases rapidly. Obviously, topographic build-up is still related to the subduction of the Nazca plate, but no longer uniformly distributed along strike. In summary, the swath profile constructed with our method shows that there are indeed characteristic topographic features perpendicular to the subduction zone and at similar distances to the trench.

4.3 Impact structures

In the analysis of impact structures, morphological features are essential to infer aspects of the impact process. The diameter and the depth of the crater are interpreted in terms of size and velocity of the impactor. The asymmetry of the crater is interpreted in terms of the impactor's shape and the obliquity

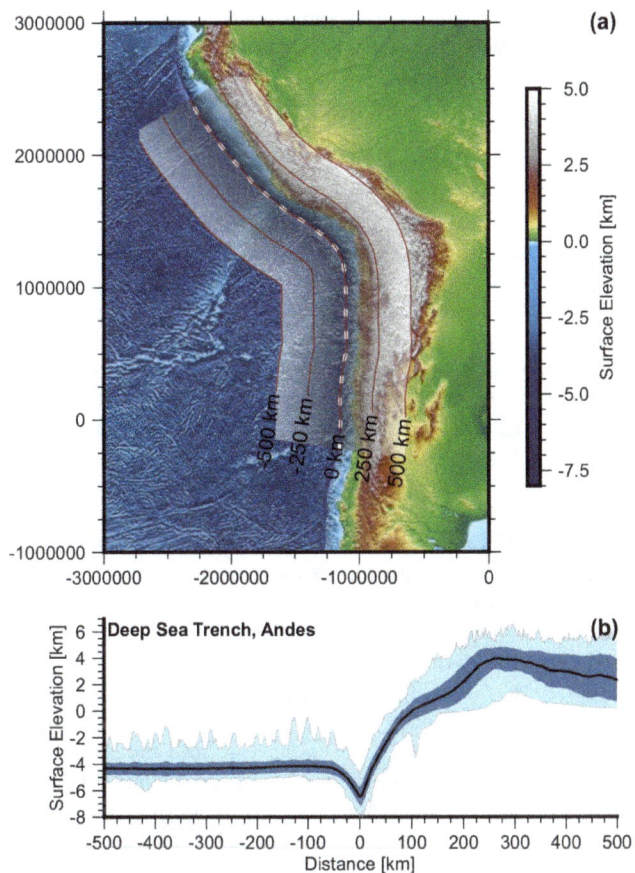

Figure 5. Topographic and bathymetric representation of the converging Nazca and South American plates as an example of a curved subduction zone. The baseline of the generalized swath profile follows the trench (white and red dashed line). Red contours indicate the signed distance to the baseline. The pale region covers all values included in the profile. The mean elevation is shown by the black line. Extreme values and mean values ±1 standard deviation are indicated by the light- and dark-blue areas, respectively. Bin width is 1 km. Map and profile were derived from the ETOPO1 digital elevation model (http://www.ngdc.noaa.gov/mgg/global/).

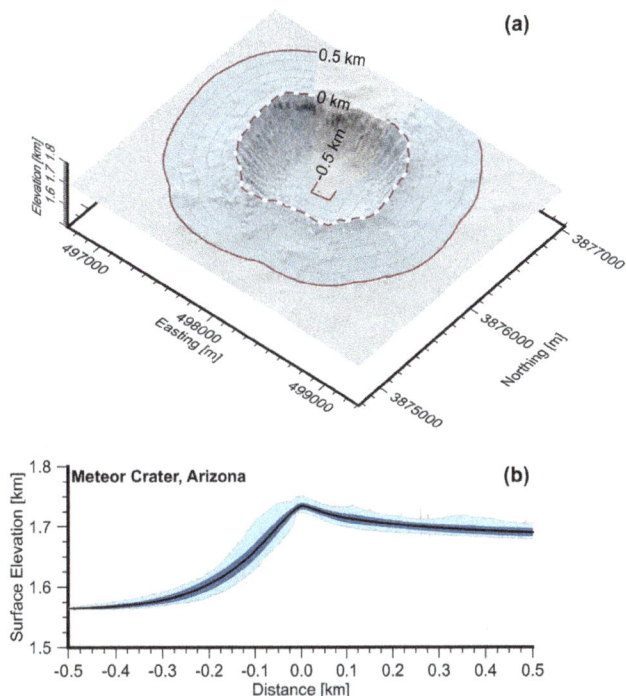

Figure 6. Generalized swath profile analysis of Meteor Crater (Arizona). The baseline (white and red dashed line) follows the crater rim, and contours indicate the signed distance to the baseline. The blue transparent region covers all values included in the profile. Mean elevation, (thick black line), extreme values (hull of the blue area), and mean elevation ±1 standard deviation (hull of the dark area) are shown. Bin width is 1 m. The map and the profile are based on the lidar point cloud processed by the National Center for Airborne Laser Mapping (http://www.ncalm.org) and interpolated onto a regular Cartesian grid (UTM coordinate system) with a spatial resolution of 2 m.

of the impact angle (Littlefield and Dawson, 2006). Fig. 6 shows our topographic analysis of Meteor Crater in Arizona, also known as Barringer Crater. It was formed by the impact of the Canyon Diablo iron meteorite about 50,000 years ago in sedimentary rocks consisting predominantly of dolomites with minor sandstones (Barringer, 1905; Nishiizumi et al., 1991; Phillips et al., 1991). The crater shows a rounded rectangular shape and deviates from a circular or elliptical form due to preexisting joints.

The baseline of our analysis follows the topographic high of the crater rim. About four-fifths of the entire crater contributes to the swath profile shown in Fig. 6b. The rest of the rim has been omitted just in order to illustrate the shape of the swath in this case. Interestingly, the surface elevation declines uniformly at increasing distance from the rim (base-line) as indicated by a very low standard deviation ranging from 1 m at the floor to 13 m at the steepest faces.

5 Conclusions

We have extended the concept of swath profiles towards curved baselines allowing the topographic analysis of river valleys, curved subduction zones, crater rims, and perhaps many other morphologic structures. Our method directly uses the signed distance of all surface points to the baseline. It immediately avoids artefacts of alternative approaches, namely the principal bias due to the curvature occurring in ordinary swath profiles and the over- and undersampling that occurs when individual, nonparallel profiles are stacked. Furthermore, the new method is robust against the roughness of the baseline and can be implemented efficiently in a numerical code. The high efficiency even persists when surfaces consisting of unstructured point clouds instead of DEMs on a regular lattice are considered. An implementation is available as an online tool (http://hergarten.at/geomorphology/swathprofile.php) and as a command line tool free of charge from the corresponding author.

Acknowledgements. The authors would like to thank I. S. Evans and M. D. Hurst for their very constructive reviews, R. Hilton for the editorial handling and several additional suggestions for the final version of the paper, and the geological survey of the federal state Salzburg for providing DEM data.

References

Barringer, D. M.: Coon Mountain and its crater, P. Acad. Nat. Sci. Phila., 57, 861–866, 1905.

Bonney, T. G.: Notes on the upper Engadine and the Italian valleys of Monte Rosa, and their relation to the glacier-erosion theory of lake-basins, Quarterly Journal of the Geological Society of London, 30, 479–489, 1874.

Bull, W. B.: Threshold of critical power in streams, Bull. Geol. Soc. Am., 90, 453–464, 1979.

Capitanio, F. A., Faccenna, C., Zlotnik, S., and Stegman, D. R.: Subduction dynamics and the origin of Andean orogeny and the Bolivian orocline, Nature, 480, 83–86, 2011.

Fielding, E., Isacks, B., Barazangi, M., and Duncan, C.: How flat is Tibet?, Geology, 22, 163–167, 1994.

Gephardt, J. W.: Topography and subduction geometry in the central Andes: Clues to the mechanics of a noncollisional orogen, J. Geophys. Res., 99, 12279–12288, 1994.

Gesch, D., Oimoen, M., Greenlee, S., Nelson, C., Steuck, M., and Tyler, D.: The National Elevation Dataset, Photogramm. Eng. Rem. S., 68, 5–11, 2002.

Grohmann, C. H.: Morphometric analysis in Geographic Information Systems: applications of free software GRASS and R, Comp. Geosci., 30, 1055–1067, 2004.

Gutenberg, B., Buwalda, J. P., and Sharp, R. P.: Seismic explorations on the floor of Yosemite Valley, California, Bull Geol. Soc. Amer., 67, 1051–1078, 1956.

Isacks, B. L.: Uplift of the central Andean plateau and bending of the Bolivian orocline, J. Geophys. Res., 93, 3211–3231, doi:10.1029/JB093iB04p03211, 1988.

Littlefield, D. L. and Dawson, A.: The role of impactor shape and obliquity on crater evolution in celestial impacts, Int. J. Impact Eng., 33, 371–371, 2006.

Mitchell, S. G. and Montgomery, D. R.: Influence of a glacial buzzsaw on the height and morphology of the central Washington Cascade Range, USA, Quaternary Res., 65, 96–107, 2006a.

Mitchell, S. G. and Montgomery, D. R.: Polygenetic topography of the Cascade Range, Washington State, USA, Am. J. Sci., 306, 736–768, 2006b.

Montgomery, D. R.: Slope distributions, hillslope thresholds and steady-state topography, Am. J. Sci., 301, 432–454, 2001.

Nishiizumi, K., Kohl, C. P., Shoemaker, E. M., Arnold, J. R., Klein, J., Fink, D., and Middleton, R.: In situ 10Be-26Al exposure ages at Meteor Crater, Arizona, Geochim. Cosmochim. Acta, 55, 2699–2703, 1991.

Pederson, J. L., Anders, M. D., Rittenhour, T. M., Sharp, W. D., Gosse, J. C., and Karlstrom, K. E.: Using fill terraces to understand incision rates and evolution of the Colorado River in eastern Grand Canyon, Arizona, J. Geophys. Res., 111, F02003, doi:10.1029/2004JF000201, 2006.

Phillips, F. M., Zreda, M. G., Smith, S. S., Elmore, D., Kubik, P. W., Dorn, R. I., and Roddy, D. J.: Age and geomorphic history of Meteor Crater, Arizona, from cosmogenic 36Cl and 14C in rock varnish, Geochim. Cosmochim. Acta, 55, 2695–2698, 1991.

Rehak, K., Strecker, M. R., and Echtler, H. P.: Morphotectonic segmentation of an active forearc, 37°–41°S, Chile, Geomorphology, 94, 98–116, 2008.

Reiners, P. W., Ehlers, T. A., Mitchell, S. G., and Montgomery, D. R.: Coupled spatial variations in precipitation and long-term erosion rates across the Washington Cascades, Nature, 426, 645–647, 2003.

Robl, J., Hergarten, S., and Stüwe, K.: Morphological analysis of the drainage system in the Eastern Alps, Tectonophysics, 460, 263–277, 2008.

Stern, R. J.: Subduction zones, Rev. Geophys., 40, 3–1–3–38, doi:10.1029/2001RG000108, 2002.

Telbisz, T., Kovács, G., Székely, B., and Szabó, J.: Topographic swath profile analysis: a generalization and sensitivity evaluation of a digital terrain analysis tool, Z. Geomorphol., 57, published online June 19, 2013, doi:10.1127/0372-8854/2013/0110, 2013.

Wernicke, B.: The California River and its role in carving Grand Canyon, Bull. Geol. Soc. Am., 123, 1288–1316, 2011.

7

Tracing the boundaries of Cenozoic volcanic edifices from Sardinia (Italy): a geomorphometric contribution

M. T. Melis[1], F. Mundula[1], F. Dessì[1], R. Cioni[2], and A. Funedda[1]

[1]Department of Chemical and Geological Sciences, University of Cagliari, Cagliari, Italy
[2]Department of Earth Sciences, University of Firenze, Firenze, Italy

Correspondence to: M. T. Melis (titimelis@unica.it)

Abstract. Unequivocal delimitation of landforms is an important issue for different purposes, from science-driven morphometric analysis to legal issues related to land conservation. This study is aimed at giving a new contribution to the morphometric approach for the delineation of the boundaries of volcanic edifices, applied to 13 monogenetic volcanoes (scoria cones) related to the Pliocene–Pleistocene volcanic cycle in Sardinia (Italy). External boundary delimitation of the edifices is discussed based on an integrated methodology using automatic elaboration of digital elevation models together with geomorphological and geological observations. Different elaborations of surface slope and profile curvature have been proposed and discussed; among them, two algorithms based on simple mathematical functions combining slope and profile curvature well fit the requirements of this study. One of theses algorithms is a modification of a function introduced by Grosse et al. (2011), which better performs for recognizing and tracing the boundary between the volcanic scoria cone and its basement. Although the geological constraints still drive the final decision, the proposed method improves the existing tools for a semi-automatic tracing of the boundaries.

1 Introduction

Unequivocal delimitation of landforms is a general objective that needs to be pursued for different purposes, both scientific, such as the morphometric analysis of a given set of landforms, and applicative, such as the conservation of the natural heritage. The continuously increasing capability of digital elaboration of the DEM (digital elevation model) data gives now the possibility of adopting tools for the automatic recognition and delimitation of specific landforms, and these tools are important for obtaining objective results that are not heavily dependent on the subjective choices of an operator. This assumes particular importance when the delimitation of a given landform is required for conservation purposes, where the delimited area is to be subject to specific regulations or laws. The improvement of objective and reproducible methods to delimitate the external perimeter of a landform, thereby reducing to a minimum the subjective choices of an operator, is very important especially if the results are to be used for applying regulations.

The morphological and morphometric characteristics of volcanic landforms and edifices have been the subject of studies since the end of the 1970s (e.g. Pike, 1978; Pike and Clow, 1981; Wood, 1980a, b), and overall reviews of the typical values of elevation, base diameter, crater diameter and average slope, and of their variability, are now available (e.g. Francis, 1993; Ollier, 1988; Cas and Wright, 1987; Thouret, 1999; Walker, 2000; Fink and Anderson, 2000). These studies have been largely used to define the main morphometric parameters of different types of volcanic edifices, and such parameters have been used to discuss and infer their origin, degradation and also age (e.g. Kereszturi et al., 2013). Conversely, to the best of our knowledge, none of the existing studies on volcanic edifice morphometrics have been used for conservation purposes up to now.

As the term volcanic landform is often used with a general meaning to indicate all the surface forms related to volcanic activity, we prefer in the following to use the more specific term "volcanic edifice" to refer to all those landforms formed by accumulation of products around the volcanic vent, whether during a single event or as the cumulated result of multiple eruptions. Volcanic edifices largely vary in type and scale: they can be constructional, such as monogenetic pyroclastic cones and lava domes at small scale, or composite volcanic massifs and shield volcanoes at large scale, or can be destructional/excavational, such as maars, craters or calderas. Consequently, the external boundary of these edifices can have largely variable sizes and shapes. In this sense, the simplest volcanic edifice is represented by a symmetric, cone-shaped pile of products built upon a planar surface. In this case the edifice boundary is circular and represented by the concave break in slope at its base (Euillades et al., 2013). Obviously, this represents an extremely simplified, somewhat theoretical, case rarely present in nature.

In their interesting review of volcanic edifice morphometrics, Grosse et al. (2012) discussed how, before the advent of DEMs, the process of delimitation of volcanic edifices, mainly pursued for studies on volcano morphometry, was based on the integrated analysis of topographic maps and air photos with field measurements. Following the large use of DEM, delimitation of volcanic edifices, mainly scoria cones, has been manually performed on the basis of shaded relief images and slope maps (e.g. Fornaciai et al., 2012); the use of dedicated algorithms for semi-automatic and automatic delimitation of volcanic edifices has been recently developed (Grosse et al., 2009, 2012; Euillades et al., 2013). Many authors have used these data for obtaining descriptive morphometric parameters (e.g. base diameter, crater elongation, cone height), or for discussing the relationships between volcanic and tectonic structures (e.g. location of depressions on the crater rim, alignment of cones, azimuths and geometry of the fracture feeding system; Tibaldi, 1995; Mazzarini et al., 2010). The studies of Hooper and Sheridan (1998), Carn (2000) and Kereszturi et al. (2012) analysed the morphometric characteristics of scoria cones in order to estimate possible relationships with their age and to evaluate the extent of erosional processes.

In the framework of the application of Article 142 of the Italian "Cultural Heritage and Landscape Code", Legislative Decree 42/2004, which states that volcanoes are areas protected by law, a detailed study aimed at identifying and delimiting Cenozoic volcanic edifices present in Sardinia was performed. In order to delimit these areas, a comprehensive morphological, volcanological and morphometrical study was undertaken, proposing a methodology based on the semi-automatic delimitation of the volcanic edifices using a DEM, largely overcoming the subjectivity related to traditional techniques.

In the first phase of this study, a landform classification was used, based on local morphometric attributes such as slope gradient and total curvature (Shary, 1995; Shary et al., 2002; Florinsky, 2012). The results, discussed below, were generally useful to recognize the landforms, but they proved to be unsatisfactory for the identification of their boundaries, as the use of a sole slope classification can hardly represent the morphologic heterogeneity of a complex set of scoria cones at the regional scale. Therefore, the study was implemented by using the methodology proposed by Grosse et al. (2012), mainly based on two additional geomorphometric parameters (slope and profile curvature), integrated with the use of a new – modified Grosse – methodology, presented and discussed in detail in the following.

2 Objectives and methods

The main objective of the research was the recognition, selection and delimitation of well-preserved volcanic edifices from the Cenozoic volcanic activity of Sardinia (Italy). The result of this process can then be used by the regional authority for adopting conservation measures and to establish development and management policies over the selected landforms. In the present section we present a short geological framework of the Cenozoic volcanism of Sardinia and the methods adopted in the study. In the following sections, we discuss in detail the morphometric approaches used to get a final delimitation of the selected areas, together with some key examples, which well illustrate the different problems encountered in the research.

Figure 1 is a flow chart showing the general approach adopted in the study.

2.1 Geological setting and volcanic landforms

During the Cenozoic, Sardinia was subject to two distinct volcanic cycles: the Oligocene–Miocene calc–alkaline cycle (32–15 Ma; Lecca et al., 1997) and the Plio–Pleistocene alkaline and tholeiitic cycle (5.1–0.1 Ma; Beccaluva et al., 1985), whose starting phase was recently reconsidered as being of the Messinian age (6.6–6.4 Ma; Lustrino et al., 2007) (Fig. 2, Table 1). The Pliocene–Pleistocene activity, whose products are largely dispersed in Sardinia, was mainly basaltic in composition, varying from mildly to strongly alkaline (mostly with sodic character) to sub-alkaline (with tholeiitic affinity), and shows a marked within-plate geochemical signature (Lustrino et al., 1996, 2000, 2002). Eruptive centres are mainly located, and in some cases aligned, on the main N–S, N60 and N90 structures. A wide range of landforms are associated with the Pliocene–Pleistocene volcanic cycle, mainly represented by extended lava fields, monogenetic volcanoes and minor compound volcanoes. Monogenetic volcanoes are mainly represented by scoria cones, distributed in the Logudoro (north-west Sardinia; Beccaluva et al., 1981), in the Orosei–Dorgali areas (central-eastern Sardinia; Lustrino et al., 2002) and in central Sardinia (Giare). Pliocene–Pleistocene scoria cones span a time interval from 3.0 ± 0.1 to 0.11 ± 0.02 Ma (Beccaluva et al., 1985). A set of well-preserved scoria cones from the Plio–Pleistocene activity is here presented in detail. They range from nearly unmodified, poorly eroded cones, characterized by about 30° sloping

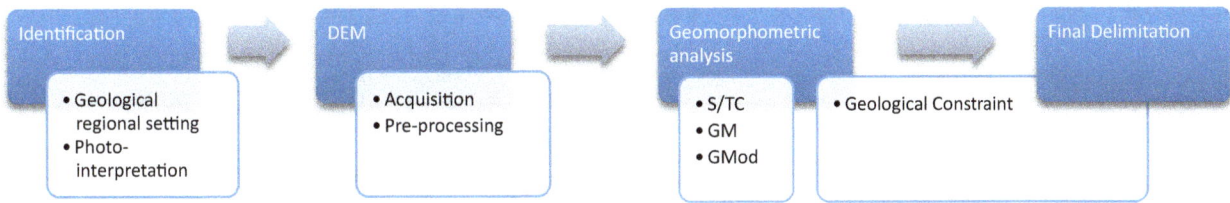

Figure 1. Schematic flow chart of the proposed methodology for the delimitation of the boundaries of the volcanic edifices with the three morphometric classifications discussed in the text.

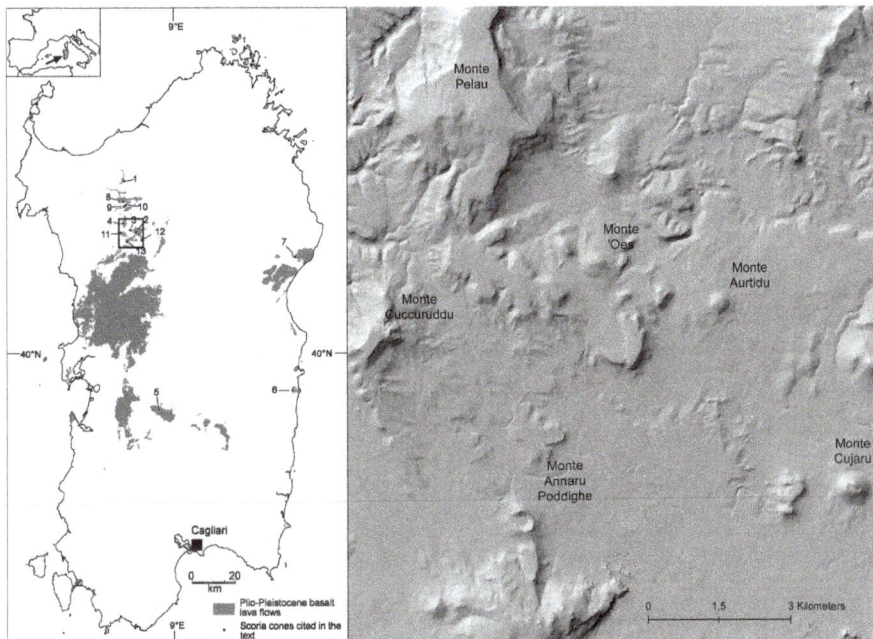

Figure 2. Distribution of Plio–Pleistocene basalt lava flows in Sardinia and localization of the scoria cones (the codes are referred to the list in Table 1). The hillshade view represents the area in the box with high concentrations of volcanic edifices.

flanks (Monte Cujaru; Fig. 3), to quite completely eroded remnants where only the conduit zone, occupied by a neck or a dike of coherent magma, is still preserved. Scoria cones were built up on different palaeo-topographic surfaces, from planar horizontal or gently dipping (as in the cases respectively of Monte Cujaru and Ibba Manna) to highly irregular surfaces (Monte Annaru Poddighe; Fig. 6), laying on scarps a few tens of metres high. Some scoria cones and the related lava fields experienced different degrees of relief inversion, occupying highlands extending a few or tens of square kilometres and elevated from tens to hundreds of metres above the surrounding valleys. Finally, all the scoria cones present different products related to a complex constructional activity, such as lava effusions from the flanks or from the base, lateral collapses related to lava effusion and variable presence of agglutinated scoria banks. These processes reflect the different amount and distribution of volcanic lithofacies in each scoria cone, determining the final morphological features.

2.2 Identification of scoria cones

The localization and classification of the volcanic edifices was achieved by integrating data from the scientific literature (e.g. Carmignani et al., 2008; Lustrino et al., 2004, and references therein) with an overview of airborne imagery (photo archives of 1968 (b/w) and 1977 (colour) respectively at the scale of 1 : 23 000 and 1 : 10 000), orthophotos (http://www.sardegnageoportale.it/navigatori/fotoaeree.html) and web 3-D digital models (Google Earth and Sardegna 3-D). This analysis allowed the identification of volcanic edifices as scoria cones, small shields or domes, and of the related relict forms and products as craters, necks, dikes, lava flows, debris flow deposits and coulees. Selection of the volcanic edifices to be studied and delimited was mainly based on this step.

2.3 DEM pre-processing

A DEM, processed in 2011, with 10 m pixel resolution and a vertical and horizontal accuracy of 2.5 m, was used for

Table 1. List of the scoria cones analysed with the basal area and the radiometric age.

Code	Name	Loc.	Lat. centroid	Long. centroid	Area (km²)	K/Ar age (Ma)
1	Monte Massa	Ploaghe	1478090	4507120	0.34	0.38 ± 0.04
2	Monte Aurtidu	Torralba	1483410	4484150	0.27	0.4 ± 0.1
3	Monte'Oes	Torralba	1480750	4484890	0.29	0.4 ± 0.2
4	Monte Pelau	Thiesi	1477390	4487950	1.22	1.9 ± 0.1
5	Zeppara Manna	Genoni	1493740	4403420	0.58	2.76 ± 0.11
6	Ibba Manna	Bari Sardo	1555870	4411920	0.17	
7	Punta Su Nurtale	Onifai	1557160	4472680	0.16	
8	Monte Pubulena	Ploaghe	1477530	4497580	0.11	0.9 ± 0.4
9	Monte Ruju	Siligo	1477910	4495680	0.10	0.6 ± 0.1
10	Monte Percia	Siligo	1478320	4496080	0.04	0.6 ± 0.1
11	Monte Cuccuruddu	Cheremule	1476520	4483540	0.12	0.11 ± 0.02
12	Monte Cujaru	Bonorva	1486150	4480380	0.51	0.8 ± 0.1
13	Monte Annaru Poddighe	Giave	1479240	4479730	0.69	> 0.2

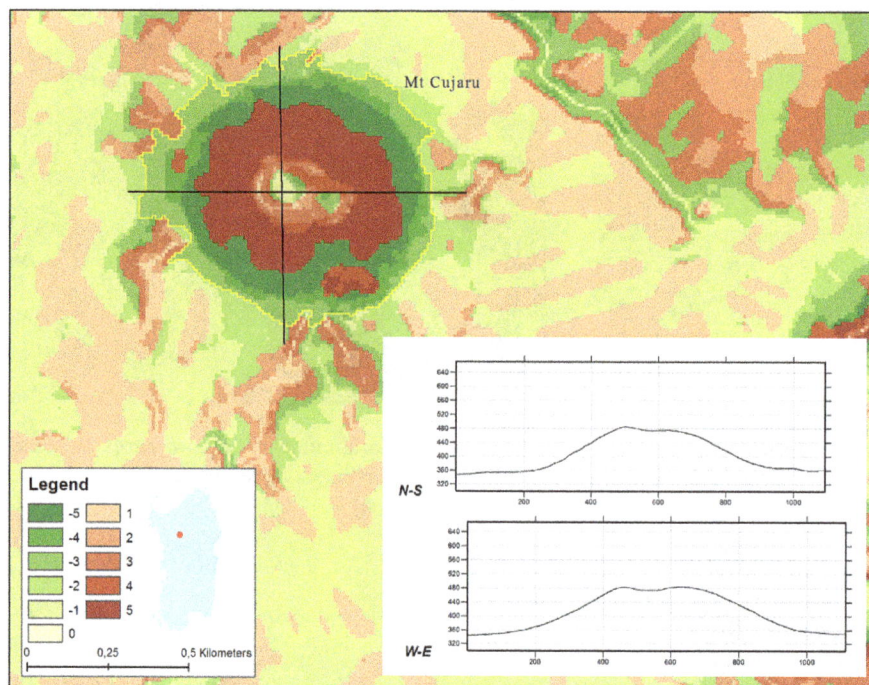

Figure 3. STC morphometric classification and topographic profiles of Mt Cujaru. The yellow line is the boundary of the volcanic edifice extracted from the transformation in polyline of the limit of the class −3. In the map, green and red colours represent respectively concave and convex areas, with the values of slope referring to the five classes indicated in Table 2. Class 0 represents plain or horizontal areas.

the study. Based on the general assumption that the smallest object to be generated on the DEM is at least twice the size of the grid (Dikau, 1989), the 10 m resolution was considered satisfactory for the requested final mapping scale of 1 : 25 000 (Melis et al., 2013).

A digital terrain model (DTM) generated by the Cartographic Service of RAS (Autonomous Region of Sardinia) using the 3-D Analyst and Spatial Analyst extensions of ArcGIS 10.0 (ESRI, Redmond, USA) was used. The DTM is georeferenced according to the WGS84 datum (EPSG code: 7030) and UTM projection system (Vacca et al., 2013). The "Contour_Lines" and "Elevation_Points" layers of the 1 : 10 000 scale geographic database (GeoDB 10k) were used as source data. A triangulated irregular network (TIN) was generated and subsequently transformed into a regular grid format. A "NoData" value was assigned to pixels of the sea surface.

This DEM was pre-processed to verify the quality and to correct the possible artefacts. As indicated by Reuter et al. (2009), the quality of a DEM determines the outcomes of geomorphometric analysis. The only DEM noises removed in this research were the artefacts due to wrongly coded elevation contours or due to missing contours. These errors were detected through a hillshade view, and the correction was done by assigning the correct value to the original data. The set of elevation points, available in the original database and used as ground control points during the photogrammetric restitution, was used to improve the quality of the roughness of the land surface with other elevation data rather than the contour lines alone.

3   The geomorphometric analysis

Different approaches based on DEM elaborations have been proposed to draw up the boundaries of volcanic edifices after their identification. The methods based on the slope–total curvature (STC) relation, as proposed for the general classification of landforms (Vacca et al., 2013), and that proposed by Grosse et al. (2012) are reviewed, and, finally, a modification of the Grosse method (GM) is proposed here, overcoming its inherent complexity and ultimately reducing the operator choices during the delimitation process.

3.1   Slope–total curvature (STC) algorithm

A first approach for a classification aimed at extracting the volcanic edifices as isolated landforms with respect to the surrounding landscape is based on slope and total curvature. These two geomorphometric parameters clearly highlight the presence of a concave upward break in slope, typical of the connection between the edifice and the basement (Euillades et al., 2013). This methodology simplifies those proposed by Iwahashi and Pike (2007) and by Gorini (2009) for landform classification. Slope is a suitable landform parameter in the case of monogenetic edifices, generally characterized by a simple shape often showing a radial symmetry, which clearly stands out with respect to the surrounding basement. The choice of appropriate classes of slope has to be related to the type of edifice, as it is well known that different types of volcanic edifices show different morphological and morphometric features (see for example Pike and Clow, 1981). In the scoria cones analysed in detail for this study, slope classes arising from average morphometric data presented in the literature for the same type of landforms were used.

There are several detailed studies on the morphometric characteristics of scoria cones. In order to define the main classes of slope, Fornaciai et al. (2012) presented a statistical analysis for more than 500 scoria cones of various nature and age, extracting some basic parameters whose values partly confirm and enlarge the results obtained by Woods (1980a, b):

Table 2. Slope classification.

Slope classes	Value expressed in degree
1	$< 3, 5$
2	$3, 5\text{–}6$
3	$6\text{–}13$
4	$13\text{–}18$
5	> 18

- The connection between the slope of the cone and the surrounding basement is typically around 3.5°.

- The minimum average gradient of a scoria cone is about 6°.

- The mean slope value for the large number of studied scoria cones is about 18°, with a standard deviation of ±5°.

Based on the values discussed in these studies, which at a first approximation well agree with the data set of the present research, five classes of slope were used in this step of the study (Table 2).

Slope S (here expressed in degrees) represents the variation of height z in the two directions x and y; it was calculated with the Horn method (Burrough and McDonell, 1998):

$$S = \arctan\left(\sqrt{(dz/dx)^2 + (dz/dy)^2}\right) \cdot 57.29578. \qquad (1)$$

The curvature K was instead calculated with respect to a 3×3 kernel using the equation of Zevenbergen and Thorne (1987):

$$K = \left(\partial^2 Z/\partial S^2\right) / \left[1 + (\partial Z/\partial S)^2\right]^{3/2}, \qquad (2)$$

where Z is the quadratic surface calculated as

$$Z = Ax^2y^2 + Bx^2y + Cxy^2 + Dx^2 + Ey^2, \\ + Fxy + Gx + Hy + I \qquad (3)$$

where the coefficients A to I are the nine elevation values of the nodes of the 3×3 matrix.

The values obtained are positive and negative for convex and concave forms respectively.

Final processing of classification involves the application of the product

$$STC = S \cdot K. \qquad (4)$$

The resulted STC is a 10-class subdivision, from -5 (strongly sloping and concave areas) to $+5$ (strongly sloping and convex areas) with 0 (horizontal or plain areas) as the central value (Table 2). The output of this process is a map representing, for each pixel, the attribution to 1 of the 10 classes; Fig. 3 shows an extract of the map related to one of the selected scoria cones (Monte Cujaru).

A preliminary delimitation of the volcanic edifice is extracted on the basis of the upper limit of the class representing a concave area with a value of slope less than 6 (class -2). As discussed above, this slope threshold is considered in the literature as the minimum average gradient of a well-preserved cone.

3.2 The Grosse algorithm and the proposed modified Grosse method

The application of the morphometric classification based on the STC algorithm described above, however, presents some limitations, mainly due to the naturally large variability, in terms of geomorphological and volcanological setting, presented by the studied volcanic edifices. In fact, a unique slope–curvature class is hardly representative of the real basal limit of the scoria cones from Sardinia, mainly due to their different nature, age, basement geology and degree of erosion.

In order to draw up the boundaries of the volcanic edifices with different morphology and age, the approach proposed by Grosse et al. (2012) was applied. The authors suggested the use of a complex function of two parameters: slope gradient and profile curvature (Wood, 1996), based on the observation that a volcano, although bounded by a major rupture of the slope that connects the flanks of the edifice with the surrounding landscape, generally presents morphological complications that often do not allow an automatic tracing of the perimeter of its base.

The function proposed by Grosse et al. (2012) for the extraction of the boundaries of volcanic edifices, the boundary delineation layer (GM), is

$$GM = PC_n f + S_n (1 - f), \qquad (5)$$

where PC_n and S_n are respectively the normalized values of profile curvature and slope and f is a weighting factor of the two functions with a suggested value of 0.7 (Grosse et al., 2012).

The normalized values of PC and S are calculated as

$$PC_n = (PC_i - PC_{min}) / PC_{range}, \qquad (6)$$

$$S_n = (S_i - S_{min})^2 / (S_{range})^2, \qquad (7)$$

and the subscripts i, min and $range$ respectively refer to the point value, the minimum calculated value and the maximum-minimum calculated range of curvature and slope. These values were calculated from a subregion related to each volcanic edifice, tracing the pixel profile that cuts the landform and extracting the minimum and maximum values for slope and curvature.

Assuming that the limit has a concave morphology, the peculiarity of this function is that it presents a relative minimum corresponding to the boundary between the volcanic

edifice and the basement. This method is referred to here as the Grosse method 9Grosse et al., 2012).

As clearly stated by the same authors, however, the method does not discriminate locally in the case of a non-concave boundary, which could be the case when the volcanic edifice is built upon a rough, irregular basement or on top of tectonic lineaments such as faults. The authors suggest applying the function and delineating the edifice boundaries by visual interpretation of the results (i.e. "manually" tracing the path along the minimum values of the generated layer), as the main problem is in general related to the unequivocal distinction of the basement from the edifice.

We propose here a modification of the GM in order to make the edifice much more distinguishable from the surrounding basement in the DEM elaboration. This result was obtained by introducing a new function of the same parameters as GM. The proposed algorithm, hereafter referred to as modified Grosse (GMod), is

$$GMod = (PC_n f) \cdot [S_n (1 - f)]. \qquad (8)$$

In this case, where the value of one of the two parameters is low (as in the case of a nearly planar or a very gentle sloping basement), it decreases the final value, thereby strongly differentiating the basement from the rest of the edifice, which is generally characterized by high values in at least one of the two parameters.

The comparison between the resulting images obtained from GM and GMod is shown in Fig. 4. The GMod images are very appropriate to calculate the threshold for masking the area around the volcanic edifice and delineate the boundary (Fig. 5). Apart from minor local irregularities, largely deviating from the smooth circular shape characterizing scoria cones, and in which manual drawing cannot be avoided, in this way the boundary is generally clearly evidenced around most of the edifice, helping in its automatic tracing.

4 Geological constraints

Parallel to the geomorphometrically based delimitation process, photogeological interpretation, geological map analysis and field survey were performed in order to check and possibly correct the final results. Some general statements can be used in this case.

– Volcanic edifice boundaries cannot cross-cut non-volcanic rocks.

– Pyroclastic edifices such as scoria cones are commonly associated to lava flows erupted from the flank or from the base of the cone.

– Scoria cone can grown on pre-existing lava fields. In this case, it is fundamental to distinguish cone-forming deposits from other volcanic products which preceded or post-dated cone formation. In general, we considered

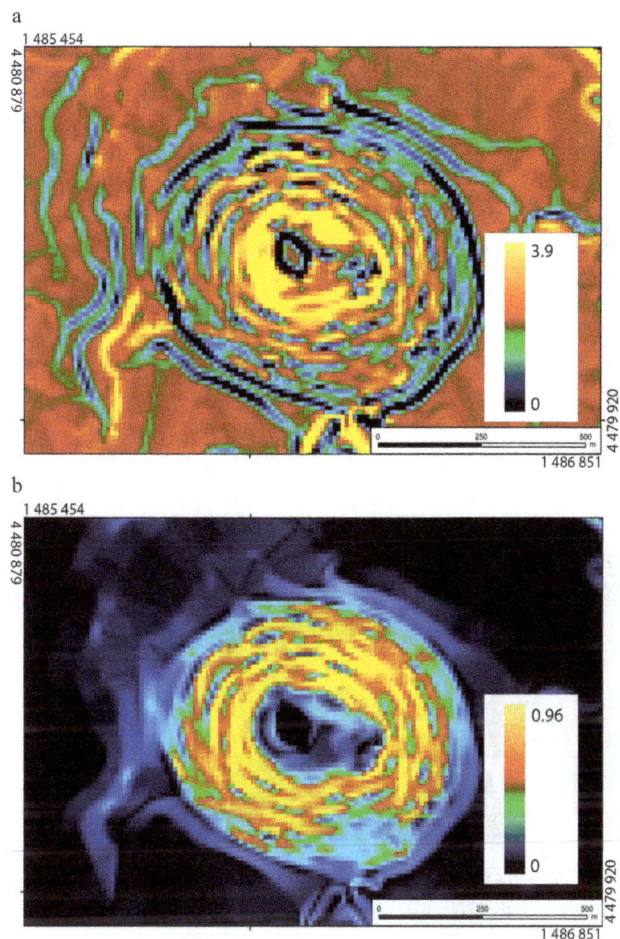

Figure 4. Comparison between the resulting image obtained with the application of the (**a**) Grosse method (GM) and (**b**) modified Grosse method (GMod) to the area of Mt Cujaru. The colour scale, from dark blue to bright yellow, indicates the range from minimum to maximum values of the resulted data. The minimum value is used for the boundary delimitation.

as an integrant part of the edifice only those deposits that lie on its flanks and are thus possibly associated with cone-related eruptive vents.

– The deposits related to gravitational processes (e.g. landslides) redistributing the primary cone deposits are here considered as part of the edifice.

– In the case of partially eroded edifices in which internal geological features and facies architecture are clearly visible, geology should represent a major constraint for delimiting the inner structure of the edifice from the partially eroded distal products.

Because the occurrence of various types of deposits can strongly reduce the efficacy of geomorphometric algorithms, we suggest that the final delimitation of the cone should always be checked against the observed geological features.

5   Results and discussion

The described methodological approach, based on DEM elaborations, was applied to 13 Plio–Pleistocene scoria cones from Sardinia, with the aim of tracing their boundaries. The results are first discussed with reference to two scoria cones, representative of two endmembers with different degrees of complexity of the landform and of their surroundings. Finally, the general results relative to the 13 delimited scoria cones are also briefly presented. The progressive refinement of edifice boundaries is obtained by the integrated use of the STC, the GM and the GMod algorithms; this methodology clearly shows how the introduction of the GMod algorithm strongly improves the final results.

Monte Cujaru scoria cone represents the simplest case (Fig. 3). It is a largely preserved, poorly eroded scoria cone built on a nearly planar surface, characterized by the presence of a summit crater depression, limited by a crater rim elevated about 130 m with respect to the base of the cone and with an average slope of the flanks of about 30°.

Conversely, Monte Annaru Poddighe scoria cone here represents the most complex case (Fig. 6). It is an asymmetric, cone-shaped relief elevated about 110–180 m above its base, mostly built on a basaltic lava plateau formed during the initial activity of the same eruptive centre. The southern flank of the scoria cone is built upon a tectonically related plateau elevated about 90–100 m above the surrounding basaltic plain and constituted by Miocene age marine sediments (Carmignani et al., 2008). The western flank of the scoria cone experienced a syn-eruptive partial collapse possibly related to the outward pushing of lava flowing from the base of the cone.

5.1   STC map

The STC map of Monte Cujaru (Fig. 3) shows an approximately concentric, sub-circular distribution of the slope classes. Deviations from circularity arise in correspondence to local irregularities, related both to minor, shallow-seated landslides and to the erosional heterogeneities of the surrounding. In the case of Monte Cujaru, the 3.5° slope limit for locating the physical junction between the cone and its basement (as suggested in Fornaciai et al., 2012) is extremely irregular, as it is mainly controlled by erosional features affecting the topographic surface around the scoria cone. Instead, the 6° threshold represents a good first approximation for the edifice boundary, although this limit cannot be considered definitive, due to the extremely irregular trend along the northern limit of the edifice and to minor irregularities unrelated to the constructional processes of the edifice.

The STC map of Monte Annaru Poddighe (Fig. 6) shows a discontinuous, approximately concentric, sub-circular distribution of the slope classes. Deviations from circularity affect almost all slope classes in the western sector, in proximity to the collapsed portion of the cone. Other deviations

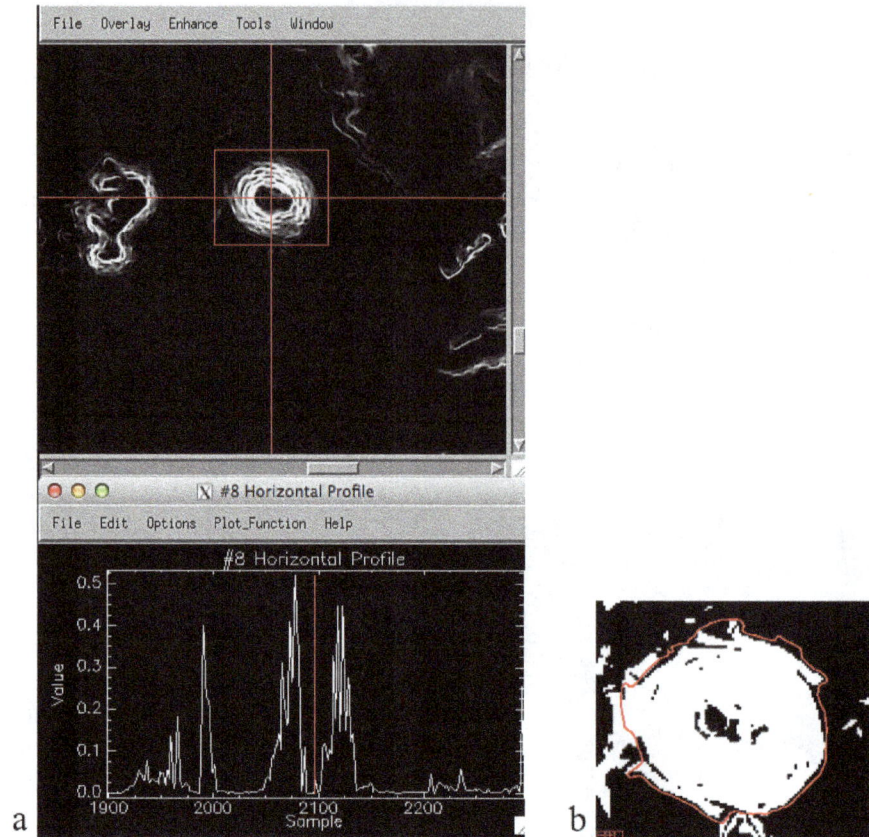

Figure 5. Application of the modified Grosse method to the DEM of Mt Cujaru: **(a)** the horizontal profile enhances the minimum and maximum values (thresholds) used to apply a mask to the image and to classify the data; **(b)** mask image (0 in black and +1 in white). In red the final boundary.

from circularity arise in correspondence to the $-2/-3$ and $-3/-4$ classes in the northern and southern flanks, in correspondence to morphological irregularities of the surrounding landscape (Fig. 6a). The $-2/-3$ classes boundary was considered in order to approximate the outer limit of the scoria cone, although this limit, due to its discontinuous trend, cannot be extended to the entire base. The 3.5° threshold in slope (here corresponding to the lower limit of the class -2) does not reflects the smooth circular morphologic nature of scoria cones (Walker, 2000).

This algorithm represents, especially in the case of a nearly regular cone-shaped edifice, a powerful tool to extract a landform from a study area, but – in both the study cases – it does not discriminate satisfactorily the volcano edifice from the basement.

5.2 GM map

On the flanks of the Monte Cujaru cone, the GM map (Fig. 4a) shows a regular concentric pattern of alternating sub-circular red and blue bands (in this representation), representing respectively the maximum and minimum of the normalized reference function (Eq. 5). At the base of the sco-

ria cone, in the eastern sector, a wide blue band, representing the relative minimum values, largely overlaps the upper limit of class -2 from the STC processing, but overall this band cross-cuts different slope classes, smoothing many irregularities which characterize the STC-based delimitation.

The GM algorithm applied to the more complex case of Monte Annaru Poddighe shows a similar concentric sub-circular pattern (Fig. 6b). A main blue band well delimitates the base of the cone in the NE sector. Conversely, an important deviation from circularity arises in the southern sector, where the blue bands outline the shape of the surrounding plateau formed by Miocene sediments, and in the NNW sector, where the main blue band marking the base of the edifice in the NE sector shows an irregular, rough trend and progressively departs from the edifice base toward the east.

As suggested by Grosse et al. (2012), the blue band can be used to manually trace the edifice boundary, but some doubts still remain due to discontinuity of this band and to the local occurrence of close multiple minima also in the simplest case. Moreover, several blue-green bands cross-cut the planar topography close to the edifice, highlighting minor topographic irregularities and making the visual interpretation of the GM map more complex.

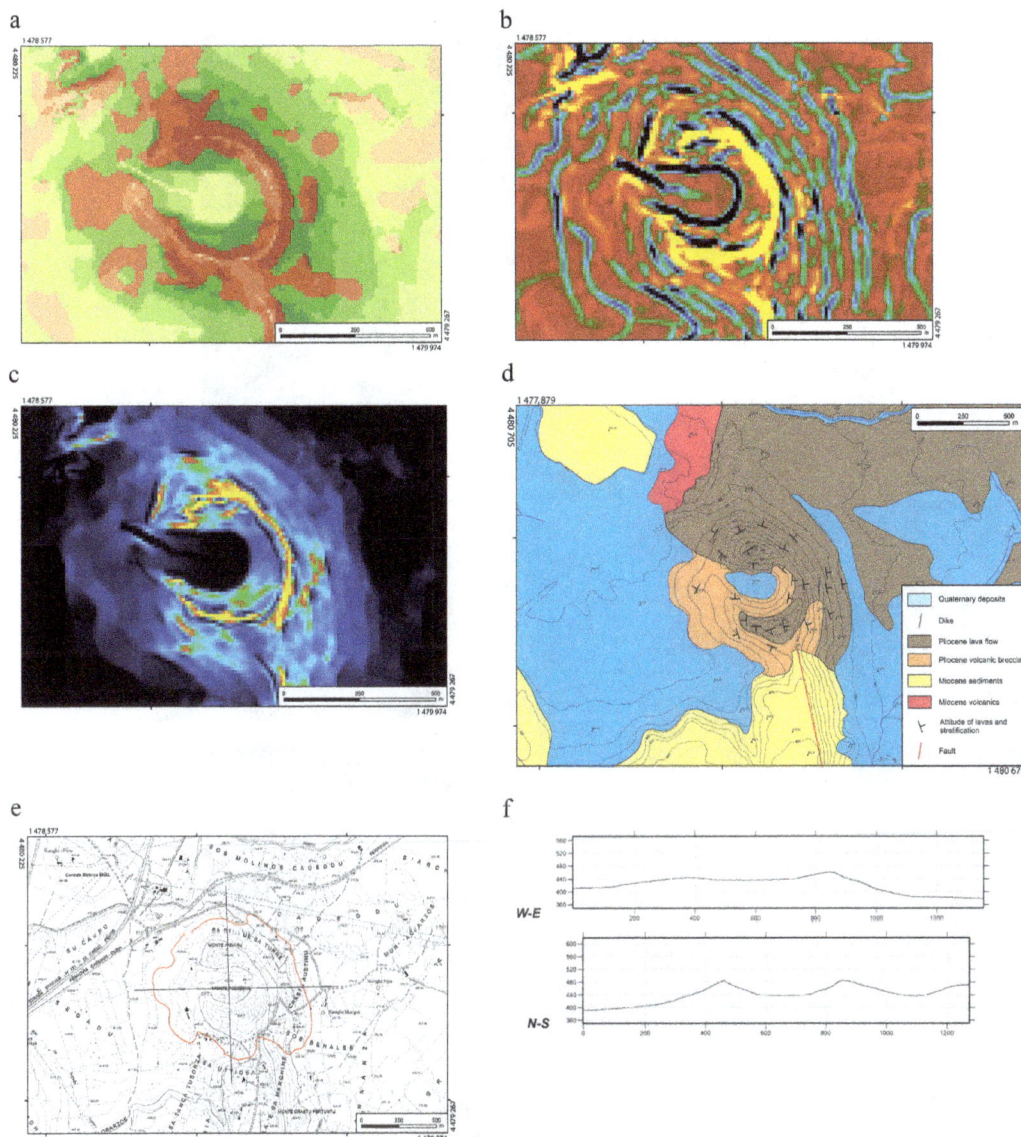

Figure 6. Mt Annaru Poddighe: **(a)** first morphometric classification (legend in Fig. 3); **(b)** Grosse method (legend in Fig. 4); **(c)** modified Grosse method (legend in Fig. 4); **(d)** geological map; **(e)** topographic map with the final boundary in red, illustrating the traces; and **(f)** topographic profiles of the volcanic edifice.

5.3 GMod map

To avoid the problems arising from the STC and GM methods and to reduce the dependency from a subjective, operator-driven selection of the edifice boundary, and to improve the accuracy of the results, the GMod algorithm was also applied.

The GMod map of the regular edifice of Monte Cujaru (Fig. 4b) shows a concentric pattern of bands on the flanks of the edifice, and blue scale near concentric irregular-shaped bands in the base zone. The boundary delineation layer in the plain surrounding the edifice has a low and poorly variable value (represented by the black colour in Fig. 4b), with only minor blue zones connecting the edifice to the plain. These transitional, blue-coloured zones are cross-cut by black or dark blue narrow bands representing the minimum values. These bands largely overlap with those of the GM map but are reduced in number. At the north-eastern sector, the GM map shows three blue bands corresponding, whereas the pattern shown by the GMod map (from the inner to the outer zone) appears largely simplified. In the same way, in the south-western sector, the two bands evidenced in the GM elaboration reduce to one band and to a steady colour transition.

In the more complex case of Monte Annaru Poddighe, the GMod elaboration proved to be very useful, as it filters out the evident multiple minima bands of the GM data between

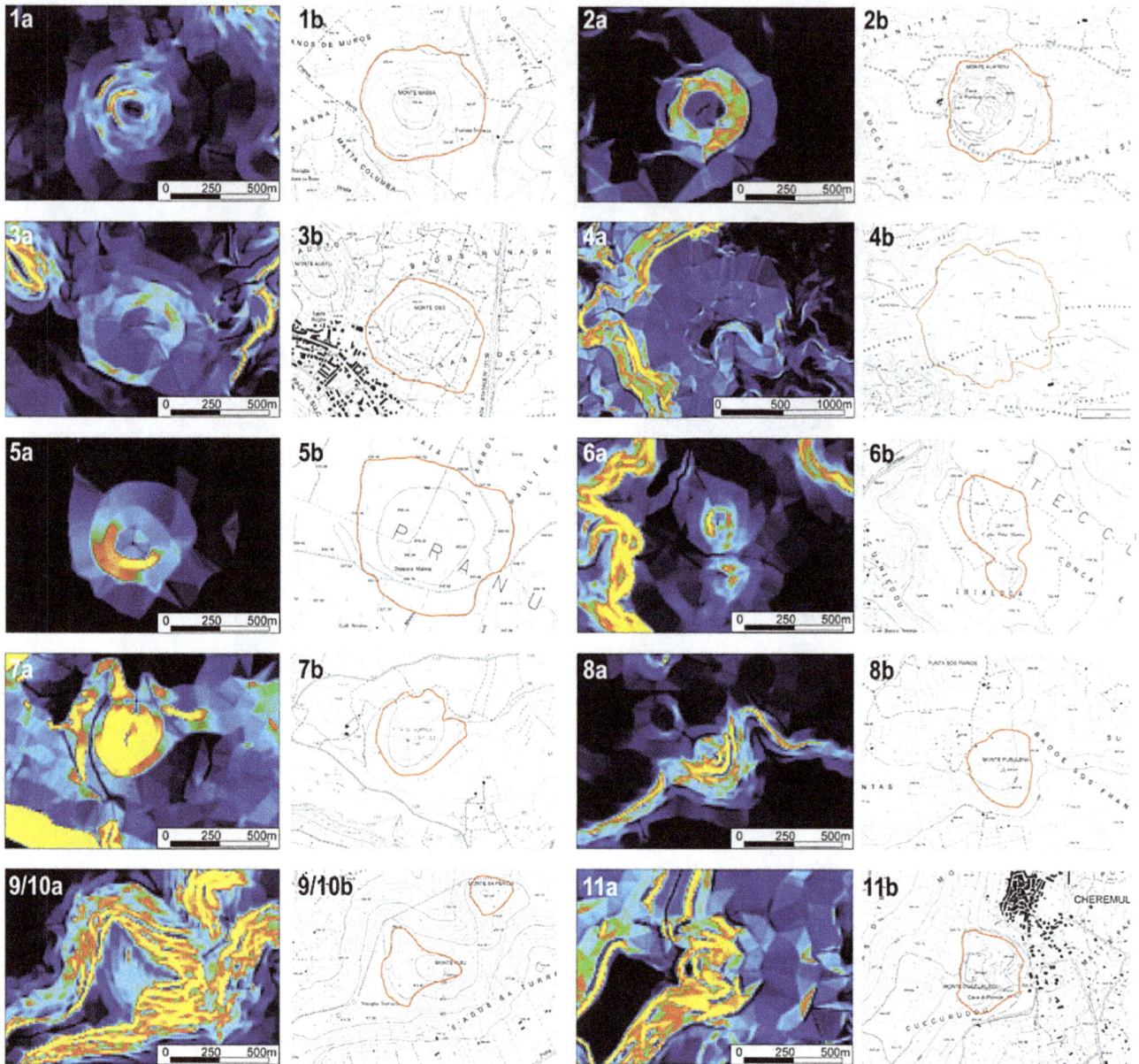

Figure 7. In this figure the applied GMod classification and the final delimitations of the boundaries are overlaid on the topographic map (see the legend in Fig. 4). The numbers of the edifices refer to Table 1.

the base of the edifice and its surroundings (Fig. 6b and c). In the southern sector, where the morphological features do not reflect the geological characteristics of the base of the cone and of its surroundings, the lithological contact between the Plio–Pleistocene volcanic and Miocene sedimentary rocks drives the delimitation process (Fig. 6d).

As a general rule, the GMod algorithm reduces the uncertainties related to the occurrence of several repeated minimum bands of the GM approach and presents the advantage of smoothing the morphologic irregularities of the plain surrounding the edifice, simplifying the visual interpretation of the DEM elaborations. It looks self-standing in simple cases

(i.e. symmetric cone shape), allowing an automatic tracing for most of the edifice perimeter. In more complex cases, where erosion is very impactful, the geological constraints still play a prominent role in the final delimitation of the edifice.

The GMod algorithm has been also applied to 11 other scoria cones, and the resulting boundaries have been extracted to compare and integrate the geological and geomorphological observations. In Fig. 7 the applied GMod classification and the final delimitations of the boundaries are presented. The results are often an intermediate step between the endmember cases discussed above. In cases 1, 2, 4, 5 and 7

(Fig. 7), the delimitation is mainly based on the results of the GMod algorithm. In these cases, minor deviations from the blue/black boundaries (specifically in the cases 5 and 7) arise considering the circular nature of scoria cones. In cases 3, 6, 8, 9, 10 and 11, the geology exerts major control in the delimitation process, and erosion and tectonics deeply modify the original shapes of the cones.

6 Conclusions

Scoria cones from Sardinia offer – due to the wide spectrum of age, erosion and constructional conditions, resulting in an extremely variable range of shapes – a suitable case study to test the efficiency of existing algorithms and for developing new algorithms aimed at the delimitation of volcanic edifices.

The obtained results, in the case of scoria cone edifices, can be summarized in the following points:

- The SCT morphometric classification well highlights the shapes of the edifices but does not precisely identify the boundary with the surrounding basement.

- The algorithm introduced by Grosse et al. (2012) performs very well in finding minima in the complex slope–curvature function but carries an intrinsic uncertainty for tracing the outer boundary of a volcanic edifice mainly due to its high sensitivity to morphological changes.

- The proposed modification of the method by Grosse (2012), based on a different function of the normalized slope and curvatures for each specific edifice, is helpful in overcoming the problem of manual final tracing of the edifice boundaries, especially in regularly shaped cones.

- In more complex shapes derived from eroded cones, geological data still play a key role in tracing the boundary of the volcanic edifices.

- As a consequence, although the presented algorithm reduces the subjectivity of tracing volcanic edifice boundary, and in agreement with that which was already stated by Evans (2012), the complete automatic delimitation of landforms still remains a research frontier.

- We suggest that the proposed method of an integrated use of opportunely processed morphometric data (STC, GM, GMod) together with detailed geological data may result in a more objective delimitation of landscape morphologies. In particular, the proposed method performs very well with a large variety of volcanic edifices, and it is probably suited for any landscape morphology clearly standing against a clearly contrasting basement.

Acknowledgements. The authors thank the Autonomous Region of Sardinia for the support in this study.

Edited by: T. Hengl

References

Beccaluva, L., Deriu, M., Macciotta, G., Savelli, C., and Venturelli, G.: Carta geopetrografica del vulcanismo Plio-Pleistocenico della Sardegna nord occidentale, Scala 1 : 50 000, Grafiche STEP cooperativa, Parma, 1981.

Beccaluva, L., Civetta, L., Macciotta, G., and Ricci, C. A.: Geochronology in Sardinia: results and problems, Rendiconti della Società Italiana di Mineralogia e Petrologia, 40, 57–72, 1985.

Burrough, P. A. and McDonell, R. A.: Principles of Geographical Information Systems, Oxford University Press, New York, 1998.

Carmignani, L., Oggiano, G., Funedda, A., Conti, P., Pasci, S., and Barca, S.: Carta Geologica della Sardegna a scala 1 : 250.000, Litografia Artistica Cartografica s.r.l., Firenze, 2008.

Carn, S.: The Lamongan volcanic field, East Java, Indonesia: physical volcanology, historic activity and hazards, J. Volcanol. Geoth. Res., 95, 81–108, 2000.

Cas, R. A. F. and Wright, J. V.: Volcanic successions, modern and ancient, Allen and Unwin, London, 1987.

Dikau, R.: The application of a digital relief model to landform analysis, in: Three dimensional Application in Geographical Information Systems, edited by: Raper, J. F., Taylor & Francis, London, 55–77, 1989.

Euillades, L., Grosse, P., and Euillades, P.: NETVOLC: An algorithm for automatic delimitation of volcano edifice boundaries using DEMs, Comput. Geosci., 56, 151–160, 2013.

Evans, I. S.: Geomorphometry and landform mapping: what is a landform?, Geomorphology, 137, 94–106, 2012.

Fink, J. H. and Anderson, S. W.: Lava Domes and coulees, in: Encyclopedia of Volcanoes, edited by: Sigurdsson, H., Houghton, B. F., McNutt, S. R., Rymer, H., and Stix, J., Academic Press, San Diego, 643–662, 2000.

Florinsky, I. V.: Digital Terrain Modeling: A Brief Historical Overview, in: Digital Terrain Analysis in Soil Science and Geology, Elsevier, Amsterdam, 1–4, 2012.

Fornaciai, A., Favalli, M., Karátson, D., Tarquini, S., and Boschi, E.: Morphometry of scoria cones, and their relation to geodynamic setting: A DEM-based analysis, J. Volcanol. Geoth. Res., 217–218, 56–72, 2012.

Francis, P.: Volcanoes: A planetary perspective, Oxford University Press, Oxford, 1993.

Gorini, M. A. V.: Physiographic classification of the ocean floor: a multi-scale geomorphometric approach, in: Proceeding of Geomorphometry Conference, edited by: Purves, R., Gruber, S., Straumann, R., and Hengl, T., University of Zurich, Zurich, Switzerland, 98–105, 2009.

Grosse, P., van Wyk de Vries, B., Euillades, P. A., Kervyn, M., and Petrinovic, I. A.: Systematic morphometric characterization of volcanic edifices using digital elevation models, Geomorphology, 136, 114–131, 2012.

Grosse, P., van Wyk de Vries, B., Petrinovic, I. A., Euillades, P. A., and Alvarado, G.: Morphometry and evolution of arc volcanoes, Geology, 37, 651–654, 2009.

Hooper, D. and Sheridan, M.: Computer-simulation models of scoria cone degradation, J. Volcanol. Geoth. Res., 83, 241–267,
1998.

Iwahashi, J. and Pike, R. J.: Automated classifications of topography from DEMs by an unsupervised nested-means algorithm and
a three-part geometric signature, Geomorphology, 86, 409–440,
2007.

Keresztauri, G., Jordan, G., Németh, K., and Doniz-Paez, J. F.: Syn-
eruptive morphometric variability of monogenetic scoria cones,
Bull. Volcanol., 74, 2171–2185, doi:10.1007/s00445-012-0658-
1, 2012.

Keresztauri, G., Geyer, A., Martí, J., Németh, K., and Dóniz-Páez,
F. J.: Evaluation of morphometry-based dating of monogenetic
volcanoes – a case study from Bandas del Sur, Tenerife (Canary
Islands), Bull. Volcanol., 75, 734–753, 2013.

Lecca, L., Lonis, R., Luxoro, S., Melis, E., Secchi, F., and Brotzu,
P.: Oligo-Miocene volcanic sequences and rifting stages in Sardinia: a review, Periodico di Mineralogia, 66, 7–61, 1997.

Lustrino, M., Melluso, L., Morra, V., and Secchi, F.: Petrology of
Plio-Quaternary volcanic rocks from central Sardinia, Periodico
di Mineralogia, 65, 275–287, 1996.

Lustrino, M., Melluso, L., and Morra, V.: The role of lower con-
tinental crust and lithospheric mantle in the genesis of Plio-
Pleistocene volcanic rocks from Sardinia (Italy), Earth Planet.
Sc. Lett., 180, 259–270, 2000.

Lustrino, M., Melluso, L., and Morra, V.: The transition from alkaline to tholeiitic magmas: a case study from the Orosei-Dorgali
Pliocene volcanic district (NE Sardinia, Italy), Lithos, 63, 83–
113, 2002.

Lustrino, M., Morra, V., Melluso, L., Brotzu, P., D'Amelio, F.,
Fedele, L., Lonis, R., Franciosi, L., and Petteruti Lieberecknect,
A. M.: The Cenozoic igneous activity in Sardinia, Periodico di
Mineralogia, 73, 105–134, 2004.

Lustrino, M., Melluso, L., and Morra, V.: The geochemical peculiarity of 'Plio-Quaternary' volcanic rocks of Sardinia in the
circum-Mediterranean Cenozoic Igneous Province, in: Cenozoic
volcanism in the Mediterranean area, edited by: Beccaluva, L.,
Bianchini, G., and Wilson, M., Geological Society of America
Special Paper 418, The Geological Society of America, Boulder,
CO, USA, 277–301, 2007.

Mazzarini, F., Ferrari, L., and Isola, I.: Self-similar clustering of cin-
der cones and crust thickness in the Michoacan–Guanajuato and
Sierra de Chichinautzin volcanic fields, Trans-Mexican Volcanic
Belt, Tectonophysics, 486, 55–64, 2010.

Melis, M. T., Loddo, S., Vacca, A., and Marrone, V. A.: Appli-
cazione di un metodo di analisi geomorfometrica a supporto della
cartografia pedologica in Sardegna, 17a Conferenza Nazionale
ASITA, Riva del Garda, Italy, 5–7, 993–1000, 2013.

Ollier, C.: Volcanoes, Blackwell, Oxford, 1988.

Pike, R. J.: Volcanoes on the inner planets-Some preliminary comparisons of gross topography, in: Lunar and Planetary Science
Conference, 9th, Houston, Tex Proceedings, 3, A79-39253 16-
91, Pergamon Press, Inc., New York, 3239–3273, 1978.

Pike, R. J. and Clow, G. D.: Revised classification of terrestrial volcanoes and a catalog of topographic dimensions with new results on edifice volume, US Geological Survey Open-File Report OF 81-1038, US Geological Survey, USA, 1981.

Reuter, H. I., Hengl, T., Gessler, P., and Soille, P.: Preparation
of DEMs for geomorphometric Analysis, in: Geomorphometry: Concepts, Software, Applications, edited by: Hengl, T. and
Reuter, H. I., Elsevier, Amsterdam, 227–254, 2009.

Shary, P. A.: Land surface in gravity points classification by a complete system of curvatures, Math. Geol., 27, 373–390, 1995.

Shary, P. A., Sharaya, L. S., and Mitusov, A. V.: Fundamental quantitative methods of land surface analysis, Geoderma, 107, 1–32,
2002.

Thouret, J. C.: Volcanic geomorphology – an overview, Earth-Sci.
Rev., 47, 95–131, 1999.

Tibaldi, A.: Morphology of pyroclastic cones and tectonics, J. Geophys. Res., 100, 24521–24535, 1995.

Vacca, A., Loddo, S., Melis, M. T., Funedda, A., Puddu, R., Verona,
M., Fanni, S., Fantola, F., Madrau, S., Marrone, V. A., Serra, G.,
Tore, C., Manca, D., Pasci, S., Puddu, M. R., and Schirru, P.: A
GIS based method for soil mapping in Sardinia, Italy: A geomatic
approach, J. Environ. Manage., 138, 87–96, 2013.

Walker, G. P. L.: Basaltic volcanoes and volcanic systems, in: Ency-
clopedia of Volcanoes, edited by: Sigurdsson, H., Houghton, B.
F., McNutt, S. R., Rymer, H., and Stix, J., Academic Press, San
Diego, 283–290, 2000.

Wood, C. A.: Morphometric evolution of cinder cones, J. Volcanol.
Geoth. Res., 7, 387–413, 1980a.

Wood, C. A.: Morphometric analysis of cinder cone degradation, J.
Volcanol. Geoth. Res., 8, 137–160, 1980b.

Wood, J.: The Geomorphological Characterization of Digital Elevation Models, Ph.D. Thesis, University of Leicester, Department
of Geography, Leicester, UK, 1996.

Zevenbergen, L. W. and Thorne, C. R.: Quantitative analysis of
the land surface topography, Earth Surf. Proc. Land., 12, 47–56,
1987.

8

Macro-roughness model of bedrock–alluvial river morphodynamics

L. Zhang[1], G. Parker[2], C. P. Stark[3], T. Inoue[4], E. Viparelli[5], X. Fu[1], and N. Izumi[6]

[1]State Key Laboratory of Hydroscience and Engineering, Tsinghua University, Beijing, China
[2]Department of Civil & Environmental Engineering and Department of Geology, Hydrosystems Laboratory, University of Illinois, Urbana, IL, USA
[3]Lamont-Doherty Earth Observatory, Columbia University, Palisades, NY, USA
[4]Civil Engineering Research Institute for Cold Regions, Hiragishi Sapporo, Japan
[5]Dept. of Civil & Environmental Engineering, University of South Carolina, Columbia, SC, USA
[6]Faculty of Engineering, Hokkaido University, Sapporo, Japan

Correspondence to: L. Zhang (zhangli10@tsinghua.edu.cn)

Abstract. The 1-D saltation–abrasion model of channel bedrock incision of Sklar and Dietrich (2004), in which the erosion rate is buffered by the surface area fraction of bedrock covered by alluvium, was a major advance over models that treat river erosion as a function of bed slope and drainage area. Their model is, however, limited because it calculates bed cover in terms of bedload sediment supply rather than local bedload transport. It implicitly assumes that as sediment supply from upstream changes, the transport rate adjusts instantaneously everywhere downstream to match. This assumption is not valid in general, and thus can give rise to unphysical consequences. Here we present a unified morphodynamic formulation of both channel incision and alluviation that specifically tracks the spatiotemporal variation in both bedload transport and alluvial thickness. It does so by relating the bedrock cover fraction to the ratio of alluvium thickness to bedrock macro-roughness, rather than to the ratio of bedload supply rate to capacity bedload transport. The new formulation (MRSAA) predicts waves of alluviation and rarification, in addition to bedrock erosion. Embedded in it are three physical processes: alluvial diffusion, fast downstream advection of alluvial disturbances, and slow upstream migration of incisional disturbances. Solutions of this formulation over a fixed bed are used to demonstrate the stripping of an initial alluvial cover, the emplacement of alluvial cover over an initially bare bed and the advection–diffusion of a sediment pulse over an alluvial bed. A solution for alluvial–incisional interaction in a channel with a basement undergoing net rock uplift shows how an impulsive increase in sediment supply can quickly and completely bury the bedrock under thick alluvium, thus blocking bedrock erosion. As the river responds to rock uplift or base level fall, the transition point separating an alluvial reach upstream from an alluvial–bedrock reach downstream migrates upstream in the form of a "hidden knickpoint". A tectonically more complex case of rock uplift subject to a localized zone of subsidence (graben) yields a steady-state solution that is not attainable with the original saltation–abrasion model. A solution for the case of bedrock–alluvial coevolution upstream of an alluviated river mouth illustrates how the bedrock surface can be progressively buried not far below the alluvium. Because the model tracks the spatiotemporal variation in both bedload transport and alluvial thickness, it is applicable to the study of the incisional response of a river subject to temporally varying sediment supply. It thus has the potential to capture the response of an alluvial–bedrock river to massive impulsive sediment inputs associated with landslides or debris flows.

1 Introduction

The pace of river-dominated landscape evolution is set by the rate of downcutting into bedrock across the channel network. The coupled process of river incision and hillslope response is both self-promoting and self-limiting (Gilbert, 1877). Although there are multiple processes that can lead to erosion into bedrock, we here focus on incision driven by abrasion of a bedrock surface as moving particles collide with it. Low rates of incision entail some sediment supply from upstream hillslopes, which provides a modicum of abrasive material in river flows that further facilitates bedrock channel erosion. Faster downcutting leads to higher rates of hillslope sediment supply, boosting the concentration of erosion "tools" and bedrock wear rates, but also leading to greater cover of the bedrock bed with sediment (Sklar and Dietrich, 2001, 2004, 2006; Turowski et al., 2007; Lamb et al., 2008; Turowski, 2009). Too much sediment supply leads to choking of the channels by alluvial cover and the retardation of further channel erosion (e.g., Stark et al., 2009). This competition between incision and sedimentation leads long-term eroding channels to typically take a mixed bedrock–alluvial form in which the pattern and depth of sediment cover fluctuate over time in apposition to the pattern of bedrock wear.

Theoretical approaches to treating the erosion of bedrock rivers have shifted over recent decades (see Turowski, 2012, for a recent review). The pioneering work of Howard and Kerby (1983) focused on bedrock channels with little sediment cover; it led to the detachment-limited model of Howard et al. (1994), in which channel erosion is treated as a power function of river slope and characteristic discharge, and the "stream-power-law" approach, in which the power-law scaling of channel slope with upstream area underpins the way in which landscapes are thought to evolve (Whipple and Tucker, 1999; Whipple, 2004; Howard, 1971, foreshadows this approach). At the other extreme, sediment flux came into play in the transport-limited treatment of mass removal from channels of, for example, Smith and Bretherton (1972), in which no bedrock is present in the channel and where the divergence of sediment flux determines the rate of lowering. Whipple and Tucker (2002) blended these approaches, and imagined a transition from detachment limitation upstream to transport-limited behavior downstream. They also discussed, in the context of the stream-power-law approach, the idea emerging at that time (Sklar and Dietrich, 1998) of a "parabolic" form of the rate of bedrock wear as a function of sediment flux normalized by transport capacity. Laboratory experiments conducted by Sklar and Dietrich (2001) corroborated this idea, and they led to the first true sediment flux-dependent model of channel erosion of Sklar and Dietrich (2004, 2006). This saltation–abrasion model was subsequently extended by Lamb et al. (2008) and Chatanantavet and Parker (2009). It was explored experimentally by Chatanantavet and Parker (2008) and Chatanantavet et al. (2013); evaluated in a field context by Johnson et al. (2009), Chatanantavet and Parker (2009), Hobley et al. (2011) and Turowski et al. (2013); adapted to treat alluvial intermittency by Lague (2010); and given a stochastic treatment by Turowski et al. (2007), Turowski (2009) and Lague (2010), the latter of whom introduced several new elements. Howard (1998) presents an alternative formulation for incision that relates bedrock wear to the thickness of alluvial cover rather than sediment supply, in a form that can be thought to be a predecessor of the present work.

At the heart of their saltation–abrasion model lies the idea of a cover factor p corresponding to the areal fraction of the bedrock bed that is covered by alluvium (Sklar and Dietrich, 2004). This bedrock bed is imagined as a flat surface on which sediment intermittently accumulates and degrades during bedload transport over it. The fraction of sediment cover is assumed to be a linear function of bedload transport relative to capacity. Bedrock wear takes place when bedload clasts strike the exposed bedrock. In the simplest form of the saltation–abrasion model, the subsequent rate of bedrock wear is treated as a linear function of the impact flux and inferred to be proportional to the bedload flux, which leads to the parabolic shape of the cover-limited abrasion curve.

The saltation–abrasion model is considerably more sophisticated and flexible (Sklar and Dietrich, 2004, 2006) than this sketch explanation can encompass. It does, however, have three major restrictions. First, it is formulated in terms of sediment supply rather than local sediment transport. The model is thus unable to capture the interaction between processes that drive evolution of an alluvial bed and those that drive the evolution of an incising of bedrock–alluvial bed. Second, for related reasons, it cannot account for bedrock topography significant enough to affect the pattern of sediment storage and rock exposure. Such a topography is illustrated in Fig. 1 for the Shimanto River, Japan. Third, it is designed for quasi-steady conditions, and thus cannot account for the effects of cyclic variation in sediment supply on channel development downstream of the point of sediment supply.

Here we address all three of these issues in a model that allows both alluvial and incisional processes to interact and coevolve. We do this by relating the cover factor geometrically to a measure of the vertical scale of elevation fluctuations of the bedrock topography, here called macro-roughness, rather than to the ratio of sediment supply rate to capacity sediment transport rate. Our model encompasses downstream-advecting alluvial behavior (e.g., waves of alluvium), diffusive alluvial behavior and upstream-advecting incisional behavior (e.g., knickpoint migration). In order to distinguish between the model of Sklar and Dietrich (2004, 2006) and the present model, we refer to the former as the CSA (Capacity-based Saltation-Abrasion) model, and the latter as the MRSAA (Macro-Roughness-based Saltation-Abrasion-Alluviation) model. We point out here that the first and third issues indicated above have also been addressed by Lague (2010), although in a substantially different way than

Figure 1. Views of the Shimanto River, a mixed alluvial–bedrock river in Shikoku, Japan. (a) Upstream view at low stage. (b) Macroscopic roughness of the bed and alluvial patches. Channel width is about 100 m.

presented here. The notation used in this paper is defined in Table A1.

2 Capacity-based Saltation-Abrasion (CSA) geomorphic incision law and its implications for channel evolution: upstream-migrating waves of incision

2.1 CSA geomorphic incision law

Sklar and Dietrich (2004, 2006) present the following model, referred to here as the Capacity-based Saltation-Abrasion (CSA) model, for bedrock incision in mixed bedrock–alluvial rivers transporting gravel. Defining E as the vertical rate of erosion into bedrock, q_a as the volume gravel transport rate per unit width (specified in their model solely in terms of a supply, or feed rate q_{af}) and q_{ac} as the capacity volume gravel transport per unit width such that $q_a < q_{ac}$,

$$E = \beta q_a \left(1 - \frac{q_a}{q_{ac}}\right), \tag{1a}$$

where β is an abrasion coefficient with dimension L^{-1}. By introducing a cover factor parameter p, this equation can be rewritten as

$$E = \beta q_{ac} p (1 - p). \tag{1b}$$

This cover factor p is defined (Sklar and Dietrich, 2006) as the areal fraction of bedrock surface covered with alluvium and is given by

$$p = \begin{cases} \frac{q_a}{q_{ac}} \cdots & 0 \le \frac{q_a}{q_{ac}} \le 1 \\ 1 \cdots & \frac{q_a}{q_{ac}} > 1, \end{cases} \tag{2}$$

where this fraction is calculated by averaging over a window larger than a characteristic macroscale of bedrock elevation variation. We refer to this formulation for cover factor p as "capacity based" because Eq. (2) dictates that p is determined in terms of the ratio of sediment supply to its capacity value in the CSA model.

In the above formulation, it is assumed that the gravel transport rate q_a over a bedrock surface can be estimated by simply multiplying the capacity rate q_{ac} by the areal cover fraction p. While this is the simplest first-order assumption, it should be recognized that the roughness of the bedrock itself can change the flow resistance, leading to a relationship that is more complex than Eq. (1b) (Inoue et al., 2014; Johnson, 2014).

Before introducing the relation of Sklar and Dietrich (2006) for abrasion coefficient β, it is of value to provide an interpretation for this parameter not originally given by Sklar and Dietrich (2004, 2006), but which plays a useful role in the analysis below. The abrasion coefficient has a physical interpretation in terms of Sternberg's law (Sternberg, 1875) for downstream diminution of grain size (Parker, 1991, 2008; Chatanantavet et al., 2010). The analysis leading to this interpretation is given in Appendix A; salient results are summarized here. Consider a clast of material that is of identical rock type to the bedrock being abraded. Sternberg's law is

$$D = D_u e^{-\alpha_d x}, \tag{3}$$

where D is gravel clast size, D_u is the upstream value of D, x is downstream distance and α_d is a diminution coefficient. If all diminution results from abrasion, α_d is related to β by

$$\alpha_d = \frac{\beta}{3}. \tag{4a}$$

In the case of constant β, and therefore constant α_d, the distance L_{half} for such a clast to halve in size is given by

$$L_{half} = \frac{\ln(2)}{\alpha_d}. \tag{4b}$$

This interpretation of abrasion coefficient β in terms of diminution coefficient α_d allows for comparison of the experimental results of Sklar and Dietrich (2001) with values of α_d previously obtained from abrasion mills (Parker, 2008: see Fig. 3-41 therein; Kodama, 1994).

The relations of Sklar and Dietrich (2004, 2006) to compute β and q_{ac} can be cast in the following form:

$$\beta = \frac{0.08\rho_g R g Y}{k\sigma_t^2}\left(\frac{\tau^*}{\tau_c^*}-1\right)^{-1/2}\left[1-\frac{\tau^*}{R_f^2}\right]^{3/2}, \qquad (5a)$$

$$R_f = \frac{v_f}{\sqrt{RgD}}, \qquad (5b)$$

$$q_{ac} = \alpha_a\sqrt{RgD}\,D\left(\tau^*-\tau_c^*\right)^{n_a}. \qquad (5c)$$

In the above relations, D corresponds to the characteristic size of the gravel clasts that are effective in abrading the bedrock; ρ_g is the material density of the grains; R is their submerged specific gravity (~ 1.65 for quartz); g is gravitational acceleration, τ^* is the dimensionless Shields number of the flow; τ_c^* is the threshold Shields number for the onset of significant bedload transport; α_a and n_a denote, respectively, a relation-specific dimensionless coefficient and an exponent; v_f is the fall (settling) velocity corresponding to grain size D, Y is the bedrock modulus of elasticity; σ_t is the rock tensile strength; and k is a dimensionless coefficient of the order of 10^{-6}. (In the above two relations and the text, several misprints in Sklar and Dietrich (2004, 2006) have been corrected on the advice of the authors.) Equation (5c) corresponds to the bedload transport relation of Fernandez Luque and van Beek (1976) when $\alpha_a = 5.7$ and $n_a = 1.5$; Sklar and Dietrich (2004, 2006) used this relation with the assumed value $\tau_c^* = 0.03$.

The relations above define a 0-D formulation. It must be augmented with other parameters and relations, including channel width, relations for hydraulics, quantification of flow discharge or flow duration curve, etc., to allow application at the river reach scale.

It is useful to cast Eq. (5a) in the form

$$\beta = \beta_{ref}\frac{\left(\frac{\tau^*}{\tau_c^*}-1\right)^{-1/2}\left[1-\frac{\tau^*}{R_f^2}\right]^{3/2}}{\left(\frac{\tau_{ref}^*}{\tau_c^*}-1\right)^{-1/2}\left[1-\frac{\tau_{ref}^*}{R_f^2}\right]^{3/2}}, \qquad (5d)$$

where β_{ref} is a reference value of β, either computed from known values of the parameters Y, k, σ_t, R_f, etc., or estimated indirectly.

2.2 Embedding of CSA into a model of bedrock surface evolution

A relation for the evolution of bedrock surface elevation η_b is obtained by substituting the CSA geomorphic law for incision of Eq. (1b) into a simplified 1-D mass conservation

equation for bedrock material subjected to piston-style rock uplift or base level fall (Sklar and Dietrich, 2006):

$$\frac{\partial\eta_b}{\partial t} = \upsilon - IE. \qquad (6a)$$

Here t denotes time, υ denotes the relative vertical velocity between the rock underlying the channel (which is assumed to undergo no deformation) and the point at which base level is maintained, and I denotes a flood intermittency factor to account for the fact that only relatively rare flow events are likely to drive incision (Chatanantavet and Parker, 2009). Also, I is assumed to be a prescribed constant; a more generalized formulation for flow hydrograph is given in Sklar and Dietrich (2006) and DiBiase and Whipple (2011). In interpreting Eq. (6a), it should be noted that υ denotes a rock uplift rate (in the sense of England and Molnar, 1990) for the case of constant base level, or equivalently a rate of base level fall for rock undergoing neither uplift nor subsidence. Below we use the term "rock uplift" as shorthand for the relative vertical velocity between the rock and the point of base level maintenance. Substituting Eq. (1b) into Eq. (6a) yields (Sklar and Dietrich, 2006)

$$\frac{\partial\eta_b}{\partial t} = \upsilon - I\beta q_{ac}p(1-p). \qquad (6b)$$

2.3 Character of the CSA model: upstream waves of incision

The MRSAA model (introduced below) has several new features as compared to CSA. These are best illustrated by first characterizing the mathematical nature of CSA in the context of Eq. (6). Let

$$S_b = -\frac{\partial\eta_b}{\partial x} \qquad (7)$$

denote the streamwise bedrock-surface slope. Reducing Eq. (6b) with Eq. (7) the CSA model of Eq. (1) reveals itself as a nonlinear kinematic wave equation with a source term:

$$\frac{\partial\eta_b}{\partial t} - c_b\frac{\partial\eta_b}{\partial x} = \upsilon, \qquad (8a)$$

$$c_b = \frac{I\beta q_{ac}p(1-p)}{S_b}. \qquad (8b)$$

Here c_b denotes the wave speed associated with bedrock incision. The form of Eq. (8a) dictates that disturbances in bedrock elevation always move upstream. We will see later that these disturbances can take the form of upstream-migrating knickpoints (e.g., Chatanantavet and Parker, 2009).

Any solution of Eqs. (8a) and (8b) subject to the cover relation of Eq. (2) requires specification of a flow model. In mountain streams, backwater effects are likely to be negligible (e.g., Parker, 2004). The normal (steady, uniform) flow assumption allows for simplification. Let Q denote water discharge during (morphodynamically active) flood flow taking place with intermittency I, and let H denote flood depth and g denote acceleration due to gravity. Momentum and mass balance take the forms

$$\tau = \rho g H S_b, \tag{9a}$$

$$Q = U B H, \tag{9b}$$

where τ is boundary shear stress at flood flow, U is the corresponding mean flow speed, B is channel width and ρ is water density. The dimensionless Shields number τ^* and dimensionless Chézy resistance coefficient Cz are defined as

$$\tau^* = \frac{\tau}{\rho R g D}, \tag{10a}$$

$$Cz = \frac{U}{\sqrt{\tau/\rho}}. \tag{10b}$$

As shown in Parker (2004) and Chatanantavet and Parker (2009), reducing Eqs. (7), (9) and (10) yields the following relations for H and τ^*:

$$H = \left(\frac{Q^2}{Cz^2 g B^2 S_b}\right)^{1/3}, \tag{11a}$$

$$\tau^* = \left(\frac{Q^2}{Cz^2 g B^2}\right)^{1/3} \frac{S_b^{2/3}}{R D}. \tag{11b}$$

A comparison of Eqs. (2), (5c) and (11b) indicates that even for constant values of other parameters, the functional forms for q_{ac} and thus p are such that c_b is in general a nonlinear function of $S_b = -\partial \eta_b / \partial x$.

2.4 Limitations of the CSA model

The CSA model (Sklar and Dietrich, 2004, 2006) was a major advance in the analysis of bedrock incision due to abrasion because it (a) accounts for the effect of alluvial cover and tool availability on the incision rate through the term $p(1 - p)$ in Eq. (1b) and (b) provides a physical basis for incision due to abrasion as gravel clasts collide with the bedrock surface. The CSA model been used, modified, adapted and extended by a number of researchers (Crosby et al., 2007; Lamb et al., 2008; Chatanantavet and Parker, 2009; Turowski, 2009; Lague, 2010).

The model does, however, have a significant limitation in that it specifically does not include either alluvial morphodynamics or the morphodynamics of transitions between bedrock and alluvial zones. Here we study this limitation, and how to overcome it, in terms of the highly simplified configuration of a reach (HSR, highly simplified reach) with constant width; fixed, non-erodible banks; constant water discharge; and sediment input only from the upstream end. For simplicity, we also neglect abrasion of the gravel itself, so that grain size D is a specified constant. (This condition, while introduced arbitrarily here, can be physically interpreted in terms of clasts that are much more resistant to

abrasion than the bedrock.) The means to relax these constraints is available (e.g., Chatanantavet et al., 2010; DiBiase and Whipple, 2011), and indeed many of them have been implemented in the SSTRIM model of Lague (2010). Such a relaxation, however, obscures the first-order physics underlying the rich patterns of interaction between completely and partially alluviated conditions illustrated herein.

In the CSA model, the bedload transport rate q_a is specified as a "supply". That is, the bedload transport rate is constrained so that it cannot change in the downstream direction, and is always equal to the bedload feed rate (supply) q_{af} at the upstream end. When the feed rate q_{af} increases, q_a must increase simultaneously everywhere. That is, a change in bedload supply is felt instantaneously throughout the entire reach, regardless of its length.

We illustrate this behavior in Fig. 2. The reach has length L. The gravel feed rate at $x = 0$ follows a cyclic "sedimentograph" (in analogy to a hydrograph) with period $T = T_h + T_l$, in which the sediment feed rate has a constant high feed rate q_{afh} for time T_h, and a subsequent constant low feed rate q_{afl} for time T_l. According to the CSA model, at $x = L$ corresponding to the downstream end of the reach, the temporal variation in bedload transport rate must precisely reflect the feed rate. That is, the model was not designed to route sediment in the downstream direction.

In a more realistic model, the effect of a change in bedload feed rate q_{af} would gradually diffuse and propagate downstream, so that the bedload transport rate at the downstream end of the reach would show more gradual temporal variation. This effect is illustrated in Fig. 2. This same diffusion and propagation can be expected in the cover fraction p, which in general should vary in both x and t. The change in cover fraction in turn should affect the incision rate as quantified in Eq. (1a). To capture this effect, however, Eq. (1b) must be coupled with an alluvial formulation that routes sediment downstream over the bedrock.

A second limitation concerns alluviation of the bedrock surface. Consider a wave of sediment moving over this surface, as shown in Fig. 3. We characterize the vertical scale of the geometric roughness of the bedrock surface (as seen in Fig. 1) in terms of a vertical macro-roughness L_{mr}. For simplicity, we assume that the bottom of the bedrock relief has a specific elevation η_b, an assumption that will be relaxed later in favor of a probabilistic formulation. The alluvial thickness above this basal elevation η_a represents an average value of local bed elevation over an appropriately defined window. It can be seen in Fig. 3 that the surface undergoes both partial ($\eta_a < L_{mr}$) and then complete ($\eta_a \geq L_{mr}$) alluviation, only to be excavated later as the wave passes through.

Bed elevation η is given as

$$\eta = \eta_b + \eta_a. \tag{12}$$

Figure 3 shows that, in the case of complete alluviation, the elevation of the bed η can be arbitrarily higher than the elevation η_b of the bedrock, the difference between the two

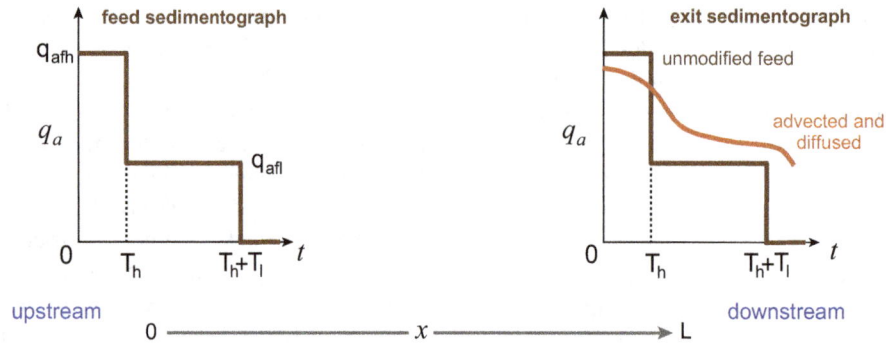

Figure 2. Schematic diagram illustrating downstream modification of a sedimentograph. At the upstream feed point ($x = 0$, left panel), the bedload transport rate q_a takes the high feed value q_{afh} for time T_h and the low feed value q_{afl} time T_l, for a total cycle time of $T = T_h + T_l$. At the downstream end ($x = L$, right panel), the discontinuous brown line represents the unaltered sedimentograph at the downstream end of the reach, assumed to have propagated instantaneously from the supply point, while the smoother red line represents the sedimentograph as modified by advective–diffusive effects.

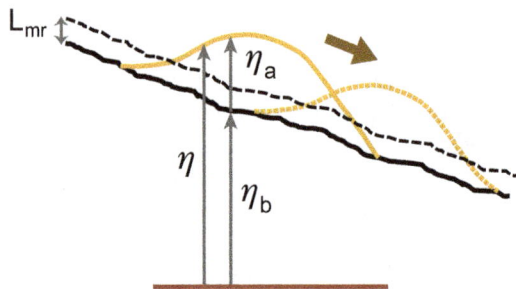

Figure 3. Schematic diagram illustrating the propagation of a wave of sediment over bedrock (orange line shifting to dashed orange line over time). Here η_b denotes the elevation of the bottom of the bedrock relief (black line), L_{mr} denotes the bedrock macro-roughness thickness (dashed black line), η_a denotes the thickness of the alluvial cover (which may be less than or greater than L_{mr}) and $\eta = \eta_b + \eta_a$ denotes the elevation of the top of the alluvium.

corresponding to the thickness η_a. The CSA model does not describe the variation in bed elevation η when the bed undergoes transitions between partial and complete alluviation; it simply infers that incision is shut down by the complete alluvial cover.

The goal of this paper is the development and implementation of a model that overcomes these limitations by (a) capturing the spatiotemporal coevolution of the sediment transport rate, alluvial cover thickness and bedrock incision rate, and (b) explicitly enabling spatiotemporally evolving transitions between bedrock–alluvial morphodynamics and purely alluvial morphodynamics. The form of the model presented here is simplified in terms of the HSR outlined above, including a constant-width channel and a single sediment source upstream.

3 Macro-Roughness-based Saltation-Abrasion-Alluviation (MRSAA) formulation and its implications for channel evolution

3.1 Formulation for alluvial sediment conservation and cover factor

The geomorphic incision law of the MRSAA model is identical to that of CSA, i.e., Eq. (1b). The essential differences are contained in (a) a formulation for the cover factor p that differs from Eq. (2) and (b) the inclusion of alluvial morphodynamics in a way that tracks the spatiotemporal evolution of the bedload transport rate, and allows for smooth spatiotemporal transitions between the bedrock–alluvial state and the purely alluvial state.

The specific case we consider here is one for which (a) the bedrock surface is rough in a hydraulic sense (as opposed to a hydraulically smooth or transitional surface; see Schlichting, 1979), and (b) the characteristic vertical scale of bedrock elevation fluctuation about a mean value based on an appropriately defined window, here denoted as the macro-roughness L_{mr} of the bedrock, is large compared to the characteristic size of the clasts constituting the alluvium. We use the term "macro-roughness" so as to clearly distinguish it from hydraulic roughness, which is specifically defined in terms the logarithmic velocity profile. Inoue et al. (2014) introduced the terms "clast-rough" and "clast-smooth", the former referring to a bedrock surface roughness that is large compared to the characteristic size of the alluvium, and the latter referring to a bedrock surface macro-roughness that is small compared to the size of the alluvium. Here we specifically consider the clast-rough case.

We formulate the problem by considering a conservation equation for the alluvium, in standard Exner form, appropriately adapted to include below-capacity transport over a non-erodible surface. The first model of this kind is due

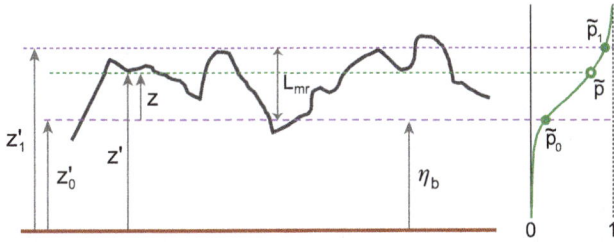

Figure 4. Illustration of the statistical structure or local hypsometry of the bedrock surface topography (dark-grey line). Here z' denotes an elevation above some arbitrary datum (dark-red line) deep in the bedrock and $\tilde{p}(z')$ is the probability (green line, marker) that a point at elevation z' is in water or alluvium rather than bedrock, i.e., above the local bedrock surface. The effective "bottom" of the bedrock relief is located at elevation $z'_0 = \eta_b$, where $\tilde{p}$ takes an appropriately selected low value $\tilde{p}_1$ (e.g., 0.05); the effective "top" of the bedrock relief is located at elevation z'_1, where $\tilde{p}$ takes an appropriately selected high value $\tilde{p}_1$ (e.g., 0.95); and the macro-roughness L_{mr} is given as $z'_1 - z'_0$. The coordinate $z = z' - z'_0$ is referenced to the effective "bottom" of the bedrock relief.

to Struiksma (1999), and further progress has been made by Parker et al. (2009, 2013), Izumi and Yokokawa (2011), Izumi et al. (2012), Tanaka and Izumi (2013) and Zhang et al. (2013). These models are expressed in continuous form; Lague (2010) presents a discrete version based on a series of reaches of finite length that allows for generalization to a continuous form.

None of the above models is specifically designed to handle the clast-rough case, in particular that shown in Fig. 1, where the elevation of the bedrock roughness has a random element. Here we handle the clast-rough case by first characterizing the statistical nature of the bedrock surface alone. As noted in Fig. 4, z' denotes elevation above an arbitrary datum deep in the bedrock, and $\tilde{p}(z')$ denotes the probability that a point located at elevation z' is located in alluvium or water rather than bedrock. Conversely, $1 - \tilde{p}$ denotes the probability that a point at elevation z' is in bedrock (rather than water or alluvium above). As seen on the right-hand side of the figure, $\tilde{p}(z') \to 0$ as $z' \to -\infty$ (pure bedrock) and $\tilde{p}(z') \to 1$ as $z' \to +\infty$. This statistical structure function (a hypsometric curve for local bedrock topography) which we use here to characterize bedrock elevation fluctuations is analogous to that used in Parker et al. (2000) for alluvial beds. It should be noted that "$-\infty$" is shorthand for "far below the bedrock surface" and "$+\infty$" is shorthand for "far above the bedrock surface". It should also be emphasized that the tilde in the parameter $\tilde{p}$ indicates it is not a cover factor, but rather a statistical parameter referring to the bedrock relief itself.

In such a statistical formulation, bedrock relief has neither a precise "bottom" nor a precise "top". Rather, the "bottom" and "top" of the bedrock topography, as well as the macro-roughness L_{mr}, are here defined in a statistical sense. This

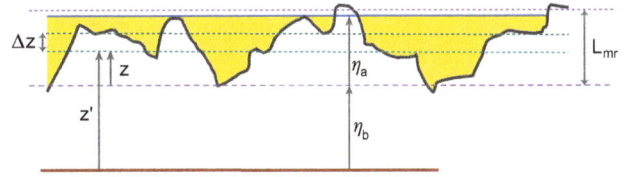

Figure 5. Schematic diagram for derivation of the Exner equation of sediment continuity over a bedrock surface (dark-grey line). As in Figs. (3) and (4), z is the elevation above the effective "bottom" of the bedrock relief, L_{mr} is macro-roughness height, η_b is elevation at base of the bedrock and η_a is the thickness of alluvium (yellow fill). The diagram shows alluvium–water interfaces (blue line) that are at spatially constant elevations. This is for illustrative purposes only; the interfaces should instead be those that result from averaging over an appropriate spatial window, i.e., alluvial fill levels are expected to vary from one pocket in the bedrock relief to another.

can be done using moments or exceedance probabilities; here we use the latter.

Let $\tilde{p}_0$ denote some low reference value of $\tilde{p}$ (e.g., $\tilde{p}_0 = 0.05$, or deep into the bedrock relief) and $\tilde{p}_1$ denote a corresponding high reference value of $\tilde{p}$ (e.g., $\tilde{p}_1 = 1 - \tilde{p}_0 = 0.95$, or near the upper portion of the bedrock relief), and z'_0 and z'_1 denote the corresponding bed elevations. An effective "base" of the bedrock relief can be set at z'_0, a macro-roughness height L_{mr} defined as

$$L_{mr} = z'_1 - z'_0 \tag{13}$$

and an effective "top" of the bedrock specified as $z'_0 + L_{mr}$. The clast-rough condition considered here satisfies the constraint that $L_{mr}/D \gg 1$.

The problem can now be rephrased in terms of a vertical coordinate z with its origin located at the effective bottom of the bedrock topography:

$$z = z' - z'_0. \tag{14}$$

As noted in Fig. 4, the statistical variable $\tilde{p}(z)$ with shifted coordinate z now has the following properties:

$$\tilde{p}(z = 0) = \tilde{p}_0, \quad \tilde{p}(z = L_{mr}) = \tilde{p}_1. \tag{15}$$

Here we define the thickness of the alluvial cover η_a as the elevation difference between the locally averaged top of the alluvium and the elevation $z = 0$. The cover fraction p associated with any alluvial thickness η_a relative to the macro-roughness height L_{mr} is then given as

$$p(\eta_a/L_{mr}) = \tilde{p}(z = \eta_a). \tag{16}$$

3.2 Exner equation of alluvial sediment conservation over a bedrock surface

The alluvial sediment is taken to have constant porosity λ. As illustrated in Fig. 5, the volume of alluvial sediment per unit area between elevations z and $z + \Delta z$ is $(1-\lambda)\,\widetilde{p}(z)\,\Delta z$, and the corresponding volume bedload transport rate per unit width q_a is estimated as $p\,q_{ac}$, where again, according to Eq. (16), $p = \widetilde{p}(\eta_a)$.

For the case of sediment of constant density, the Exner equation for mass balance of alluvial sediment can be expressed as

$$(1-\lambda)\frac{\partial}{\partial t}\int_0^{\eta_a}\widetilde{p}\,\mathrm{d}z = -I\frac{\partial p q_{ac}}{\partial x}, \tag{17}$$

where the factor I accounts for the fact that morphodynamics are active only during floods. Reducing Eq. (17) using Leibniz's rule,

$$(1-\lambda)p\frac{\partial \eta_a}{\partial t} = -I\frac{\partial p q_{ac}}{\partial x}. \tag{18}$$

The above formulation for the conservation of alluvial sediment over a bedrock surface differs in one essential way from the earlier forms due to Struiksma (1999), Parker et al. (2009, 2013), Izumi and Yokokawa (2011), Izumi et al. (2012), Tanaka and Izumi (2013) and Zhang et al. (2013). Specifically, in Eq. (18), the cover fraction p is present on the left-hand side of the equation as well as the right-hand side. It is shown below that this feature dictates a strong nonlinearity in the speed of propagation of alluvial waves over a bedrock surface, such that wave speed increases with decreasing wave amplitude. This feature is specifically captured by means of Leibniz's rule, as implemented between Eqs. (17) and (18).

The combination of Eqs. (6b) and (18) delineates a formulation encompassing both mixed bedrock–alluvial rivers and alluvial rivers.

3.3　Closure model for cover relation

In the present formulation, the cover fraction p is free to vary in both x and t, i.e., $p = p(x, t)$. In order to complete the problem, however, it is necessary to specify a closure model for p. We characterize the local variation in bedrock elevation in terms of the macro-roughness, i.e., the vertical length scale L_{mr} of Fig. 4. Here we seek a formulation that averages over a window capturing a statistically relevant sample of this local variation. In general, we assume a cover relation that characterizes to what extent the alluvial cover "drowns" the bedrock roughness elements. More specifically, we assume the form

$$p = f(\chi), \tag{19a}$$
$$\chi = \frac{\eta_a}{L_{mr}} \geq 0, \tag{19b}$$
$$f(\chi = 0) = \widetilde{p}_0, \tag{19c}$$
$$f(\chi = 1) = \widetilde{p}_1. \tag{19d}$$

The precise details of the relation can be expected to vary from case to case, but the overall characteristics that we hypothesize are illustrated in Fig. 6a–c. It is seen therein that p

takes the residual value $p = \widetilde{p}_0$ (e.g., 0.05) at $\chi = 0$, increases monotonically to $p = \widetilde{p}_1$ (e.g., 0.95) at $\chi = 1$, and then takes the asymptotic value $p \to 1$ as χ becomes sufficiently large. The first of these conditions corresponds to a bedrock surface that is bare of alluvium except in deep pockets of the macro-roughness elements, the second to a bedrock surface that is nearly completely alluviated but with some parts of the macro-roughness elements exposed, and the third to a bedrock surface that is deeply alluviated.

Note that the cover relation of Fig. 6 and Eq. (19) is based on the macro-roughness height scale L_{mr} rather than the transport capacity q_{ac} of Eq. (2). This is the motivation for referring to the new model presented here as the Macro-Roughness-based Saltation-Abrasion-Alluviation (MRSAA) model.

In applying the MRSAA model to general cases, it is useful to delineate the simplest functional form for the closure relation for cover fraction that satisfies the constraints of Eq. (19) and Fig. 6. This relation is the piecewise-linear form

$$p = f(\chi) = \begin{cases} \widetilde{p}_0 + (\widetilde{p}_1 - \widetilde{p}_0)\chi & \dots 0 \leq \chi \leq \frac{1-\widetilde{p}_0}{\widetilde{p}_1-\widetilde{p}_0} \\ 1 & \dots \chi > \frac{1-\widetilde{p}_0}{\widetilde{p}_1-\widetilde{p}_0} \end{cases} \tag{20a}$$

or rephrasing to emphasize the dependence of p on the thickness of alluvial cover η_a,

$$p = f(\eta_a/L_{mr}) = \begin{cases} \widetilde{p}_0 + (\widetilde{p}_1 - \widetilde{p}_0)\frac{\eta_a}{L_{mr}} & \dots 0 \leq \frac{\eta_a}{L_{mr}} \leq \frac{1-\widetilde{p}_0}{\widetilde{p}_1-\widetilde{p}_0} \\ 1 & \dots \frac{\eta_a}{L_{mr}} > \frac{1-\widetilde{p}_0}{\widetilde{p}_1-\widetilde{p}_0} \end{cases} \tag{20b}$$

The above relation is illustrated in Fig. 7a with the sample evaluations

$$\widetilde{p}_0 = 0.05, \quad \widetilde{p}_1 = 0.95. \tag{21}$$

Equations (20) and (21) are used in implementations of the MRSAA below. One way to develop forms of Eq. (19) that can show a wider variety of behavior than Eq. (20) would be through the performance of experiments similar to those of Chatanantavet and Parker (2008), but with a specific focus on various forms of macro-roughness that mimic those in the field.

The form for the derivative of Eq. (20) with respect to χ, which is given below, will prove useful in succeeding analysis.

$$\frac{\mathrm{d}f}{\mathrm{d}\chi} = \begin{cases} \widetilde{p}_1 - \widetilde{p}_0 & \dots 0 \leq \chi \leq \frac{1-\widetilde{p}_0}{\widetilde{p}_1-\widetilde{p}_0} \\ 0 & \dots \chi > \frac{1-\widetilde{p}_0}{\widetilde{p}_1-\widetilde{p}_0} \end{cases} \tag{22}$$

Our model is specifically meant to apply to the case for which the characteristic size of the roughness elements is large compared to the size of clasts transported as bedload. With this in mind, it should be noted that in Fig. 6, no bed elevation variations are shown over parts of the bed that are covered with alluvium. This is done only for simplicity, and reflects

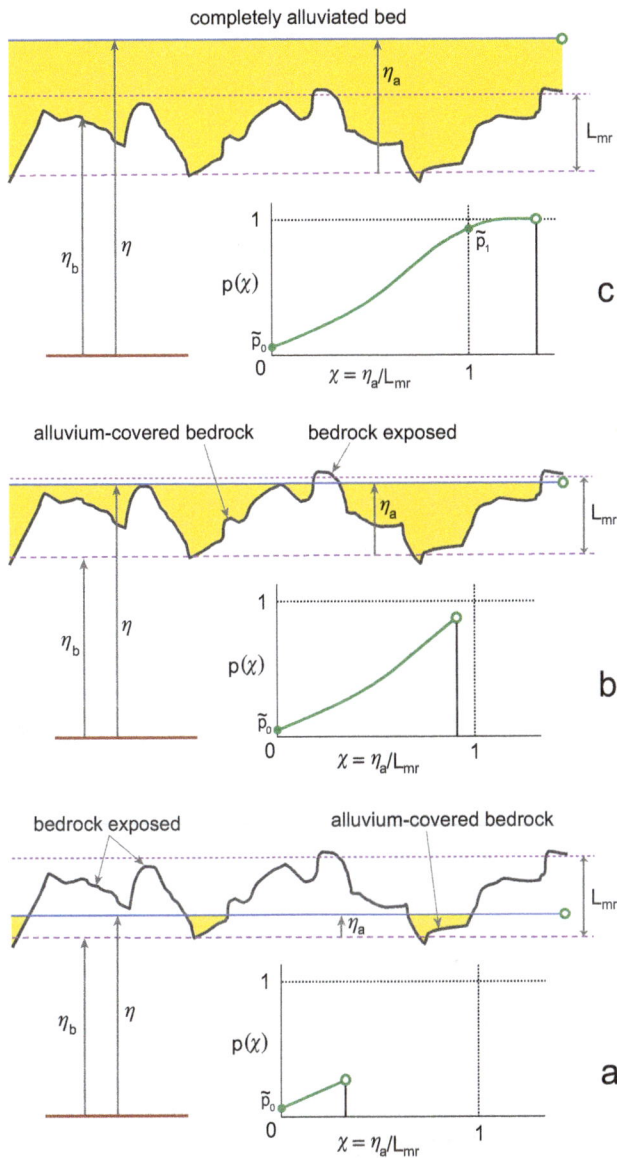

Figure 6. Illustration of the MRSAA model relation between areal fraction of alluvial cover of bedrock $p(\chi)$ (green curves) and $\chi = \eta_a / L_{mr}$, where η_a is the thickness of alluvium (yellow fill) and L_{mr} is the macro-roughness height, for **(a)** low cover, **(b)** intermediate cover **(c)** complete alluviation above the top of the bedrock. The diagrams show alluvium–water interfaces at spatially constant elevations (blue lines). This is for illustrative purposes only: the interfaces should instead be those that result from averaging over an appropriate spatial window.

the condition that in the clast-rough case considered here, grain size is small compared to macro-roughness height. Figure 6 also contains another simplification, in that all pockets are assumed to be filled to the same level by alluvium. While this condition is not likely to be true at the local scale, it is a reasonable first approximation when averaging over an appropriately defined window.

The formulation presented here has an obvious limitation. Since it is a 1-D expression of sediment conservation over a bedrock surface, it cannot capture 2-D variation, which will result in a more complex pattern than that shown in Fig. 6, and in particular will provide more connectivity between adjacent pockets. This two-dimensionality is known to have an effect on the pattern of incision, as illustrated by Johnson and Whipple (2007). The extension of the formulation to the 2-D case represents a future goal; some relevant comments can be found in the "Discussion" section (Sect. 8).

Sections 3.4 and 3.5, immediately below focus, on the mathematical interpretation of the MRSAA problem in terms of diffusion and wave characteristics. The reader whose primary interest is in applications may jump directly to Sect. 3.6, with the two exceptions of Eqs. (27) and (28) in Sect. 3.5. Equation (27) is a version of Eq. (6b) in which incisional morphodynamics are recast into a kinematic wave equation, revealing upstream-migrating waves of incision with wave speed c_b. Equation (28) is a version of Eq. (18) in which alluvial morphodynamics are recast in terms of an advection–diffusion equation, with downstream-migrating waves of alluviation with speed c_a, and alluvial diffusion with kinematic coefficient of diffusion κ_a.

3.4 Character of the alluvial part of the MRSAA problem: alluvial diffusion and downstream-migrating waves of alluviation

Equation (18) may be reduced to reveal the presence of an alluvial wave speed as follows. The derivative on the right-hand side of the equation is expanded using the chain rule, the derivative $\partial p / \partial x$ is reduced in accordance with the general closure form of Eq. (19), and both sides of Eq. (18) are then divided by p to yield

$$\frac{\partial \eta_a}{\partial t} + c_a \frac{\partial \eta_a}{\partial x} = -\frac{1}{(1-\lambda)} I \frac{\partial q_{ac}}{\partial x}, \tag{23a}$$

where

$$c_a = \frac{I}{(1-\lambda)} \frac{q_{ac}}{L_{mr} p} \frac{\mathrm{d}f}{\mathrm{d}\chi}. \tag{23b}$$

The left-hand side of Eq. (23a) thus takes a kinematic wave form, such that c_a is the wave speed of downstream-directed alluviation.

It is important to realize that alluvial wave speed c_a is a nonlinear function of alluvial thickness η_a. Using the example functions of Eqs. (20) and (22), the speed c_{ai} of an alluvial wave of infinitesimal height $\eta_a = 0$ is given from Eq. (23b) as

$$c_{ai} = \frac{I}{(1-\lambda)} \frac{q_{ac}}{L_{mr}} \frac{(\tilde{p}_1 - \tilde{p}_0)}{\tilde{p}_0}. \tag{24a}$$

The ratio c_a / c_{ai} of the wave speed c_a at height η_a to the corresponding value for an infinitesimal wave is then found to be

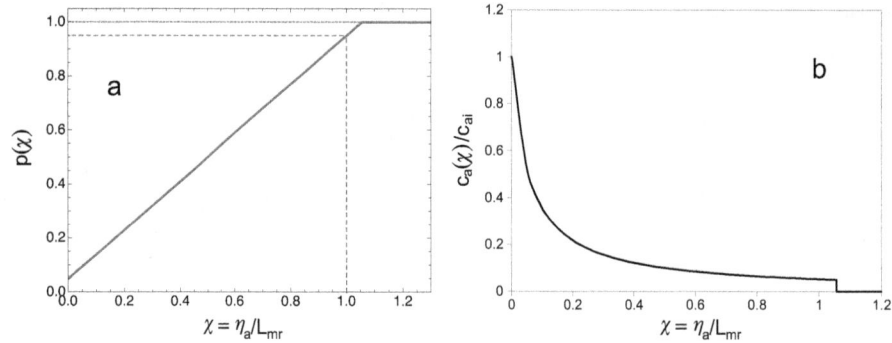

Figure 7. (a) Simplest modified form the MRSAA cover function (green curve) satisfying the conditions $p(0) = r$, $p(1) = 1 - r$ and $p(\infty) = 1$, where in this case $r = 0.05$. The parameter on the vertical axis $p(\chi)$ denotes the cover fraction, and the parameter on the horizontal axis is $\chi = \eta_a/L_{mr}$, where η_a is the thickness of alluvium and L_{mr} is the macro-roughness height. **(b)** Plot of the ratio of alluvial wave speed $c_a(\chi)$ at a finite value of $\chi = \eta_a/L_{mr}$ to the corresponding speed c_{ai} for an alluvial wave of infinitesimal height versus χ. The plot shows that alluvial wave speed declines with increasing alluvial thickness, and that it vanishes under complete alluviation.

$$\frac{c_a}{c_{ai}} = \frac{(\tilde{p}_1 - \tilde{p}_0)}{\tilde{p}_0} \frac{1}{p} \frac{df}{d\chi}. \tag{24b}$$

In Fig. 7b, the ratio c_a/c_{ai} is plotted against $\chi = \eta_a/L_{mr}$ in accordance with the specifications of Eqs. (20), (21), and (22). It is seen therein that alluvial wave speed takes its maximum value for the limit $\eta_a \to 0$, and decreases to a vanishing value as the bed becomes completely alluviated ($\eta_a/L_{mr} = 1.056$). That is, waves of alluvium run fastest over a nearly bare bed, and wave-like behavior ceases to exist under conditions of complete alluviation. This latter result is in accordance with Lisle et al. (2001) and Cui et al. (2003a, b).

It is of interest to inquire as to how the model would behave if the clast-rough condition, i.e., $L_{mr}/D \gg 1$, were not satisfied here. The limiting case of clast-smooth conditions would correspond to a probability distribution $\tilde{p}(z')$ in Fig. 4 that obeys a step function; $\tilde{p}(z')$ would be vanishing up to a smooth, horizontal bedrock surface, and would take the value unity above it. Consequently, the cover fraction p would converge precisely to zero as $\eta_a \to 0$, thus resulting, according to Eq. (23b), in an infinite speed of propagation of an alluvial wave of infinitesimal height. This is not entirely unrealistic: the physical realization would consist of clasts rolling rapidly over a smooth bed with no alluviation (Inoue et al., 2014). The presence of such a singularity, would, however, preclude the modeling of the migration of a pulse of alluvium of finite extent over the otherwise bare bed schematized in Fig. 3. In the present clast-rough formulation, the presence of deep pockets within the bedrock relief where alluvium can be stored without transport ensures that the wave speed of alluvium never displays a singularity.

The form of Eq. (23a) can be further clarified by rewriting it as

$$\frac{\partial \eta_a}{\partial t} + c_a \frac{\partial \eta_a}{\partial x} - \frac{\partial}{\partial x}\left(\kappa_a \frac{\partial \eta_a}{\partial x}\right) = \frac{\partial}{\partial x}\left(\kappa_a \frac{\partial \eta_b}{\partial x}\right), \tag{25}$$

where

$$\kappa_a = \frac{I q_{ac}}{(1 - \lambda) S}, \tag{26a}$$

$$S = -\frac{\partial \eta}{\partial x} = -\frac{\partial \eta_b}{\partial x} - \frac{\partial \eta_a}{\partial x}. \tag{26b}$$

In the above relation, κ_a has the physical meaning of a kinematic diffusivity. In general, q_{ac}, q_{ac}/S and thus κ_a are nonlinear functions of S. The alluvial problem thus takes the form of a nonlinear advective–diffusive problem with a source term arising from a bedrock term.

3.5 Full MRSAA formulation: alluvial diffusion, upstream-migration waves of incision, downstream-migrating waves of alluviation

The full MRSAA model consists of the kinematic wave equation with a source term Eq. (8a) for the bedrock part, Eqs. (23b), (25) and (26) for the alluvial part, and the linkage between the two embodied in the cover relation of Eq. (19). Restating these equations for emphasis, Eq. (6a) can be recast as

$$\frac{\partial \eta_b}{\partial t} - c_b \frac{\partial \eta_b}{\partial x} = \upsilon, \tag{27a}$$

$$c_b = I\beta q_{ac} p(1 - p)\left(-\frac{\partial \eta_b}{\partial x}\right)^{-1}, \tag{27b}$$

where c_b denotes the speed of upstream-migrating incisional waves. Equation (18) can be cast in conjunction with Eqs. (19) and (20) as

$$\frac{\partial \eta_a}{\partial t}+c_a\frac{\partial \eta_a}{\partial x}-\frac{\partial}{\partial x}\left(\kappa_a\frac{\partial \eta_a}{\partial x}\right)=\frac{\partial}{\partial x}\left(\kappa_a\frac{\partial \eta_b}{\partial x}\right), \quad (28a)$$

$$c_a=\frac{I}{(1-\lambda)}\frac{q_{ac}}{L_{mr}p}\frac{df}{d\chi}, \quad (28b)$$

$$\frac{df}{d\chi}=\frac{dp}{d(\eta_a/L_{mr})}, \quad (28c)$$

$$\kappa_a=I(1-\lambda)q_{ac}\left[-\frac{\partial}{\partial x}(\eta_a+\eta_b)\right]^{-1} \quad (28d)$$

and

$$p=f(\chi)=\begin{cases}\tilde{p}_0+(\tilde{p}_1-\tilde{p}_0)\chi & \ldots 0\le\chi\le\frac{1-\tilde{p}_0}{\tilde{p}_1-\tilde{p}_0}\\ 1 & \ldots \chi>\frac{1-\tilde{p}_0}{\tilde{p}_1-\tilde{p}_0},\end{cases} \quad (29)$$

where c_a denotes the speed of downstream-migrating alluvial waves, and κ_a is the kinematic diffusivity of alluvium. In this way, upstream-migrating incisional waves are combined with downstream-migrating alluvial waves and alluvial diffusion.

In MRSAA, then, the spatiotemporal variation in the cover fraction $p(x,t)$ is specifically tied to the corresponding variation in η_a through Eq. (19), e.g., the specific example of Eq. (29) above. This variation then affects incision through Eq. (27). Consider the simplified case of a wave of alluvium of finite extent illustrated in Fig. 3. There is no incision upstream of the wave because $p=0$ and there is no sediment in motion over the bed. At the peak of the wave, $\eta_a>L_{mr}$, so $p=1$; the bed is entirely covered with sediment, and again there is no incision. Incision can only occur on the rising and falling parts of the wave, where bedrock is partially exposed and sediment is in motion over it, i.e., $0<p<1$. It can thus be expected that the spatiotemporal variation in cover thickness η_a will affect the evolution of the long profile of an incising river that undergoes transitions between alluvial and mixed bedrock–alluvial states.

3.6 Amendment of the flow component of the MRSAA model

The flow model, and in particular Eqs. (9a) and (11), must be modified to include the alluvial formulation, so that bedrock slope S_b is replaced with slope S of the top of the bed, where

$$S=-\frac{\partial \eta}{\partial x}=S_b+S_a, \quad (30a)$$

$$S_b=-\frac{\partial \eta_b}{\partial x}, \quad (30b)$$

$$S_a=-\frac{\partial \eta_a}{\partial x}. \quad (30c)$$

Thus Eqs. (9a) and (11a, 11b) are amended to

$$\tau=\rho g H S \quad (31)$$

and

$$H=\left(\frac{Q^2}{Cz^2gB^2S}\right)^{1/3}, \quad (32a)$$

$$\tau^*=\left(\frac{Q^2}{Cz^2gB^2}\right)^{1/3}\frac{S^{2/3}}{RD}. \quad (32b)$$

The purely alluvial case, i.e., $p=1$, $df/d\chi=0$ and $\eta_b=\text{const}<\eta_a$ in Eq. (28), results in the purely diffusional relation

$$\frac{\partial \eta_a}{\partial t}=\frac{\partial}{\partial x}\left(\kappa_a\frac{\partial \eta_a}{\partial x}\right), \quad (33)$$

in which the diffusivity κ_a is a function of $S_a=-\partial\eta_a/\partial x$.

3.7 How the governing equations connect to each other

In the numerical analysis below, the actual equations used to solve for morphodynamic evolution are not those of Sects. 3.4 and 3.5, but rather the primitive forms presented earlier. The unknowns to be solved are η_b, η_a, η, p, q_{ac}, τ^*, S, S_b and S_a. These nine parameters are connected to each other via nine equations, i.e., Eqs. (5c), (6b), (12), (18), (20), (30a, b, c) and (32b).

3.8 Equivalence of the MRSAA and CSA models at steady state

In the restricted case of the highly simplified reach (HSR) configuration constrained by (a) temporally constant, below-capacity sediment feed (supply) rate q_{af}, (b) bedload transport rate q_a everywhere equal to the feed rate q_{af}, and (c) a steady-state balance between incision and rock uplift, η_a, p and S_b become constant and S_a vanishes, so that Eq. (28) is satisfied exactly. Equation (18) integrates to give

$$p=\frac{q_{af}}{q_{ac}}, \quad (34)$$

so that η_a can then be back-calculated from Eq. (20). In this case, then, the MRSAA model reduces to Eqs. (27) and (34), i.e., the CSA model.

4 The below-capacity steady-state case common to the CSA and MRSAA models

The steady-state form of Eq. (6) under below-capacity conditions ($p<1$) can be expressed with the aid of Eq. (2) in the form

$$p_{ss} = 1 - \varphi, \tag{35a}$$

$$\varphi = \frac{\upsilon}{I\beta_{ss}q_{af}}, \tag{35b}$$

$$q_{acss} = \frac{q_{af}}{p_{ss}}, \tag{35c}$$

where p_{ss}, β_{ss} and q_{acss} denote steady-state values of p, β and q_{ac}, respectively. Equations (35a)–(35c) describe a balance between the incision rate and relative vertical rock velocity (e.g., constant rock uplift rate at constant base level or constant rock elevation with constant rate of base level fall). CSA and MRSAA yield the same solution for this case, which must be characterized before showing how the models differ.

Equation (35a) has an interesting character. When the value of the dimensionless number φ exceeds unity, p falls below zero and no steady-state solution exists. Equation (35b) reveals that φ can be interpreted as a dimensionless rock uplift rate. Thus when the rock uplift rate is sufficiently large for φ to exceed unity, incision cannot keep pace with rock uplift. The model thus implicitly predicts the formation of a hanging valley. This issue was earlier discussed in Crosby et al. (2007).

In solving for this steady state, and in subsequent calculations, we use the bedload transport relation of Wong and Parker (2006a), a development and correction of the semi-empirical relation of Meyer-Peter and Müller (1948), rather than the similar formulation of Fernandez Luque and van Beek (1976); in the case of the former, $\alpha_a = 4$, $n_a = 1.5$ and $\tau_c^* = 0.0495$. We consider two cases: one for which $\beta_{ss} = \beta$ is a specified constant, and one for which only a reference value β_{ref} is specified, and β_{ss} is computed from Eq. (5d).

In the case of a specified constant abrasion coefficient β, specification of υ, I and q_{af} allow computation of φ, p_{ss} and q_{acss} from Eqs. (35a)–(35c). Further specification of R (here chosen to be 1.65, the standard value for quartz) and D allows the steady-state Shields number τ_{ss}^* to be computed from Eq. (5c). Steady-state bedrock slope S_{bss} can then be computed from Eq. (11b) upon specification of flood discharge Q, Chézy resistance coefficient Cz and channel width B. In the case of β_{ss} calculated according to Eq. (5d) using a specified reference value β_{ref}, the problem can again be solved with Eqs. (35), (5c) and (11b), but the solution is implicit.

We performed calculations for conditions loosely based on (a) field estimates for a reach of the bedrock Shimanto River near Tokawa, Japan (Fig. 1), for which bed slope S is about 0.002 and channel width is about 100 m and (b) estimates using relations in Parker et al. (2007) for alluvial gravel-bed rivers with similar slopes, and reasonable choices for otherwise poorly constrained parameters. The input parameters, $Cz = 10$, $Q = 300\,\mathrm{m^3\,s^{-1}}$, $B = 100\,\mathrm{m}$, are loosely justified in terms of bankfull characteristics of alluvial gravel-bed rivers of the same slope (Parker et al., 2007;

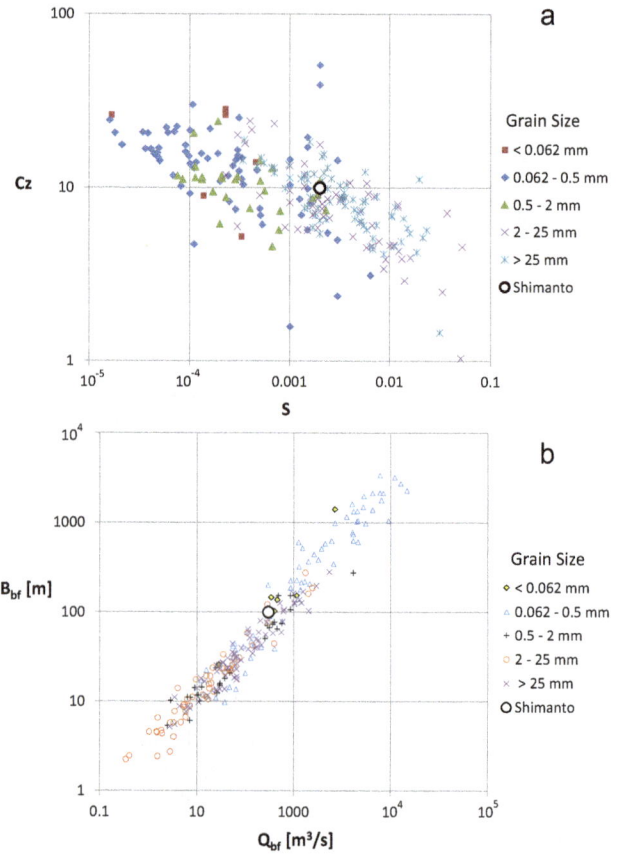

Figure 8. (a) Chézy resistance coefficient Cz plotted against bed slope S for alluvial rivers. Also included are estimated values for the Shimanto River, Japan. (b) Bankfull width B_{bf} versus bankfull discharge Q_{bf} for alluvial rivers. Also included are estimates for bank-to-bank width and characteristic flood discharge in the Shimanto River, Japan. The ranges for characteristic bed material size of the alluvial rivers are denoted in the legends.

Wilkerson and Parker, 2011) as shown in Fig. 8a and b. The value $D = 20\,\mathrm{mm}$ represents a reasonable characteristic size of the substrate (and thus the bedload) for gravel-bed rivers; a typical size for surface pavement is 2 to 3 times this (e.g., Parker et al., 1982). Flood intermittency I is estimated at 0.05, i.e., 18 days per year, and thus a reasonable estimate for a river subject to frequent heavy storm rainfall. Alluvial porosity is $\lambda = 0.35$.

Two sediment feed rates were considered. The high feed rate was set at $3.5 \times 10^5\,\mathrm{t\,yr^{-1}}$, which corresponds to the following steady-state parameters at capacity conditions: Shields number $\tau^* = 0.12$, depth $H = 1.5\,\mathrm{m}$, steady-state alluvial bed slope $S_{ass} = 0.0026$ and Froude number $Fr = 0.51$, where

$$Fr = \frac{Q}{BH\sqrt{gH}}. \tag{36}$$

The low feed rate was set at $3.5 \times 10^4\,\mathrm{t\,yr^{-1}}$, corresponding to the following parameters at capacity conditions: Shields number $\tau^* = 0.064$, depth $H = 2.1$ m, steady-state alluvial bed slope $S_{ass} = 0.0010$ and Froude number $Fr = 0.32$.

The value $\beta_{ss} = 0.05$ km^{-1} was used for the case of a constant steady-state abrasion coefficient. This corresponds to a value of α_d of 0.017 km^{-1}, which falls in the middle of the range measured by Kodama (1994) for chert, quartz and andesite (see Fig. 3-41 of Parker, 2008). For the case of a variable abrasion coefficient, Eq. (1a) was used with β_{ref} set to 0.05 km^{-1} and τ^*_{ref} set to 0.12, i.e., the value for the high feed rate. This value of τ^*_{ref} is about 2.5 times the threshold value of Wong and Parker (2006a).

For the high feed, predicted relations for a steady-state abrasion coefficient β_{ss} versus rock uplift rate v are shown in Fig. 9a; the corresponding predictions for S_{bss} versus v are shown in Fig. 9b; the corresponding predictions for p_{ss} and φ are shown in Fig. 9c. Both the cases of constant and variable β_{ss} are shown. There are five notable aspects of these figures: (a) in Fig. 9a, the predictions for variable β_{ss} are very similar to the case of constant, specified β_{ss}, and indeed are nearly identical for $v \leq 3.3$ mm yr^{-1} (corresponding to $\varphi \leq 0.05$ in Fig. 9c). (b) In Fig. 9b and c, the predictions for S_{bss}, p_{ss} and φ for variable β_{ss} are again nearly identical to those for constant β_{ss}, and again essentially independent of v for $v \leq 3.3$ mm yr^{-1}. (c) In Fig. 9c, p_{ss} is only slightly below unity (i.e., $p_{ss} \geq 0.95$), and $\varphi \leq 0.05$ for $v \leq 3.3$ mm yr^{-1}). (d) For $v > 3.3$ mm yr^{-1}, the predictions for S_{bss} and p_{ss} become dependent on v, such that S_{bss} increases, and p_{ss} decreases, with increasing v. The values for constant β_{ss} diverge from those for variable β_{ss}, but are nevertheless close to each other up to some limiting value. (e) This limiting value corresponds to $\varphi = 1$ and thus $p_{ss} = 0$ from Eq. (35a); larger values of φ lead to hanging valley formation. Here $\varphi = 1$ for the very high values $v = 65$ mm yr^{-1} for constant β_{ss} and $v = 30$ mm for variable β_{ss}.

These results require interpretation. It can be seen from Eqs. (35a–35c) that when $v/(I \beta_{ss} q_{af}) = \varphi \ll 1$, p becomes nearly equal to unity (very little exposed bedrock), in which case q_{af} is constrained to be only slightly smaller than q_{ac}. From Eqs. (5c) and (11), then, S_{bss} is only slightly above the steady-state alluvial bed slope S_{ass}. Note that the steady-state bedrock slope decouples from rock uplift rate under these conditions: the predictions for $v = 0.2$ mm yr^{-1} are nearly identical to this for $v = 3.3$ mm yr^{-1}. This behavior is a specific consequence of the condition $\varphi \ll 1$ corresponding to a low ratio of uplift rate to reference incision rate $E_{ref} = I \beta_{ss} q_{af}$. They imply a wide range of conditions for which (a) very little bedrock is exposed, and (b) bedrock slope is independent of uplift rate.

The results for the low feed rate are very similar. The values for variable steady-state abrasion coefficient β_{ss} differ from the constant value β_{ss} in Fig. 10a, but this is because the constant value $\beta_{ss} = 0.05$ was set based on the high feed rate. The results in Fig. 10b and c are qualitatively the same for Fig. 9b and c; the uplift rate below which $\varphi < 0.05$ is $v < 0.33$ mm yr^{-1} for the case of constant β_{ss},

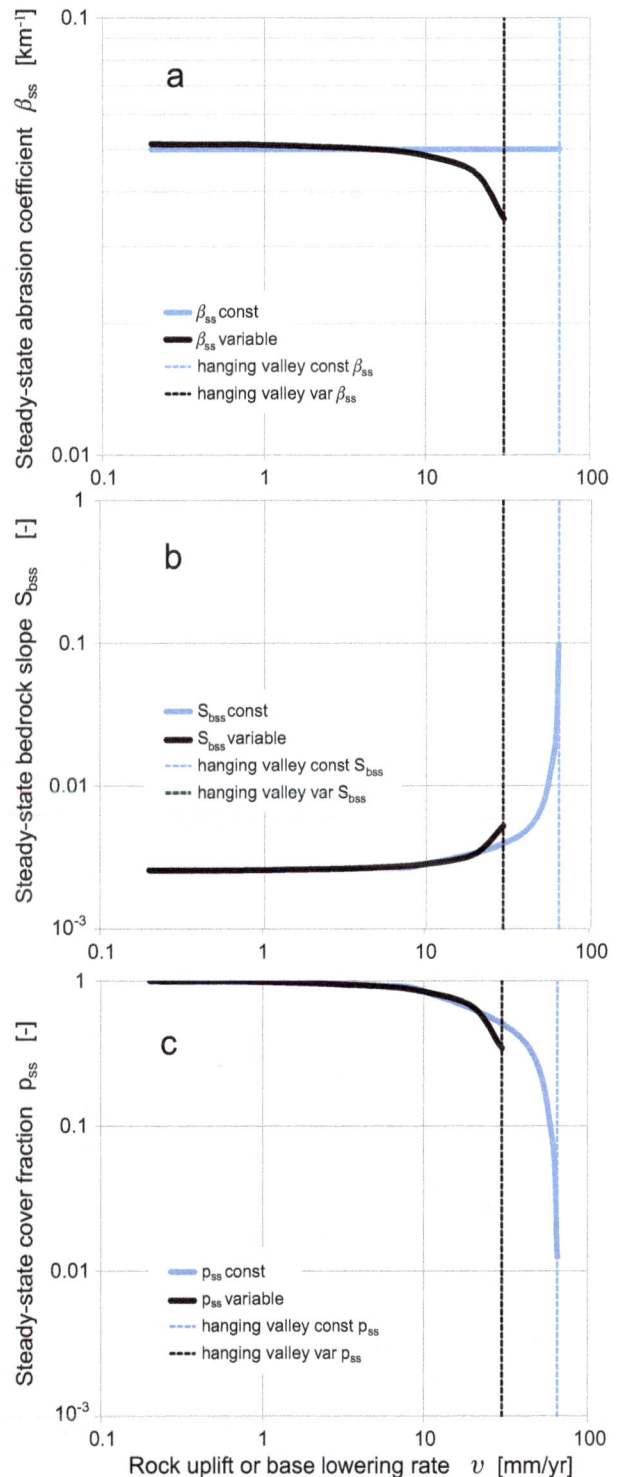

Figure 9. Variation at steady state (black curves) of (a) abrasion coefficient β_{ss}, (b) bedrock slope S_{bss} and (c) cover fraction p_{ss} and parameter φ on rock uplift or base lowering rate v, for a high bedload feed rate of 3.5×10^5 t yr^{-1}. The cases of constant, specified β_{ss} and S_{bss} varying according to Eq. (34) are shown as blue curves. The vertical dashed lines denote the incipient conditions for the formation of a hanging valley. The predictions are the same for the CSA and MRSAA models.

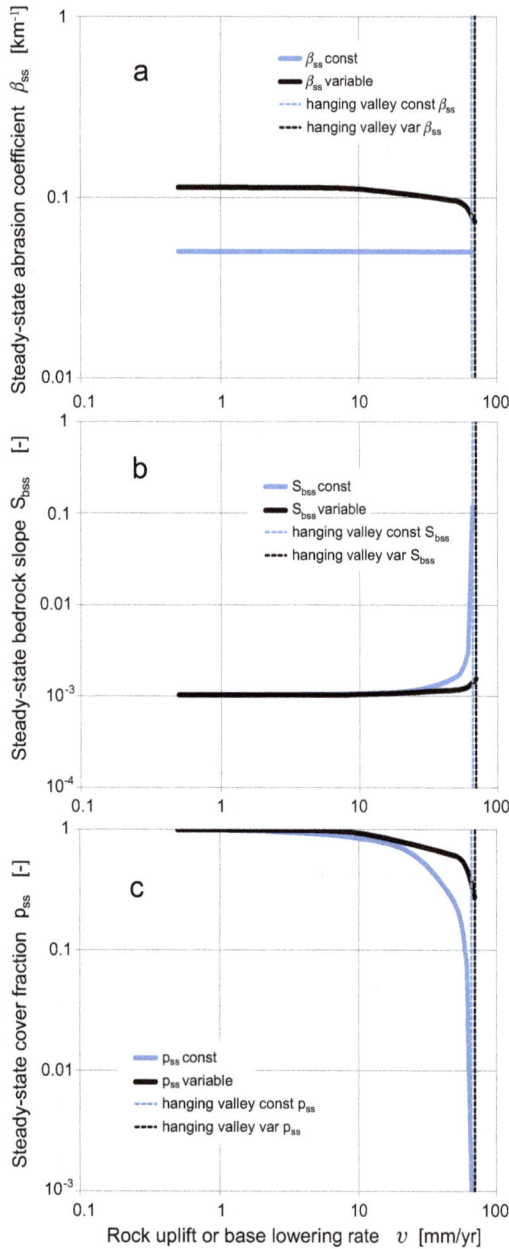

Figure 10. Variation at steady state (black curves) of (a) abrasion coefficient β_{ss}, (b) bedrock slope S_{bss} and (c) cover fraction p_{ss} and parameter φ with rock uplift or base lowering rate υ, for a low bedload feed rate of $3.5 \times 10^4\,t\,yr^{-1}$. The cases of constant, specified β_{ss} and S_{bss} varying according to Eq. (34) are shown as blue curves. The vertical dashed lines denote the incipient conditions for the formation of a hanging valley. The predictions are the same for the CSA and MRSAA models.

The lack of dependence of steady-state bedrock slope S_{bss} on rock uplift rate υ below a threshold value for the steady-state solutions of the CSA model (and thus the MRSAA model as well) is in stark contrast to earlier work for which the incision rate E is assumed to have the following dependence on slope S_b and drainage area A ("slope–area" formulation, Howard and Kerby, 1983):

$$E = K S_b^n A^m, \tag{37}$$

where A denotes drainage area, n and m are specified exponents, and K is a constant assumed to decrease with increasing rock hardness.

In order to compare the steady-state predictions of the slope–area relation in Eq. (37) for constant υ with CSA, drainage area A must be taken to be a constant value A_o so as to correspond to the HSR configuration used here. The steady-state slope S_{bss} corresponding to a balance between incision and rock uplift is found from Eq. (37) to be

$$S_{bss} = \frac{\upsilon^{1/n}}{K^{1/n} A_o^{m/n}}. \tag{38}$$

The issue as to the values of m and n has been considered by many researchers, including Whipple and Tucker (2002) and Lague (2014).

In their Table 1, Whipple and Tucker (2002) quote a range of values of n, but their most quoted value is $n = 2$. We compare the results for CSA for S_{bss} with the predictions from Eq. (38) with $n = 2$ by normalizing against a reference value S_{bref} that corresponds to a reference rock uplift rate υ_{ref} of $0.2\,mm\,yr^{-1}$. Equation (38) yields

$$\frac{S_{bss}}{S_{bref}} = \left(\frac{\upsilon}{\upsilon_{ref}}\right)^{1/2}. \tag{39}$$

In Fig. 11, Eq. (39) is compared against the CSA predictions of Figs. 9b and 10b (high and low feed rate, respectively) for both constant and variable β_{ss}. In order to keep the plot within a realistic range, only values of υ between 0.2 and $10\,mm\,yr^{-1}$ (the upper limit corresponding to Dadson et al., 2003) have been used in the CSA results. The remarkable insensitivity of the CSA predictions for steady-state slope S_{bss} on rock uplift rate is readily apparent from the figure.

One more difference between the CSA and slope–area formulations is worth noting. If the slope–area relation is installed into Eq. (6) in place of CSA, it is readily shown that bedrock slope gradually relaxes to zero in the absence of rock uplift. CSA does not obey the same behavior under the constraint of constant sediment feed rate: Figs. 9b and 10b indicate that bedrock slope converges to a constant, nonzero value as rock uplift declines to zero. This is not necessarily a shortcoming of CSA; the sediment feed rate can be expected to decline as relief declines.

and $\upsilon < 0.73\,mm\,yr^{-1}$ for the case of variable β_{ss}. The critical value of υ beyond which a hanging valley forms is $\upsilon \geq 6.8\,mm\,yr^{-1}$ for constant β_{ss} and $\upsilon \geq 7.1\,mm\,yr^{-1}$ for variable β_{ss}.

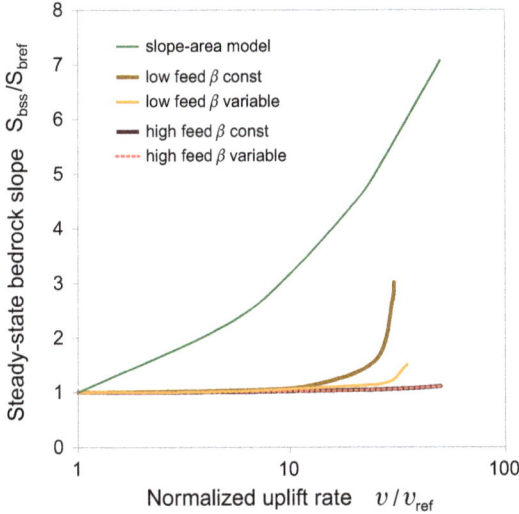

Figure 11. Normalized steady-state bedrock slope $S_{\mathrm{bss}}/S_{\mathrm{bref}}$ versus normalized rock uplift rate v as predicted by the CSA model for a low feed rate (orange-brown curves) and a high feed rate (black curve and dashed red curve), and for constant and variable abrasion coefficient. The results are the same for the MRSAA model. Also shown is the prediction of a model for which the incision rate is specified in terms of bedrock slope and upstream drainage area (green curve). Note that the predictions for steady-state bedrock slope of the CSA model are insensitive to the rock uplift rate over a wide range.

5 Boundary conditions and parameters for numerical solutions of the MRSAA model

Having conducted a fairly thorough analysis of the steady state common to the CSA and MRSAA models, it is now appropriate to move on to examples of behavior that can be captured by the MRSAA model, but are not captured by models that assume a relation for cover based on the ratio of sediment supply to capacity transport rate, i.e., Eq. (2). Before doing so, however, it is necessary to delineate the boundary conditions and other assumptions used in the MRSAA model.

Let L denote the length of the reach. Equation (27a) indicates the formulation for bedrock incision is first order in x and so requires only one boundary condition. The example considered here is that of a downstream bedrock elevation, i.e., base level, set to zero:

$$\eta_{\mathrm{b}}|_{x=L} = 0. \tag{40}$$

According to Eq. (18), or alternatively Eq. (28a), the alluvial formulation is second order in x and thus requires two boundary conditions. The following boundary condition applies at the upstream end of the reach, where $q_{\mathrm{af}}(t)$ denotes a feed rate that may vary in time,

$$q_{\mathrm{a}}|_{x=0} = q_{\mathrm{af}}(t). \tag{41}$$

At the downstream end, a free boundary condition is applied for $\eta_{\mathrm{a}}/L_{\mathrm{mr}} < 1$, and a fixed boundary condition is applied for $\eta_{\mathrm{a}}/L_{\mathrm{mr}} \geq 1$ as follows:

$$\left[(1-\lambda)p\frac{\partial \eta_{\mathrm{a}}}{\partial t} + I\frac{\partial pq_{\mathrm{ac}}}{\partial x}\right]_{x=L} = 0 \quad \dots \text{if } \left[\frac{\eta_{\mathrm{a}}}{L_{\mathrm{mr}}}\right]_{x=L} < 1, \tag{42a}$$

$$\eta_{\mathrm{a}}|_{x=L} = L_{\mathrm{mr}} \quad \dots \text{if } \left[\frac{\eta_{\mathrm{a}}}{L_{\mathrm{mr}}}\right]_{x=L} \geq 1. \tag{42b}$$

Here, Eq. (42a) specifies a free boundary in the case of partial alluviation, thus allowing below-capacity sediment waves to exit the reach. Equation (42b), on the other hand, fixes the maximum downstream elevation at $\eta = \eta_{\mathrm{a}} = L_{\mathrm{mr}}$.

In order to illustrate the essential features of the new formulation of the MRSAA model for the morphodynamics of mixed bedrock–alluvial rivers, it is useful to consider the most simplified case that illustrates its expanded capabilities compared to the CSA model. Here we implement the HSR simplification. In addition, based on the results of the previous section, we approximate β_{ss} as a prescribed constant. Finally, we assume that the clasts of the abrading bedload are sufficiently hard compared to the bedrock so that grain size D can be approximated as a constant. These constraints are easily relaxed.

In the numerical solution of the differential Eqs. (6b) and (18), spatial derivatives have been computed using an upwinding scheme for short timescales (so as to capture downstream-migrating alluvial waves) and a downwinding scheme for long timescales (so as to capture upstream-migrating incisional waves). Time derivatives have been computed using the Euler step method.

6 Sediment waves over a fixed bed: stripping and emplacement of alluvial layer and advection–diffusion of a sediment pulse

Three numerical solutions of the MRSAA model are studied here: (a) stripping of an alluvial cover to bare bed, (b) emplacement of an alluvial cover over a bare bed and (c) advection–diffusion of an alluvial pulse over a bare bed. Reach length L is 20 km. As the time for alluvial response is short compared to incisional response, β_{ss} and v are set equal to zero for these calculations. In addition, flood intermittency I is set to unity so as to illustrate the migration from the feed point to the end of the reach under the condition of continuous flow. The macro-roughness L_{mr} is set to 1 m based on observation of the Shimanto River near Tokawa, Japan. The values for Cz, Q, B, D and λ are the same as in Sect. 5, i.e., $Cz = 10$, $Q = 300\,\mathrm{m}^3\,\mathrm{s}^{-1}$, $B = 100\,\mathrm{m}$, $D = 20\,\mathrm{mm}$ and $\lambda = 0.35$. Bedrock slope S_{b}, which is constant due to the absence of abrasion, is set to 0.004. The above numbers combined with Eqs (5c) (using the constants of the formulation of Wong and Parker, 2006a), (32a) and (32b) yield the following values: depth $H = 1.32\,\mathrm{m}$, Froude number $Fr = 0.63$, Shields number $\tau^* = 0.016$ and capacity bedload transport rate $q_{\mathrm{ac}} = 0.0017\,\mathrm{m}^2\,\mathrm{s}^{-1}$.

None of these three cases can be treated using models that assume a relation for cover based on the ratio of sediment supply to capacity transport rate, i.e., Eq. (2). They thus illustrate capabilities unique to MRSAA.

6.1 Alluvial stripping

The case of stripping of an initial alluvial layer to bare bedrock is considered here. In this simulation, the bedload feed rate $q_{af} = 0$ and the initial thickness of alluvial cover η_a is set to 0.8 m, i.e., 80 % of the macro-roughness length L_{mr}. To drive stripping of the alluvial layer, the feed rate is set equal to zero. Figure 12a shows how the alluvial cover is progressively stripped off from upstream to downstream as a wave of alluvial rarification migrates downstream. The alluvial layer is completely removed (except for residual sediment in deep pockets, as specified by Eq. 21) after a little more than 0.12 years.

Of interest in Fig. 12a is the fact that the wave of stripping maintains constant form in spite of the diffusive term in Eq. (28a) which should cause the wave to spread. The reason the wave does not spread is the nonlinearity of the wave speed c_a in Eq. (28b): since p enters into the denominator on the right-hand side of the equation, wave speed is seen to increase as p decreases, and thus η_a decreases. As a result, the lower portion of the wave tends to migrate faster than the higher portion, sharpening the wave and opposing diffusion.

6.2 Emplacement of an alluvial layer over an initially bare bed

In this simulation, the initial thickness of alluvium η_a is set to zero and the sediment feed rate is set to 0.0013 m^2 s^{-1}, i.e., 80 % of the capacity value. The result of the calculation is shown in Fig. 12b. Here nonlinear advection and diffusion act in concert to cause the wave of alluviation to spread. The steady-state thickness of alluvium is 0.83 m; by 0.1 years it has been emplaced only down to about 5 km from the source. This steady-state condition, and only this condition, corresponds to a convergence of results from MRSAA and CSA.

6.3 Propagation of a pulse of alluvium over an initially bare bed

In this example the initial bed is bare of sediment. The sediment feed rate is set equal to 0.0012 m^2 s^{-1}, i.e., 70 % of the capacity value for 0.05 years from the start of the run, and then dropped to zero for the rest of the run. Figure 12c shows the propagation of a damped alluvial pulse through the reach, with complete evacuation of the pulse in a little more than 0.15 years. Nonlinear advection acts against diffusion to suppress the spreading of the upstream side of the pulse, but advection acts together with diffusion to drive spreading of the downstream side of the pulse.

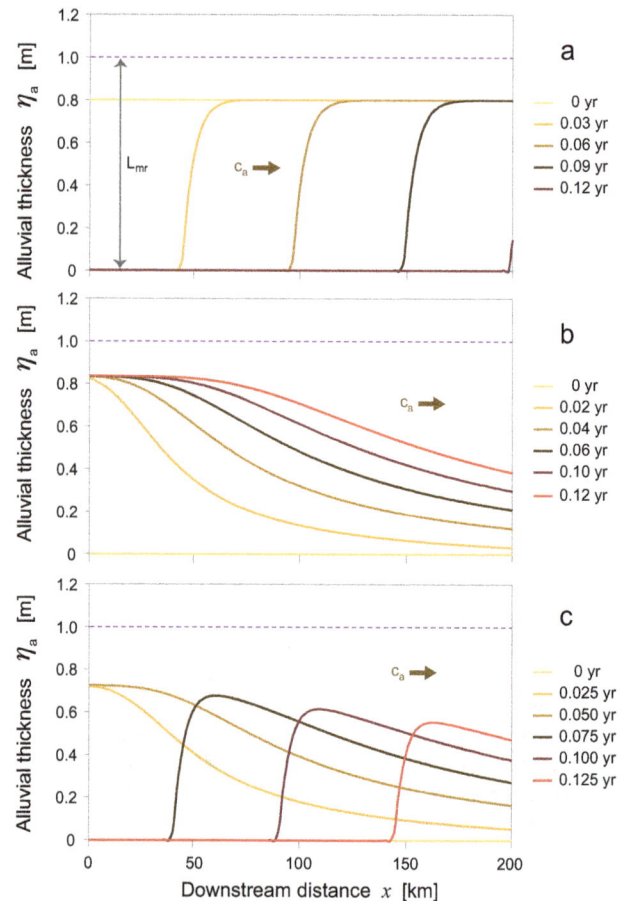

Figure 12. MRSAA model solutions for (a) stripping of an alluvial layer to bare bedrock, (b) emplacement of an alluvial cover over initially bare bedrock and (c) evolution of a pulse of sediment over bare bedrock. Numerical simulations of the evolution of alluvial thickness $\eta_a(x, t)$ over streamwise distance x are shown for a series of time steps t as indicated in the legends.

7 Comparison of evolution to uplift-driven steady state for the CSA and MRSAA models

Here we consider three cases of channel profile evolution to steady state that include both rock uplift and incision. In the first case, the initial bedrock slope is set to a value below the steady-state value, and the sediment feed rate is set to a value that is well above the steady-state value for the initial bedrock slope, causing early-stage massive alluviation. The configuration for the second case is a simplified version of a graben with a horst upstream and a horst downstream. The configuration for the third case is such that there is an alluviated river mouth downstream and a bedrock–alluvial transition upstream. In all cases, MRSAA predicts evolution that cannot be predicted by models that assume a relation for cover based on the ratio of sediment supply to capacity transport rate, i.e., Eq. (2).

7.1 Evolution of bedrock profile with early-stage massive alluviation

Here we set Q, B, Cz, D and λ to the same values as Sect. 6. The reach length L is 20 km, the flood intermittency I is set to 0.05, macro-roughness L_{mr} is set to 1 m, initial alluvial thickness $\eta_a|_{t=0} = 0.5$ m, downstream bed elevation $\eta_b|_{x=L} = 0$ and the abrasion coefficient β_{ss} is 0.05 km^{-1}. The initial bed slope is 0.004. The feed rate is set to twice the capacity rate for this slope, i.e., $q_{af} = 0.0033$ m^2 s^{-1}. The uplift rate is set to the very large value of 5 mm yr^{-1}. It should be noted, however, that as shown in Fig. 10b, the steady-state bedrock slope for this feed rate is independent of the uplift rate for $\upsilon \leq 5$ m yr^{-1}. This is because the steady-state value of φ is 0.019, i.e., $\varphi \ll 1$.

The results for the CSA model are shown in Fig. 13a. The bed slope evolves from the initial value of 0.004 to a final steady-state value of 0.0068. Evolution is achieved solely by means of an upstream-migrating knickpoint. Only the first 4000 years of evolution are shown in the figure.

Figure 13b shows the results of the first 400 years of the calculation with MRSAA. By 100 years, the bed is completely alluviated, and by 400 years, the thickness of the alluvial layer at the upstream end of the reach is 52 m. This massive alluviation is, unsurprisingly, not predicted by CSA, which was designed to treat incision only. Figure 13c shows the results of the first 4000 years of evolution. The upstream-migrating knickpoint takes the same form as CSA, but it is nearly completely hidden by the alluvial layer. The knickpoint gradually migrates upstream, driving the completely alluviated layer out of the domain, but this process is not complete by 4000 years. A comparison of Fig. 13a and c show that a knickpoint that is exposed in CSA is hidden in MRSAA. Models that assume a relation for cover based on the ratio of sediment supply to capacity transport rate cannot predict the presence of a hidden knickpoint.

7.2 Evolution of horst–graben configuration

In this example, Cz, Q, B, D, λ, I, β_{ss}, L_{mr}, L, $\eta_a|_{t=0}$ and $\eta_b|_{x=L}$ are set to the values used in Sect. 7.1. The sediment feed rate $q_{af} = 0.00083$ m^2 s^{-1}, and the initial bedrock slope S_b is set to the steady-state value for a rock uplift rate of 1 mm yr^{-1}, i.e., 0.0027. The model is then run for a rock uplift rate of 1 mm yr^{-1} for the domains $0 \leq x \leq 8$ km and 12 km $\leq x \leq 20$ km and a rock subsidence rate of 1 mm yr^{-1} for the domain 8 km $< x < 12$ km. This configuration corresponds to a simplified 1-D configuration of a graben bounded by two horsts, one upstream and one downstream.

This case cannot be implemented in models that assume a relation for cover based on the ratio of sediment supply to capacity transport rate, i.e., Eq. (2). This is because such models are not designed to handle the case of alluvial fill of accommodation space created by subsidence. The results for MRSAA are shown in Fig. 14. By 15 kyr, the uplifting

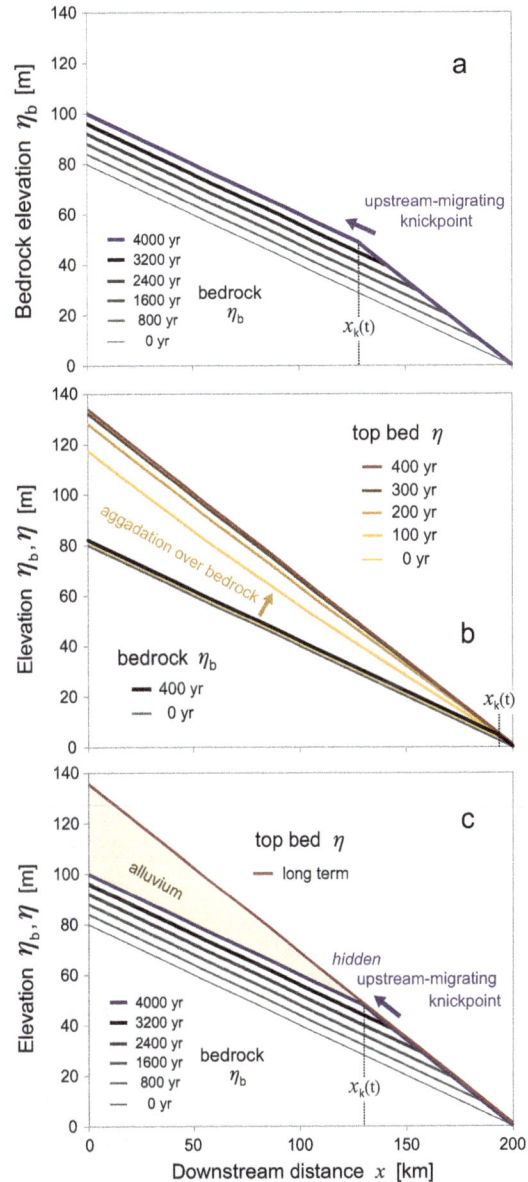

Figure 13. Progression to steady state after an impulsive increase in sediment supply: (**a**) CSA model; (**b**) MRSAA model, early stage; and (**c**) MRSAA model, late stage. Note the knickpoints $x_k(t)$ in bedrock in (**a**) and (**b**) and hidden in a migrating alluvial–bedrock transition in (**c**). Here η denotes elevation of the top bed surface, η_b denotes elevation of the base of bedrock relief, and x denotes streamwise distance. The retreating alluvial wedge is shaded in (**c**) for emphasis.

domains evolve to a steady state in terms of both bedrock elevation and alluvial cover. The bedrock elevation of the subsiding domain never reaches steady state because it is completely alluviated. The profile at the top of the alluvium in this domain has indeed reached steady state by 15 000 years, with a bed slope that deviates only modestly from the steady-state bedrock slope driven by a uniform $\upsilon = 1$ mm yr^{-1}.

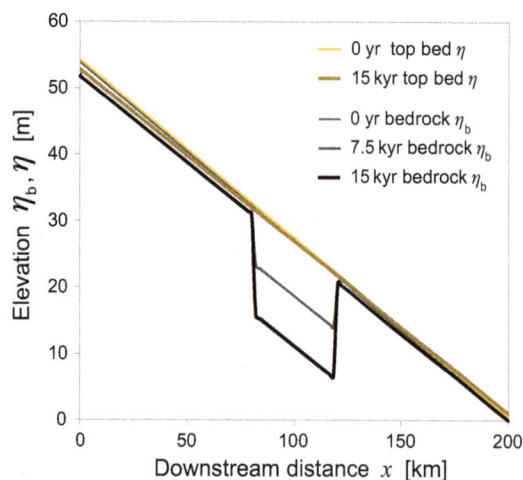

Figure 14. Evolution predicted by the MRSAA model for localized subsidence at a narrow graben superimposed on broader uplift. Note the bedrock–alluvial and alluvial–bedrock transitions at the margins of the graben. By 15 kyr, the bed top has reached steady state, even though the bedrock surface in the graben continues to subside. The regional rock uplift rate and graben subsidence rate are assumed constant for simplicity. Here η denotes elevation at the bed top, η_b denotes elevation of the bottom of bedrock relief, and x denotes streamwise distance.

7.3 Evolution of river profile with alluviated zone at river mouth

In this example Cz, Q, B, D, λ, β_{ss}, L_{mr}, L and $\eta_a|_{t=0}$ are again set to the values chosen in Sect. 7.1. The bedload feed rate is $0.00083\ \mathrm{m^2\,s^{-1}}$; the steady-state bedrock slope S_b associated with this feed rate is 0.0026 for $\upsilon < 5\ \mathrm{mm\,yr^{-1}}$ (Fig. 10b). The initial bedrock slope is set, however, to the higher value of 0.004. The rock uplift rate υ for this case is set to zero, for which the steady-state slope is again 0.0026.

The result of CSA for this case, with base level $\eta_b|_{x=L}$ pinned at zero elevation, is shown in Fig. 15. As in the case of Sect. 7.1, the bedrock slope evolves from the initial value of 0.004 to the steady-state value 0.0026 by means of an upstream-migrating knickpoint. Only 4000 years of evolution are shown in the figure, by which time the knickpoint is 4.8 km from the feed point.

MRSAA is implemented with somewhat different initial and downstream boundary conditions in order to model the case of a bed that remains alluviated at the downstream end. This condition thus corresponds to an alluviated river mouth. The initial bedrock slope is again 0.004, and the downstream bedrock elevation $\eta_b|_{x=L}$ is again 0 m. The downstream alluvial elevation $\eta_a|_{x=L}$, however, is held at 10 m, so that the downstream end is completely alluviated. The initial slope S for the top of the bed is 0.0021, a value chosen so that the bed elevation equals the bedrock elevation at the upstream end. Results of the MRSAA simulation are shown in Fig. 16.

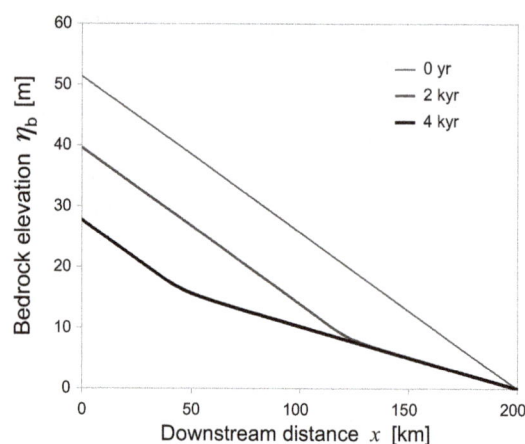

Figure 15. CSA model evolution of an initial bedrock profile towards a steady-state profile. Compare with the MRSAA model behavior in Fig. 16. Here η_b denotes elevation of the bottom of bedrock relief, and x denotes streamwise distance.

Figure 16a–c show the early-stage evolution, i.e., at $t = 0$, 10 and 100 years. Over this period, a bedrock–alluvial transition (from mixed bedrock–alluvial to purely alluvial) migrates downstream from the feed point to $x = 13.6\ \mathrm{km}$, i.e., 6.4 km upstream of the terminus. Bedrock incision is negligible over this period.

Figure 16d–f show the bedrock and top bed profiles for 1000, 2000 and 4000 years. Over this period, the bedrock–alluvial transition migrates upstream. As it does so, the bedrock slope downstream of $x = 13.6\ \mathrm{km}$ remains alluviated and does not change. The bedrock slope between the transition and $x = 13.6\ \mathrm{km}$ evolves to the steady-state value of the case in Sect. 7.2, and the top bed slope downstream of the transition evolves to the same slope as the steady-state bedrock slope (because with $\upsilon = 0$, φ vanishes). The figures show that the upstream-migrating bedrock knickpoint is located at the bedrock–alluvial transition. By 4000 years, the transition has migrated out of the domain and the bed is completely alluviated. The thickness of the alluvial cover upstream of $x = 13.6\ \mathrm{km}$ is, however, only 1.05 m, i.e., only slightly larger than the macro-roughness height of 1 m. This means that although the reach is everywhere alluvial at 4000 years, the bedrock is only barely covered.

8 Discussion

The MRSAA model is a direct descendant of the model of Sklar and Dietrich (2004) in terms of the formulation for bedrock incision, and the model of Struiksma (1999) in terms of the formulation of the conservation alluvium over a partly covered bedrock surface. In terms of its capabilities, however, it shares much in common with the previous work of Lague (2010), and in particular with his SSTRIM model. These include (a) the melding of incision and allu-

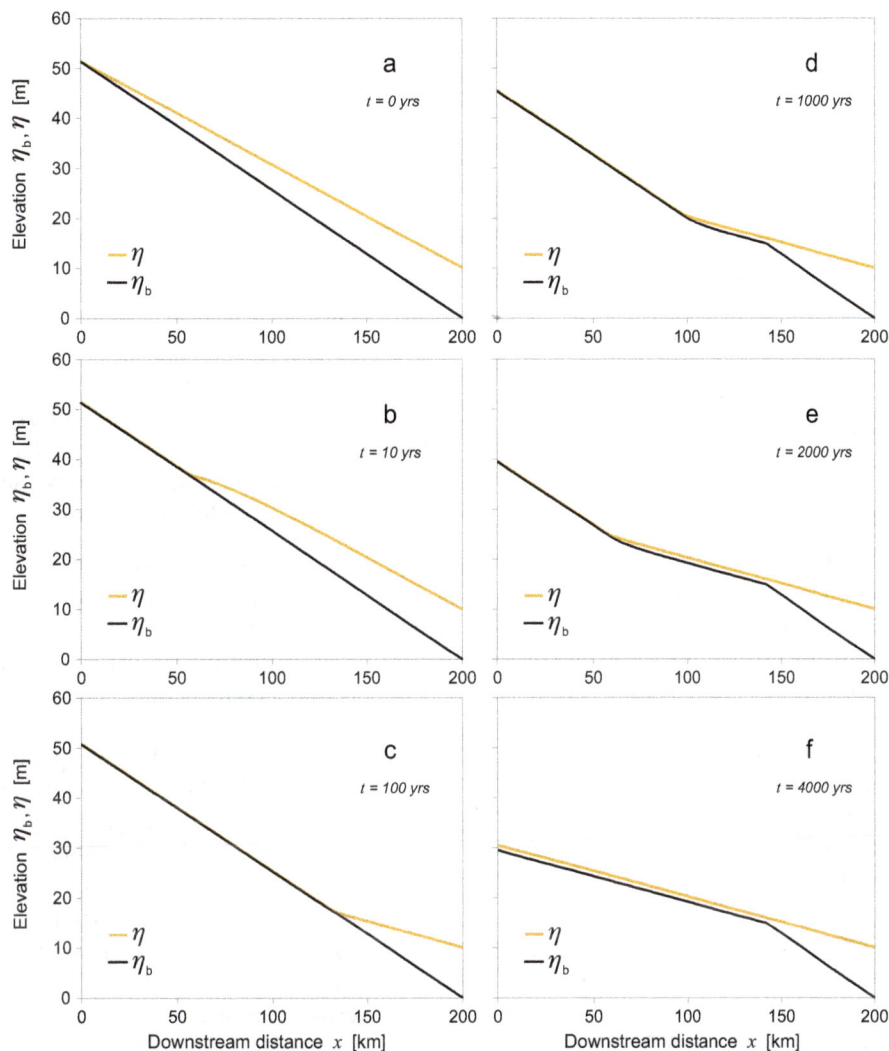

Figure 16. MRSAA model evolution of bed top and bedrock profiles with an imposed alluvial river mouth at the downstream end and an upstream-migrating bedrock–alluvial transition. The results are for (a) $t = 0$ years, (b) $t = 10$ years, (c) $t = 100$ years, (d) $t = 1000$ years, (e) $t = 2000$ years and (f) $t = 4000$ years (steady state). Here (a), (b) and (c) show the early response, and (d), (e) and (f) show the late response. In these plots, η denotes elevation of the bed surface, η_b denotes elevation of the bottom of bedrock relief, and x denotes streamwise distance. Compare with the CSA model solution in Fig. 15.

viation into a single model, (b) the inclusion of a cover relation that is based on geometric bed structure, and (c) the ability to track simultaneously the spatiotemporal variation in both incision rate and alluvial cover. Priority should accrue to Lague (2010) in regard to these features. The present model has the following advantages: (a) the Exner equation of sediment conservation is specifically based on a formulation of the statistics of partial and complete cover over a rough bedrock surface; (b) the formulation yields a specific relation for alluvial wave velocity as a function of cover, ranging to a maximum value for minimum cover to 0 for complete alluviation; and (c) it allows for explicit description of the nonlinear advective–diffusive physics of the problem in terms of an alluvial diffusivity and two wave celerities, one

directed upstream and associated with bedrock incision, and one directed downstream and associated with alluviation.

The form of the MRSAA model presented here has been simplified as much as possible, i.e., to treat a HSR (highly simplified reach) with constant grain size D. This has been done to allow for a precise and complete characterization of the behavior of the governing equations. It can relatively easily be extended to: (a) abrasion of the clasts that abrade the bed, so abrasional downstream fining is captured (Parker, 1991); (b) size mixtures of sediment (Wilcock and Crowe, 2003); (c) multiple sediment sources (Lague, 2010; Yanites et al., 2010); (d) channels with width variation downstream (Lague, 2010); (e) discharge varying according to a flow duration curve (Sklar and Dietrich, 2006; Lague, 2010), or fully unsteady flow (An et al., 2014); and (f) cyclically varying hydrographs (Wong and Parker, 2006b) or "sedimentographs", the latter corresponding to events for which the sediment supply rate first increases, and then decreases cyclically (Zhang et al., 2013). In addition, the model can and should be extended to include the stochasticity emphasized by Lague (2010). Sections 6 and 7 illustrate features captured by MRSAA but not by models that assume a cover relation based on the ratio of sediment supply to capacity transport rate, i.e., Eq. (2).

The MRSAA model presented here is applied to several 1-D cases with spatiotemporal variation. The model can easily be generalized to 2-D simply by expressing Eq. (18) in 2-D form. In any such implementation, however, the effect of 2-D connectivity between deep holes should be considered in the relation to the cover factor.

The MRSAA model in the form presented here has a weakness in that the flow resistance coefficient Cz is a prescribed constant. The recent models of bedrock incision of Inoue et al. (2014) and Johnson (2014) provide a much more detailed description of flow resistance. In addition to characterizing macro-roughness, their models use two micro-roughnesses, one characterizing the hydraulic roughness of the alluvium and the other characterizing the hydraulic roughness of the bedrock surface. Their models can thus discriminate between (a) "clast-smooth" beds, for which bedrock roughness is lower than clast roughness and (b) "clast-rough" beds, for which bedrock roughness is greater than clast roughness. This characterization allows for two innovative features: (a) both bed resistance and fractional cover become dependent on the ratio of bedrock micro-roughness to alluvial micro-roughness and (b) incision can result from throughput sediment passing over a purely bedrock surface with no alluvial deposit. The models of Inoue et al. (2014) and Johnson (2014) use modified forms of the capacity-based form for cover of Eq. (2) in order to capture these phenomena. Their models are thus unable to capture the coevolution of bedrock–alluvial and purely alluvial processes of MRSAA. Amalgamation of their models and the one presented here, however, appears to be feasible and is an attractive future goal.

Because MRSAA tracks the spatiotemporal variation in both bedload transport and alluvial thickness, it is applicable to the study of the incisional response of a river subject to temporally varying sediment supply. It thus has the potential to capture the response of an alluvial–bedrock river to massive impulsive sediment inputs associated with landslides or debris flows. A preliminary example of such an extension is given in Zhang et al. (2013). When extended to multiple sediment sources, it can encompass both the short- and long-term responses of a bedrock–alluvial river to intermittent massive sediment supply due to landslides and debris flows. As such, it has the potential to be integrated into a framework for managing sediment disturbance in mountain rivers systems such as those affected by the 2008 Wenchuan earthquake in Sichuan, China. Over 200 landslide dams formed during that event (Xu et al., 2009; Fu et al., 2011). A similar potential application is the case of drastic sediment supply to, and evacuation from, rivers in Taiwan due to typhoon-induced or earthquake-induced landsliding (e.g., Yanites et al., 2010).

9 Conclusions

We present a 1-D model of alluvial transport and bedrock erosion in a river channel whose bed may be purely alluvial, or mixed bedrock–alluvial, or may transition freely between the two morphologies. Our model, which we call the Macro-Roughness-based Saltation-Abrasion-Alluviation (MRSAA) model, specifically tracks not only large-scale bedrock morphodynamics but also the morphodynamics of the alluvium over it. The key results are as follows:

1. The transport of alluvium over a bedrock surface cannot in general be described simply by a supply rate that instantaneously affects the entire river reach downstream as it is varied in time. Here we track the alluvium in terms of a spatiotemporally varying alluvial thickness.

2. The area fraction of cover p enters into both incisional and alluvial evolution. The alluvial part allows for the downstream propagation and diffusion of sediment waves, so that at any given time the alluvial bed can be above or below the top of the bedrock. The model thus allows for spatiotemporal transitions between complete cover, under which no incision occurs; partial cover, for which incision may occur; and no cover, for which no incision occurs.

3. The MRSAA model captures three processes: downstream alluvial advection at a fast timescale, alluvial diffusion, and upstream incisional advection at a slow timescale. Only the third of these processes is captured by models that assume a relation for cover based on the ratio of sediment supply to capacity transport rate rather than a measure of the thickness of alluvial cover itself. The CSA model can be thought of as a 0-D model that applies locally. The MRSAA model lends itself more directly to application to long 1-D reaches because it embeds the elements necessary to route sediment down the reach.

4. The MRSAA model reduces to the CSA model under the conditions of steady-state incision in balance with rock uplift and below-capacity cover. The steady-state bedrock slope predicted by both models is insensitive to the rock uplift rate over a wide range of conditions. This insensitivity is in marked contrast to the commonly used incision model in which the incision rate is a power function of bedrock slope and drainage area upstream. The two models can differ substantially under transient

conditions, particularly under those that include migrating transitions between the bedrock–alluvial and purely alluvial state.

5. In the MRSAA model, inclusion of alluvial advection and diffusion lead to the following phenomena: (a) a wave-like stripping of antecedent alluvium over a bedrock surface in response to cessation of sediment supply, (b) advection–diffusional emplacement of a sediment cover over initially bare bedrock and (c) the propagation and deformation of a sediment pulse over a bedrock surface.

6. In the case of transient imbalance between rock uplift and incision with a massive increase in sediment feed, MRSAA captures an upstream-migrating transition between a purely alluvial reach upstream and a bedrock–alluvial reach downstream (here abbreviated as a alluvial–bedrock transition). The bedrock profile shows an upstream-migrating knickpoint, but this knickpoint is hidden under alluvium. Models that assume a relation for cover based on the ratio of sediment supply to capacity transport rate, i.e., Eq. (2), capture only the knickpoint, which is completely exposed, and thus miss the thick alluvial cover predicted by MRSAA.

7. MRSAA captures the mixed incisional–alluvial evolution for the case of a simplified 1-D subsiding graben bounded by two uplifting horsts. It captures alluvial filling of the graben, and thus converges to a steady-state top-bed profile with a bedrock–alluvial transition at the upstream end of the graben and an alluvial–bedrock transition at the downstream end.

8. In the case studied here of an uplifting bedrock profile with an alluviated bed at the downstream end modeling a river mouth, MRSAA predicts an upstream-migrating bedrock–alluvial transition at which the bedrock undergoes a sharp transition from a higher to a lower slope. MRSAA further predicts a bedrock long profile under the alluvium that has the same slope as the top bed. It also predicts that the cover is thin, so that the purely alluvial reach is only barely so. The steady state for this case is purely alluvial.

9. The new MRSAA model provides an entry point for the study of how bedrock–alluvial rivers respond to occasional large, impulsive supplies of sediment from landslides and debris flows. It thus can provide a tool for forecasting river-sedimentation disasters associated with such events. An example application would be treatment of the aftereffects of the 2008 Wenchuan earthquake, which triggered massive alluviation and the formation of over 200 landslide dams.

Appendix A: Interpretation of the abrasion coefficient β

Consider a clast or grain of size D and volume $V_g \sim D^3$ causing abrasion over an exposed bedrock surface. The bedload transport rate q_a is given as

$$q_a = E_g L_g, \tag{A1}$$

where E_g denotes the volume rate per unit time per unit area at which clasts are ejected from the bed into saltation, and L_g denotes the saltation length. Each clast ejected into saltation collides with the bed a distance L_g later; therefore, the number of clasts that collide with the bedrock (rather than with other bed particles) per unit time per unit area is found using Eq. (A1),

$$(1-p)\frac{E_g}{V_g} = (1-p)\frac{q_a}{V_g L_g}. \tag{A2}$$

The volume lost from the striking clast per strike is defined as $\beta_g^* V_g$, and the volume lost from the stricken bedrock per strike is similarly defined as $\beta^* V_g$. The parameters β_g^* and β^* could be expected to be approximately equal if the striking grain is the same rock type as the bedrock.

The rate at which a grain strikes the bed per unit distance moved is $1/L_g$; hence the rate at which grain volume decreases downstream is given by

$$\frac{dV_g}{dx} = -\beta_g V_g, \tag{A3a}$$

$$\beta_g = \frac{\beta_g^*}{L_g}. \tag{A3b}$$

Using $V_g \sim D^3$, Eq. (A3) reduces to

$$\frac{dD}{dx} = -\alpha_d D, \tag{A4a}$$

$$\alpha_d = \frac{1}{3}\beta_g. \tag{A4b}$$

Equation (A4a) is the differential form of Sternberg's law; α_d is a diminution coefficient with units L^{-1}. The exponential form of Eq. (3) corresponds to the case of spatially constant α_d.

The incision rate of the bedrock E is the number of grains that collide with bedrock per unit area per unit time (Eq. A2) multiplied by the volume lost per strike $\beta^* V_g$, which gives the relation

$$E = (1-p)\frac{q_a}{V_g L_g}\beta^* V_g = \beta q_a(1-p) = \beta q_{ac} p(1-p), \tag{A5}$$

$$\beta = \frac{\beta^*}{L_g}. \tag{A6}$$

The above relation is identical to Eq. (1b).

Table A1. Notation.

A	upstream drainage area [L^2]
B	channel width [L]
CSA	acronym for Capacity-based Saltation-Abrasion model
Cz	dimensionless Chézy resistance coefficient [$-$]
c_a	speed of propagation of an alluvial disturbance (positive downstream) [$L\,T^{-1}$]
c_{ai}	speed of propagation of an alluvial disturbance of infinitesimal height [$L\,T^{-1}$]
c_b	speed of propagation of an incisional disturbance (positive downstream) [$L\,T^{-1}$]
D, D_u	characteristic grain size of clasts effective in abrading the bed; upstream value of D [L]
E	bedrock incision rate [$L\,T^{-1}$]
Fr	Froude number $= Q/(B\,H\,\sqrt{g\,H})$ [$-$]
f	function of χ describing cover fraction [$-$]
g	gravitational acceleration [$L\,T^{-2}$]
H	flow depth [$L\,T^{-1}$]
HSR	acronym for highly simplified reach
I	flood intermittency, i.e., fraction of time the river is in flood [$-$]
k	coefficient in Eq. (5a) [$-$]
L	reach length [L]
L_g	grain saltation length [L]
L_{half}	distance a clast travels to lose half its size (diameter) by abrasion [L]
L_{mr}	height of macro-roughness height [L]
MRSAA	acronym for Macro-Roughness-based Saltation-Abrasion-Alluviation model
$K, m, n; A_o$	symbols used in slope–area relation; reference drainage area [$-$]
n_a	exponent in bedload transport relation [$-$]
p	areal fraction of bed that is covered by alluvium [$-$]
p_{ss}	steady-state value of p [$-$]
p_0	lower reference cover fraction (0.05 herein) [$-$]
p_1	upper reference cover fraction (0.95 herein) [$-$]
Q	flood discharge [$L^3\,T^{-1}$]
q_a, q_{ac}, q_{acss}	volume bedload transport rate per unit width; capacity value of q_a; steady-state value of q_{ac} [$L^2\,T^{-1}$]
q_{af}	feed, or supply value of q_a [$L^2\,T^{-1}$]
q_{ak}	value of q_a at knickpoint [$L^2\,T^{-1}$]
R	submerged specific gravity of sediment clasts [$-$]
R_f	$= v_f/(R\,g\,D)^{1/2}$ [$-$]
S, S_b, S_a	bed slope; slope of bedrock; slope of alluvial thickness, $-\partial\eta_a/\partial x$ [$-$]
$S_{bi}, S_{bss}, S_{bref}$	initial bedrock slope; steady-state bedrock slope; reference bedrock slope [$-$]
S_{bu}, S_{bl}	bedrock slope upstream of a knickpoint; bedrock slope downstream of a knickpoint [$-$]
S_{ass}	steady-state alluvial bed slope at capacity [$-$]
T, T_h, T_l	period of cycled hydrograph; duration of high flow; duration of low flow [T]
t	time [T]
U	flow velocity during floods [$L\,T^{-1}$]
u_*	shear velocity $= (\tau/\rho)^{1/2}$ [$L\,T^{-1}$]
v_f	fall velocity of a bedload grain [$L\,T^{-1}$]
V_g	single bedload grain volume [L^3]
x	streamwise distance [L]
$\hat{x}$	x/L [$-$]
x_k	distance to knickpoint [L]
Y	bedrock modulus of elasticity [$M\,L^{-1}\,T^2$]
z, z'	vertical coordinates (relative to bedrock base and arbitrary vertical datum, respectively) [L]
z'_0	bed elevation such that cover fraction $p = p_0$ [L]
z'_1	bed elevation such that cover fraction $p = p_1$ [L]
α_a	coefficient in bedload transport relation [$-$]
α_d	diminution coefficient for an abrading clast [L^{-1}]
$\beta, \beta_{ref}, \beta_{ss}$	coefficient of wear (abrasion); reference value of β; steady-state value of β [L^{-1}]
χ	$= \eta_a/L_{mr}$ [$-$]
η, η_a, η_b	bed elevation; thickness of alluvial layer; bedrock elevation [L]

Table A1. Continued.

κ_a	alluvial diffusivity defined in Eq. (26a) $[\mathrm{L}^2\,\mathrm{T}^{-1}]$
φ	$= \upsilon/(I\,\beta_{ss}\,q_{af})$ $[-]$
λ	porosity of alluvial deposit $[-]$
ρ	density of water $[\mathrm{M}\,\mathrm{L}^{-3}]$
ρ_g	density of a bedload grain $[\mathrm{M}\,\mathrm{L}^{-3}]$
σ_t	rock tensile strength $[\mathrm{M}\,\mathrm{L}^{-1}\,\mathrm{T}^{-2}]$
τ^*, τ_c^*	Shields number $= u_*^2/(R\,g\,D)$; critical value of τ^* at threshold of motion $[-]$
τ	bed shear stress $[\mathrm{M}\,\mathrm{L}^{-1}\,\mathrm{T}^{-2}]$
υ, υ_{ref}	relative vertical speed between the (nondeforming) rock underlying the channel and the point at which base level is maintained, e.g., rock uplift rate or base level fall rate; reference uplift rate $[\mathrm{L}\,\mathrm{T}^{-1}]$

Acknowledgements. The participation of L. Zhang and X. Fu in this work was made possible by the National Natural Science Foundation of China (grant nos. 51379100 and 51039003). The participation of G. Parker was made possible in part by a grant from the US National Science Foundation (grant no. EAR-1124482) The participation of C. P. Stark was made possible in part by grants from the US National Science Foundation (grant nos. EAR-1148176, EAR-1124114 and CMMI-1331499). The participation of T. Inoue was made possible by support from the Hokkaido Regional Development Bureau.

Edited by: T. Coulthard

References

An, C. G., Fu, X. D., and Parker, G.: River morphological evolution in earthquake-hit region: effects of floods and pulsed sediment supply, in: River Flow 2014 Conference, 3–5 September 2014, Lausanne, Switzerland, p. 7, 2014.

Chatanantavet, P. and Parker, G.: Experimental study of bedrock channel alluviation under varied sediment supply and hydraulic conditions, Water Resour. Res., W12446, doi:10.1029/2007WR006581, 2008.

Chatanantavet, P. and Parker, G.: Physically based modeling of bedrock incision by abrasion, plucking, and macroabrasion, J. Geophys. Res., 114, F04018, doi:10.1029/2008JF001044, 2009.

Chatanantavet, P., Lajeunesse, E., Parker, G., Malverti, L., and Meunier, P.: Physically-based model of downstream fining in bedrock streams with lateral input, Water Resour. Res., 46, W02518, doi:10.1029/2008WR007208, 2010.

Chatanantavet, P., Whipple, K. X., Adams, M. A., and Lamb, M. P.: Experimental study on coarse grain saltation dynamics in bedrock channels, J. Geophys. Res., 118, 1–16, doi:10.1002/jgrf.20053, 2013.

Crosby, B. T., Whipple, K. X., Gasparini, N. M., and Wobus, C. W.: Formation of fluvial hanging valleys: theory and simulation, J. Geophys. Res., 112, F03S10, doi:10.1029/2006JF000566, 2007.

Cui, Y., Parker, G., Lisle, T., Gott, J., Hansler, M., Pizzuto, J. E., Allmendinger, N. E., and Reed, J. M.: Sediment pulses in mountain rivers, Part 1. Experiments, Water Resour, Res., 39, 1239, doi:10.1029/2002WR001803, 2003a.

Cui, Y., Parker, G., Pizzuto, J. E., and Lisle, T. E.: Sediment pulses in mountain rivers, Part 2. Comparison between experiments and numerical predictions, Water Resour. Res., 39, 1240, doi:10.1029/2002WR001805, 2003b.

Dadson, S. J., Hovius, N., Chen, H., Dade, W. B. Hsieh, M. L., Willett, S. D., Hu, J. C., Horng, M. J., Chen, M. C., Stark, C. P., Lague, D., and Lin, J. C.: Links between erosion, runoff variability and seismicity in the Taiwan orogeny, Nature, 426, 648–651, 2003.

DiBiase, R. A. and Whipple, K. X.: The influence of erosion thresholds and runoff variability on the relationships among topography, climate, and erosion rate, J. Geophys. Res., 116, F04036, doi:10.1029/2011JF002095, 2011.

England, P. and Molnar, P.: Surface uplift, uplift of rocks and exhumation rate of rocks, Geology, 18, 1173–1177, 1990.

Fernandez Luque, R. and van Beek, R.: Erosion and transport of bedload sediment, J. Hydraul. Res., 14, 127–144, 1976.

Fu, X. D., Liu, F., Wang, G. Q., Xu, W. J., and Zhang, J. X.: Necessity of integrated methodology for hazard mitigation of quake lakes: case study of the Wenchuan Earthquake, China, Front. Arch. Civ. Eng. China, 5, 1–10, 2011.

Gilbert, G. K.: Report on the geology of the Henry Mountains: geographical and geological survey of the Rocky Mountain region, Government Printing Office, Washington, D.C., 106 pp., 1877.

Hobley, D. E. J., Sinclair, H. D., Mudd, S. M., and Cowie, P. A.: Field calibration of sediment flux dependent river incision, J. Geophys. Res., 116, F04017, doi:10.1029/2010JF001935, 2011.

Howard, A. D.: Simulation of stream capture, Geol. Soc. Am. Bull., 82, 1355–1376, 1971.

Howard, A. D.: Long profile development of bedrock channels: interaction of weathering, mass wasting, bed erosion, and sediment transport, Am. Geophys. Un. Monogr., 107, 297–319, 1998.

Howard, A. D. and Kerby, G.: Channel changes in badlands, Bull. Geol. Soc. Am., 94, 739–752, 1983.

Howard, A. D., Seidl, M. A., and Dietrich, W. E.: Modeling fluvial erosion on regional to continental scales, J. Geophys. Res., 99, 13971–13986, 1994.

Inoue, T., Izumi, N., Shimizu, Y., and Parker, G.: Interaction between alluvial cover, bed roughness and incision rate in purely bedrock and alluvial-bedrock channel, in press, J. Geophys. Res., 119, 2123–2146, doi:10.1002/2014JF003133, 2014.

Izumi, N. and Yokokawa, M.: Cyclic steps formed in bedrock rivers, Proceedings of River, Coastal and Estuarine Morphodynamics, RCEM 2011, Tsinghua University Press, Beijing, 2084–2090, 2011.

Izumi, N., Yokokawa, M., and Parker, G.: Cyclic step morphology formed on bedrock, J. JSCE Div. B, 68, I_955–I_960, 2012.

Johnson, J. P. L.: A surface roughness model for predicting alluvial cover and bedload transport rate in bedrock channels, J. Geophys. Res., 119, 2147–2173, doi:10.1002/2013JF003000, 2014.

Johnson, J. P. L. and Whipple, K. X.: Feedbacks between erosion and sediment transport in experimental bedrock channels, Earth Surf. Proc. Land., 32, 1048–1062, doi:10.1002/esp.1471, 2007.

Johnson, J. P. L., Whipple, K. X., Sklar, L. S., and Hanks, T. C.: Transport slopes, sediment cover, and bedrock channel incision in the Henry Mountains, Utah, J. Geophys. Res., 114, F02014, doi:10.1029/2007JF000862, 2009.

Kodama, Y.: Experimental study of abrasion and its role in producing downstream fining in gravel-bed rivers, J. Sed. Res. A, 64, 76–85, 1994.

Lague, D.: Reduction of long-term bedrock incision efficiency by short-term alluvial cover intermittency, J. Geophys. Res., 115, F02011, doi:10.1029/2008JF001210, 2010.

Lague, D.: The stream power river incision model: evidence, theory and beyond, Earth Surf. Proc. Land., 39, 38-61, doi:10.1002/esp.3462, 2014.

Lamb, M. P., Dietrich, W. E., and Sklar, L. S.: A model for fluvial bedrock incision by impacting suspended and bedload sediment, J. Geophys. Res., 113, F03025, doi:10.1029/2007JF000915, 2008.

Lisle, T. E., Cui, Y., Parker, G., Pizzuto, J. E., and Dodd, A. M.: The dominance of dispersion in the evolution of bed material waves in gravel-bed rivers, Earth Surf. Proc. Land., 26, 1409–1420, doi:10.1002/esp.300, 2001.

Meyer-Peter, E. and Müller, R.: Formulas for bed-load transport, in: Proceeding of the 2nd IAHR Meeting, International Association for Hydraulic Research, 7–9 June 1948, Stockholm, Sweden, 39–64, 1948.

Parker, G.: Selective sorting and abrasion of river gravel: theory, J. Hydraul. Eng., 117, 131–149, 1991.

Parker, G.: 1-D Sediment Transport Morphodynamics with Applications to Rivers and Turbidity Currents, available at: http://vtchl.uiuc.edu/people/parkerg/ (last access: 18 January 2015), 2004.

Parker, G.: Transport of Gravel and Sediment Mixtures, Sedimentation Engineering: Processes, Measurements, Modeling and Practice, ASCE Manual of Practice 110, ch. 3, edited by: Garcia, M. H., American Society of Civil Engineers, Reston, USA, 165–252, 2008.

Parker, G., Klingeman, P., and McLean, D.: Bedload and size distribution in natural paved gravel bed streams, J. Hydraul. Eng., 108, 544–571, 1982.

Parker, G., Paola, C., and Leclair, S.: Probabilistic form of Exner equation of sediment continuity for mixtures with no active layer, J. Hydraul. Eng., 126, 818–826, 2000.

Parker, G., Wilcock, P., Paola, C., Dietrich, W. E., and Pitlick, J.: Quasi-universal relations for bankfull hydraulic geometry of single-thread gravel-bed rivers, J. Geophys. Res., 112, F04005, doi:10.1029/2006JF000549, 2007.

Parker, G., Nittrouer, J. A., Mohrig, D., Allison, M. A., Dietrich, W. E., and Voller, V. R.: Modeling the morphodynamics of the lower Mississippi River as a quasi-bedrock river, Eos Trans. AGU 90, Fall Meet., John Wiley and Sons, Hoboken, NJ, USA, Suppl., Abstract EP32A-01, 2009.

Parker, G., Viparelli, E., Stark, C. P., Zhang, L., Fu, X., Inoue, T., Izumi, N., and Shimizu, Y.: Interaction between waves of alluviation and incision in mixed bedrock-alluvial rivers, in: Advances in River Sediment Research, published by CRC Press/Balkema, Proc. of the 12th Inter. Symp. on River Sedimentation, Kyoto, p. 8, 2013.

Schlichting, H.: Boundary layer theory, 7th Edn. 1979, McGraw-Hill, New York, 1979.

Sklar, L. S. and Dietrich, W. E.: River longitudinal profiles and bedrock incision models: stream power and the influence of sediment supply, in: Rivers over Rock: Fluvial Processes in Bedrock Channels, edited by: Tinkler, K. and Wohl, E. E., Am. Geophys. Un. Geophys. Monogr., 107, 237–260, 1998.

Sklar, L. S. and Dietrich, W. E.: Sediment and rock strength controls on river incision into bedrock, Geology, 29, 1087–1090, 2001.

Sklar, L. S. and Dietrich, W. E.: A mechanistic model for river incision into bedrock by saltating bed load, Water. Resour. Res., 40, W06301, doi:10.1029/2003WR002496, 2004.

Sklar. L. S. and Dietrich, W. E.: The role of sediment in controlling steady-state bedrock channel slope: implications of the saltation–abrasion incision model, Geomorphology, 82, 58–83, 2006.

Smith, T. R. and Bretherton, F. P.: Stability and the conservation of mass in drainage basin evolution, Water Resour. Res., 8, 1506–1529, 1972.

Stark, C. P., Foufoula-Georgiou, E., and Ganti, V.: A nonlocal theory of sediment buffering and bedrock channel evolution, J. Geophys. Res., 114, F01029, doi:10.1029/2008JF000981, 2009.

Sternberg, H.: Untersuchungen über Längen- und Querprofil geschiebeführender Flüsse, Z. Bauwesen, 25, 483–506, 1875.

Struiksma, N.: Mathematical modelling of bedload transport over non-erodible layers, in: Vol. 1, Proceeding of IAHR Conference on River, Coastal and Estuary Morphodynamics, Genova, Italy, 89–98, 1999.

Tanaka, G. and Izumi, N., The bedload transport rate and hydraulic resistance in bedrock channels partly covered with gravel, J. JSCE Div. B, 69, I_1033–I_1038, 2013.

Turowski, J. M.: Stochastic modeling of the cover effect and bedrock erosion, Water Resour. Res., 45, W03422, doi:10.1029/2008WR007262, 2009.

Turowski, J. M.: Semi-alluvial channels and sediment-flux-driven bedrock erosion, in: Gravel Bed Rivers: Processes, Tools, Environments, 1st Edn., ch. 29, edited by: Church, M., Biron, P. M., and Roy, A., John Wiley & Sons, Chichester, UK, 401–416, 2012.

Turowski, J. M., Lague, D., and Hovius, N.: Cover effect in bedrock abrasion: a new derivation and its implications for the modeling of channel morphology, J. Geophys. Res., 112, F04006, doi:10.1029/2006JF000697, 2007.

Turowski, J. M., Badoux, A., Leuzinger, J., and Hegglin, R.: Large floods, alluvial overprint, and bedrock erosion, Earth Surf. Proc. Land., 38, 947–958, doi:10.1002/esp.3341, 2013.

Whipple, K. X.: Bedrock rivers and the geomorphology of active orogens, Ann. Rev. Earth Pl. Sci., 32, 151–185, 2004.

Whipple, K. X. and Tucker, G. E.: Dynamics of the stream-power river incision model: implications for height limits of mountain ranges, landscape response timescales, and research needs, J. Geophys. Res., 104, 17661–17674, 1999.

Whipple, K. X. and Tucker, G. E.: Implications of sediment-flux-dependent river incision models for landscape evolution, J. Geophys. Res., 107, 2039, doi:10.1029/2000JB000044, 2002.

Wilcock, P. R. and Crowe, J. C.: Surface-based transport model for mixed-size sediment, J. Hydraul. Eng., 129, 120–128, 2003.

Wilkerson, G. V. and Parker, G.: Physical basis for quasi-universal relations describing bankfull hydraulic geometry of sand-bed rivers, J. Hydraul. Eng., 137, 739–753, 2011.

Wong, M. and Parker, G.: Reanalysis and correction of bed-load relation of Meyer-Peter and Müller using their own database, J. Hydraul. Eng., 132, 1159–1168, 2006a.

Wong, M. and Parker, G.: One-dimensional modeling of bed evolution in a gravel bed river subject to a cycled flood hydrograph, J. Geophys Res., 111, F3018, doi:10.1029/2006JF000478, 2006b.

Xu, Q., Fan, X. M., Huang, R. Q., and Van Westen, C.: Landslide dams triggered by the Wenchuan Earthquake, Sichuan Province, south west China, Bull. Eng. Geol. Environ., 68, 373–386, 2009.

Yanites, B. J., Tucker, G. E., Mueller, K. J., and Chen, Y. G.: How rivers react to large earthquakes: Evidence from central Taiwan, Geology, 38, 639–642, doi:10.1130/G30883.1, 2010.

Zhang, L., Fu, X. D., Stark, C. P., Fernandez, R., and Parker, G.: Modeling of incision of bedrock rivers subject to temporally varying sediment supply, Proceedings of the 2013 IAHR World Congress, Chengdu, Tsinghua University Press, China, p. 11, 2013.

9

Seasonal logging, process response, and geomorphic work

C. H. Mohr[1], A. Zimmermann[1], O. Korup[1], A. Iroumé[2], T. Francke[1], and A. Bronstert[1]

[1]Institute of Earth and Environmental Science, University of Potsdam, Potsdam, Germany
[2]Faculty of Forest Sciences and Natural Resources, Universidad Austral de Chile, Valdivia, Chile

Correspondence to: C. H. Mohr (cmohr@uni-potsdam.de)

Abstract. Deforestation is a prominent anthropogenic cause of erosive overland flow and slope instability, boosting rates of soil erosion and concomitant sediment flux. Conventional methods of gauging or estimating post-logging sediment flux often focus on annual timescales but overlook potentially important process response on shorter intervals immediately following timber harvest. We resolve such dynamics with non-parametric quantile regression forests (QRF) based on high-frequency (3 min) discharge measurements and sediment concentration data sampled every 30–60 min in similar-sized ($\sim 0.1\,\mathrm{km}^2$) forested Chilean catchments that were logged during either the rainy or the dry season. The method of QRF builds on the random forest algorithm, and combines quantile regression with repeated random sub-sampling of both cases and predictors. The algorithm belongs to the family of decision-tree classifiers, which allow quantifying relevant predictors in high-dimensional parameter space. We find that, where no logging occurred, $\sim 80\,\%$ of the total sediment load was transported during extremely variable runoff events during only 5 % of the monitoring period. In particular, dry-season logging dampened the relative role of these rare, extreme sediment-transport events by increasing load efficiency during more efficient moderate events. We show that QRFs outperform traditional sediment rating curves (SRCs) in terms of accurately simulating short-term dynamics of sediment flux, and conclude that QRF may reliably support forest management recommendations by providing robust simulations of post-logging response of water and sediment fluxes at high temporal resolution.

1 Introduction

Increased soil erosion ranks among the least disputed geomorphic consequences of timber harvesting (Gomi et al., 2005; Sidle et al., 2006). The major impacts occur during and a few years after harvesting operations, before the vegetation re-establishes and the road surfaces and embankments stabilize. Clear cutting may intensify erosive overland flow (Malmer and Grip, 1990) and cause debris flows and riverbank erosion (Gomi et al., 2004), eventually releasing infrequent sediment pulses into the drainage network. Landsliding may further intensify due to modified drainage areas following the construction of timber roads (Montgomery et al., 2000). As a result, boosted erosion and re-deposition of soil promote the long-term degradation of soil and water resources not only on harvest patches but also often in downstream areas (Sidle et al., 2006).

Clear cutting is the most common technique for harvesting timber in the plantation forests of Chile. The nation is currently intensifying and extending its forestry sector, and recent projections point to increasing growth rates of timber and cellulose production (FAO, 2010), as well as an exacerbation of soil erosion in the future. At the same time, the forestry sector provides a major income source and thus a comprehensive assessment of the economic, social, and ecological benefits of forestry is required. Reliable knowledge of pre- and post-disturbance sediment fluxes is vital in this regard, and may be acquired by physics-based modelling or statistical treatment of field data. This holds particularly true for Chile, where law mandates immediate reforestation after

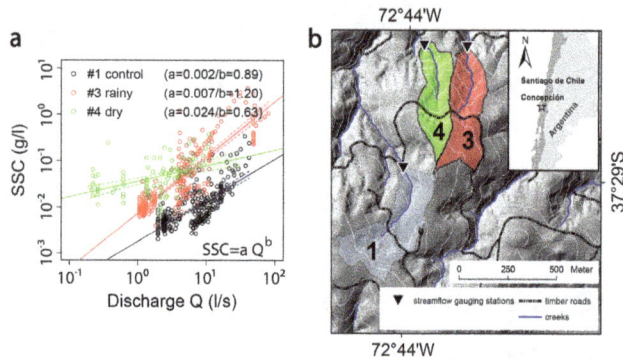

Figure 1. (a) Sediment rating curves for the catchments with fitted power-law intercepts a ($gs^b\,L^{-(b+1)}$), slopes b, and 95 % confidence intervals about regression lines. **(b)** Location of study catchments (star in inset) including stream gauges, nearest rain gauge, and un-paved timber roads. Topography derived from lidar survey; contour spacing is 20 m. Numbers (consistent with previous work) refer to catchments. See Huber et al. (2010), Mohr et al. (2012), and Mohr et al. (2013) for detailed descriptions of these catchments.

clear cuttings. However, in many situations, sample size for a robust assessment remains limited, because both time and resources for sampling hydro-geomorphic impacts are often tightly constrained; hence the acquired field data may not represent the full range of water and sediment fluxes. This limited data availability requires an analysis technique capable of dealing with few samples of high variance under changing environmental conditions (Fig. 1a).

Conventional sediment rating curves (SRC) rely on an empirical relationship between water discharge and suspended sediment concentration (SSC), but are prone to high uncertainty where SSC-discharge dynamics are subject to disturbances or nonlinear effects. Recent work revealed that antecedent rainfall, intra-event discharge dynamics (Francke et al., 2008a; Zimmermann et al., 2012), and disturbances due to clear cutting (Mohr et al., 2013) strongly bias SSC prediction based on SRC. This calls for methods capable of reliably simulating antecedent and changing environmental conditions, and predicting SSC following clear cuts. Ideally, such methods should not only sufficiently capture the high rates of sediment transport immediately following timber harvest (e.g. Walsh et al., 2011) but also the underlying process dynamics. Except for very few studies, e.g. Webb et al. (2012), most work set out to quantify erosion response to logging has largely neglected high-frequency time series of water and sediment flux. Here we use quantile regression forests (QRF), a robust multivariate and non-parametric regression technique (Meinshausen, 2006), as a viable and more robust alternative to the traditional SRC approach. We are motivated by the successful application of QRF to model multiple SSC peak events and hysteresis loops between stream flow and suspended sediment discharge (Francke et al., 2008a, b; Zimmermann et al., 2012).

In this study we apply QRF to predict from a high-frequency (3 min) time series of stream discharge and discrete SSC samples the impacts of different seasonal logging on the frequency–magnitude distribution of catchment sediment flux. We show that this technique allows for resolving changes in the distribution of geomorphic work at hitherto unprecedented detail, thus providing unique insights into hydro-geomorphic process dynamics following forestry operations.

2 Study sites

We focus on three small ($\sim 0.1\,km^2$) headwater catchments that are part of a network of 11 experimental catchments in the coastal mountains of south-central Chile, close to the city of Nacimiento in the Biobío River basin (Fig. 1b). The catchments have largely similar size, geology, soils, hydrogeology, and vegetation, but differing forestry practices. All catchments are comparable in terms of topography; for example, catchment slopes range between $14.2° \pm 8.6°$ and $20.4° \pm 10.8°$. A more detailed description and discussion of the catchments' morphometric features is stated in Huber et al. (2010) and Mohr et al. (2012). The dominant soil type is a clayey to loamy Luvisol that is locally disturbed by forestry operations and underlain by a deeply weathered saprolite on top of schist bedrock (Mohr et al., 2012). The climate is Mediterranean, and rainfall intensities are low and only rarely exceed $10\,mm\,h^{-1}$ during single events. Intense convective storms are extremely rare. Previous work shows that only 5 % of the registered rainfall events exceed $23\,mm\,h^{-1}$ (Mohr et al., 2013).

Two catchments previously planted with *Pinus radiata* were logged by the same clear-cutting technique during different seasons: catchment 3 was clear-cut during the winter rainy season (July–August 2009), and remained bare for ~ 1 year, whereas catchment 4 was harvested during the end of the dry summer season (March–April 2010), and replanted in early spring 2010 (September–October 2010) (Fig. 2a). Both catchments were reforested by *Eucalyptus globulus* (Schuller et al., 2013). Although clear cutting is permitted under the Chilean standards, the forest companies are requested to adopt best management practices in accordance with Forest Stewardship Council certification agreements. Among other practices, these include cable harvesting on slopes > 30 %, the use of ground skidders in areas of lower slopes, the maintenance of riparian buffer strips (which in the study sites are ~ 7.5 m wide both sides of the channel network), and piling up forestry residues along contour lines at the end of the harvesting operations. The logging of catchment 4 severely damaged the riparian buffer strip, whereas the buffer strip in catchment 3 remained unaffected by the timber harvest. Overall, ~ 88 % of the area of catchment 3 were logged, and more in catchment 4. The clear cut was done using heavy rubber-tired skidders to drag logs uphill to landings, whereas

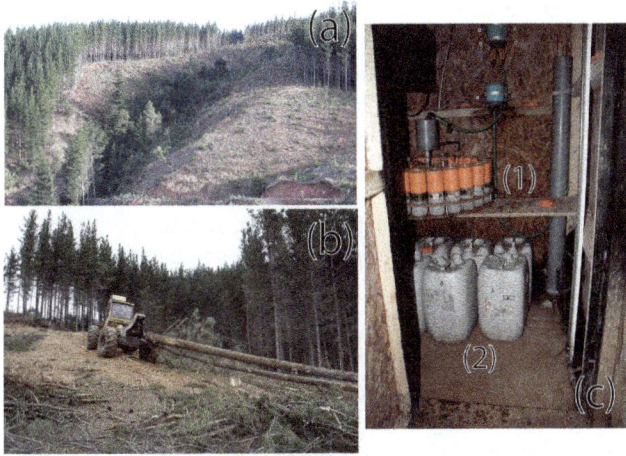

Figure 2. Pictures showing the experimental catchments, the logging procedure, and the suspended sediment monitoring devices. (a) Rainy season logged watershed (watershed 3) in the subsequent dry season (March 2010). (b) Skidder dragging logged stems uphill to the next landing in watershed 3. (c) Custom-built sediment sampling system: (1) horizontal rotating table used to sample suspended sediment per event and (2) recipient used to collect bulk sample of suspended sediment on weekly basis.

cable logging was limited to steep slopes (Mohr et al., 2013) (Fig. 2b). The loggings covered the entire catchment area including their ridges. Catchment 1 remained unlogged and covered with *P. radiata*, and served as a control catchment. On 27 February 2010, the study area was hit by the M_w 8.8 Maule earthquake that caused ground shaking for 2.5 min at ground accelerations of $\sim 0.3\,g$. The regional hydrological response featured an abrupt drop in stream discharge followed by a rapid increase (Mohr et al., 2012).

3 Methods

3.1 Field sampling

We measured stream discharge with V-notch Thompson weirs equipped with custom-built water-stage recorders at a frequency of 3 min, and a water-level accuracy of 2 mm (Huber et al., 2010; Mohr et al., 2012, 2013). Rainfall was recorded by a Hobo tipping bucket rain gauge with an accuracy of 0.2 mm. A Wilcox rank sum test was used to assess whether hourly rainfall intensities were statistically different ($p \leq 0.05$) between the years. Total sediment yields estimated by bulk samples from June 2008 to September 2009 in these and adjacent catchments indicate that pine plantations were more prone to soil erosion than eucalyptus plantations (Huber et al., 2010). With bed load being negligible in the coastal mountains (Iroumé, 1992), i.e. < 1 % of the total load (Huber and Mohr, unpublished data), we acquired high-frequency data on instantaneous SSC from June 2009 to August 2010 in order to quantify sediment flux in response to logging activi-

Table 1. Number of total samples for each catchment. Sample size of pre-logging period given in parentheses.

Catchment	Sample number n	Start date	End date
1	278	06/27/2009	08/15/2010
3	276 ($n_{pre} = 89$)	06/27/2009	08/29/2010
4	100 ($n_{pre} = 24$)	02/19/2010	08/28/2010

ties. We sampled SSC on an event basis with an electric pump attached to a floating device submerging the pump aperture at a constant depth of 5 cm below the water surface in the weirs. We took instantaneous SSC samples on an event basis at 30 to 60 min intervals (Fig. 2c). In the absence of significant rainfall events, we took at least one complementary daily sample during February–March and August 2010 for characterizing low-flow conditions (Table 1). All SSC samples were then rounded to the next 3 min interval to synchronize with discharge measurements. SSCs were determined gravimetrically with an accuracy of 0.5 mg after filtering the runoff samples (Mohr et al., 2013). We obtained sediment yields by multiplying the SSC with the runoff volume summed over the respective time intervals

$$\text{SSY} = \int_{t1}^{t2} Q(t)\,\text{SSC},(t)\,\mathrm{d}t \tag{1}$$

where SSY is suspended sediment yield ($g\,s^{-1}$), $Q(t)$ is instantaneous discharge ($L\,s^{-1}$), and $\text{SSC}(t)$ is instantaneous sediment concentration ($g\,L^{-1}$).

We complemented this event-based sampling by monitoring suspended sediment flux with weekly volume-weighted bulk sampling (Huber et al., 2010) (Fig. 2c). Despite longer sampling intervals, this alternative monitoring scheme provided data without the need to interpolate SSC. We considered these data as first-order benchmarks for the modelled sediment fluxes. Any given bulk sample merged four samples each day over a period of one week (Huber et al., 2010).

3.2 Quantile regression forests (QRF)

Quantile regression forests (QRF) is a robust non-parametric regression technique (Meinshausen, 2006) that builds on Random Forest (RF) regression tree ensembles, a data mining method based on the repeated random selection of both training data and predictors (Breiman, 2001). The method considers the full distribution of tree predictions, thus quantifying inherent uncertainties of each model (Zimmermann et al., 2012). The QRF approach also helps to incorporate effects of variable interaction while at the same time offering means to quantify relative variable importance (Francke et al., 2008a; Zimmermann et al., 2012) by assessing the decline in model performance due to randomizing single predictor variables while keeping all other predictors unchanged.

3.3 QRF model

We set up individual QRF models for each catchment to predict SSC from the (a) rainfall and discharge time series, (b) day of year to account for possible seasonality effects, and (c) change in discharge to capture dynamics between events (Francke et al., 2008a). We quantified antecedent hydrometeorological conditions by computing predictor variables that integrated antecedent rainfall and discharge values over multiples of the sampling interval. Time interval and number of aggregation levels were set to 3 and 6, respectively. These settings describe the successive increase of aggregation windows into the past and their total number in the generation of the aggregated predictors. In order to prevent co-linearity, overlaps between each window were avoided (Zimmermann et al., 2012). For example, P_{28-81} refers to the rainfall accumulated between 28 and 81 min prior to a given SSC sample (Zimmermann et al., 2012). In order to account for changing environmental conditions, we added variables accounting for elapsed time after clear cutting to capture possible effects of timber harvest and vegetation recovery over time. We further defined a switch variable that stratified the data into pre- and post-seismic periods to identify potential earthquake impacts (Supplement Table 1).

We assessed the relative predictor importance based on permutation (Strobl et al., 2008). This measure accounts for multi-collinearity and associated overestimation of variable importance due to spurious correlation artefacts (Liaw and Wiener, 2002). We validated model performance applying the root-mean-square error

$$\text{RMSE} = \sqrt{\frac{1}{N} \Sigma_{i=1}^{N} (x_i - \hat{x}_i)^2} \qquad (2)$$

for N measurements x_i and predictions $\hat{x}_i$. In order to avoid arbitrary decisions during the validation procedure, e.g. size and location of the test data set, we applied a 20-fold cross validation leaving out continuous data blocks of 5% of the data to test the models, (Zimmermann et al., 2012). We defined 10% of the SSC range (g L^{-1}) as a threshold range for acceptable model performance accounting for the inherent erosion model limitations (Nearing, 1998) and the distinct parameter range of measured SSC. Finally, we estimated suspended sediment yields for each 3 min time step applying a Monte Carlo simulation (Francke et al., 2008a). To this end we randomly drew a SSC prediction from the distribution realized by the QRF model for each time step. Based on these samples, we estimated event dynamics and both monthly and annual sediment yields by summing up the products of Q and SSC at each time step over each target period. By repeating this procedure 250 times, we obtained a distribution of suspended sediment yields (SSY) estimates which was then inspected to determine whether the data followed a Gaussian distribution. The latter allowed us to calculate their mean value and standard deviation to assess the spread of the predicted sediment yields (Zimmermann et al., 2012).

3.4 Sediment rating curve (SRC)

For each catchment, we fitted sediment rating curves to a power-law function relating SSC values to the correlate discharge Q (e.g. Gomi et al., 2005)

$$\text{SSC} = aQ^b, \qquad (3)$$

where a (g s^b L$^{-(b+1)}$) and b are empirical fitting parameters of log-transformed data. Based on the SRC, we predicted SSC during the study period and performed the same 20-fold cross-validation procedure as described for QRF.

4 Results

We find that QRF predicted SSC with high accuracy under both low- and high-flow regimes, as well as unlogged and logged conditions. Fig. 3 illustrates the predictive accuracy for high SSC under disturbed conditions, and the additional advantage of QRF to compensate for poor, or impute missing, rainfall and discharge data (Fig. 7a–c). The method also reproduced hysteresis loops and the occurrence of multiple peak events (Fig. 3). Treating errors < 10% of the measured SSC range as acceptable, both QRF and SRC met this criterion across all catchments (Supplement Table 2). Yet QRF generally outperformed SRC, except for rainy-season logging, where the large range of measured SSC values shrunk relative differences in model performance to < 1%.

We compared monthly and annual specific SSY predicted from both QRF and SRC with the bulk data, using a Monte Carlo simulation (Francke et al., 2008a) (Table 2, Supplement Tables 3–4). Specific sediment yield averaged for the first two years following rainy-season logging was 3.27 ± 0.09 t ha^{-1} or ~ 20 times the SSY predicted for unlogged conditions (0.19 ± 0.004 t ha^{-1}, $\pm 1\sigma$).

However, monthly SSY from the catchment planned to be harvested during rainy season exceeded that in the unlogged control catchment by a factor of ~ 5 even before logging commenced (Fig. 4a). Similarly, the catchment that was subjected to dry-season logging yielded ~ 4 times the SSY of the control catchment before it was clear-cut (Fig. 4b). The decreasing slope of the double-mass curve after dry-season logging indicates that soil erosion decreased. In contrast, sediment flux intensified over undisturbed conditions after rainy-season logging.

When normalized to the increase under unlogged control conditions, SSYs increased from 2009 to 2010 by ~ 125% following rainy-season logging, but decreased by ~ 40% after dry-season logging. This finding is in line with our bulk data measurements (Table 2). Overall, QRF predicted substantially higher sediment yields than the SRC approach.

Table 2. Bulk data and modelled annual suspended sediment yields in 2009 and 2010 at the catchment outlets of control (1), rainy-season clear cutting (3), and dry-season clear cutting (4). Annual rainfalls from local rain gauge (Mohr et al., 2012); errors are ± 1 standard deviation.

	Year	Catchment 1 (t ha^{-1} yr)	Catchment 3 (t ha^{-1} yr)	Catchment 4 (t ha^{-1} yr)	Rainfall (mm yr^{-1})
Measured	2009	0.15	0.28	0.22	1463.9
bulk data	2010	0.56	0.88	0.55	1120.8
QRF	2009	0.06 ± 0.00	0.83 ± 0.01	0.39 ± 0.00	
	2010	0.14 ± 0.01	2.43 ± 0.08	0.48 ± 0.01	
SRC	2009	0.02	0.08	0.12	
	2010	0.02	0.12	0.14	

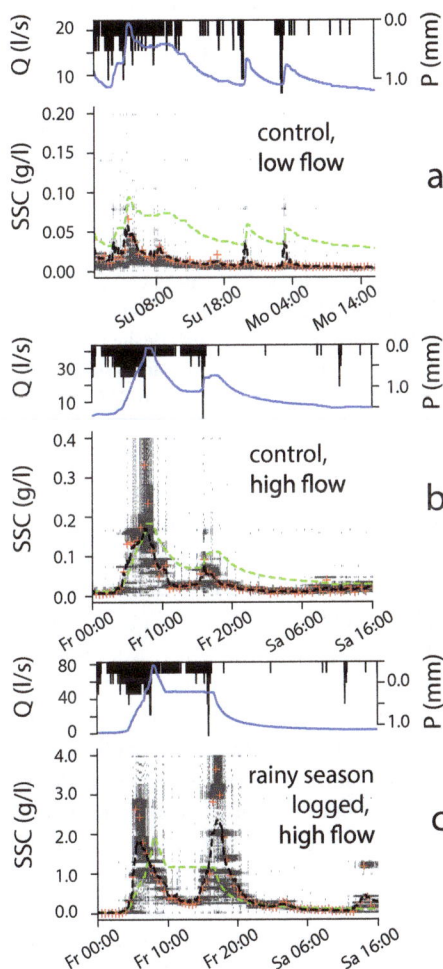

Figure 3. Water discharge and SSC dynamics under unlogged and logged conditions during two rainfall events: (**a**) 27–29 June 2009, catchment 1; (**b–c**) 14–15 August 2009, catchments 1 and 3. Density of SSC predictions of the QRF model for each time step encoded by SSC in coloured histograms: black dashed lines are means of these predictions, red crosses are measured SSC, and green dashed lines are SSC predictions of the SRC. Data are from calibration period; that is, periods are covered with SSC samples used for model building (see Fig. 7 for limits to model predictions).

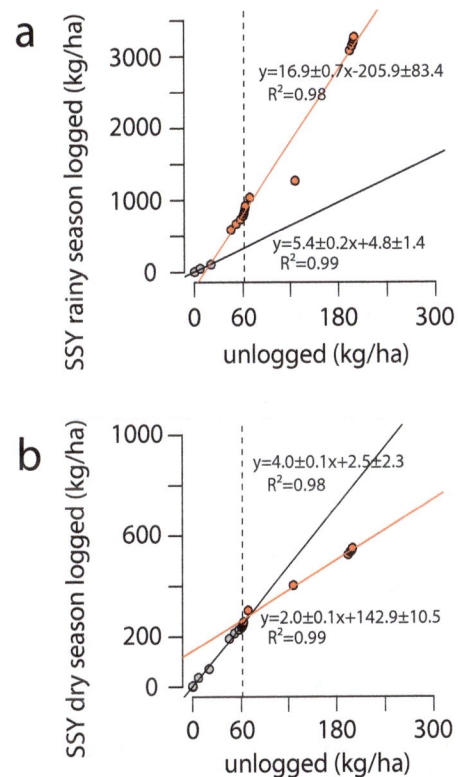

Figure 4. Monthly double-mass-curve analysis between the sediment yields (SSY) of catchments logged during (**a**) rainy and (**b**) dry season, and the unlogged control catchment. Black vertical dashed line separates 2009 and 2010 study periods; grey and red circles are pre- and post-logging sediment yields, respectively; lines are best-fit linear regression models. Uncertainties are ±1 standard deviation.

Only for undisturbed conditions and dry-season logging were SRC predictions within the same order of magnitude. Based on bulk data, SRC underestimates annual SSY by a factor of 2–28 (Supplement Table 3) despite overestimating sediment flux during individual peak runoff events (Fig. 3, Supplement Fig. 1).

Our QRF-derived estimates show that, under unlogged conditions, ~ 80 % of the total sediment load carried during the monitoring period was transported during only ~ 5 % of

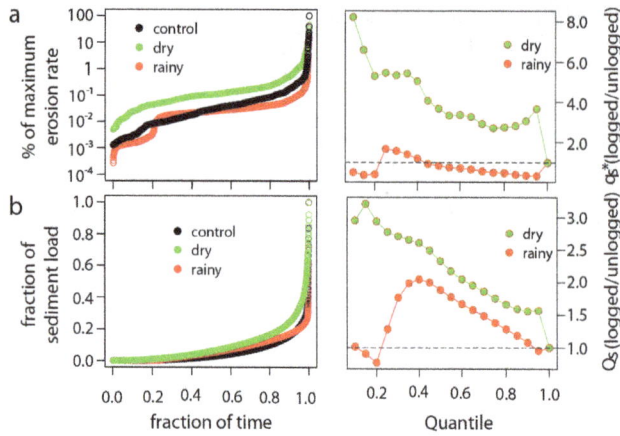

Figure 5. (a) Fraction of instantaneous sediment transport rates normalized to catchment maximum as a function of monitoring time during which these rates were not exceeded. (b) Fraction of total sediment load normalized per catchment as a function of the fraction of total monitoring period for unlogged conditions as well as rainy- and dry-season logging. Right-hand panels show resulting ratios of instantaneous transport rates q_s^* and total sediment loads Q_s per quantile for logged versus unlogged conditions. Black horizontal dashed lines are 1 : 1 ratio. Empirical cumulative distribution functions differ significantly ($p < 0.01$, Kolmogorov–Smirnov test).

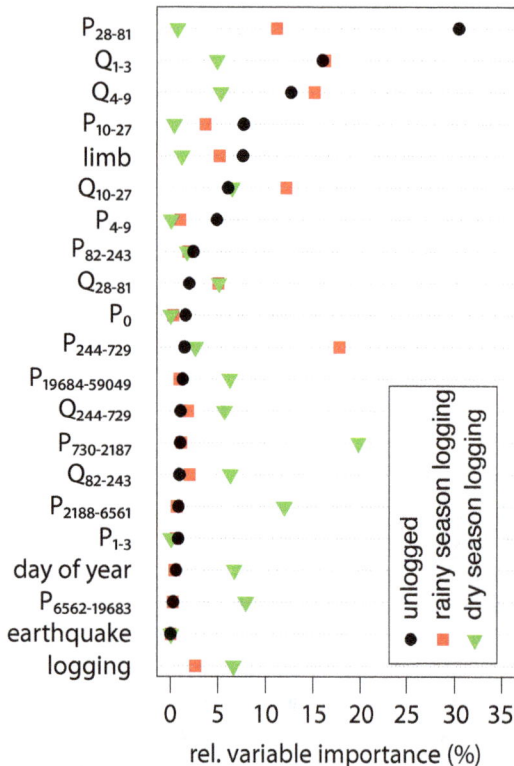

Figure 6. Variable importance of the quantile regression models for each catchment scaled to 100 % in order to facilitate inter-catchment comparison. See Supplement Table 1 for predictor variables.

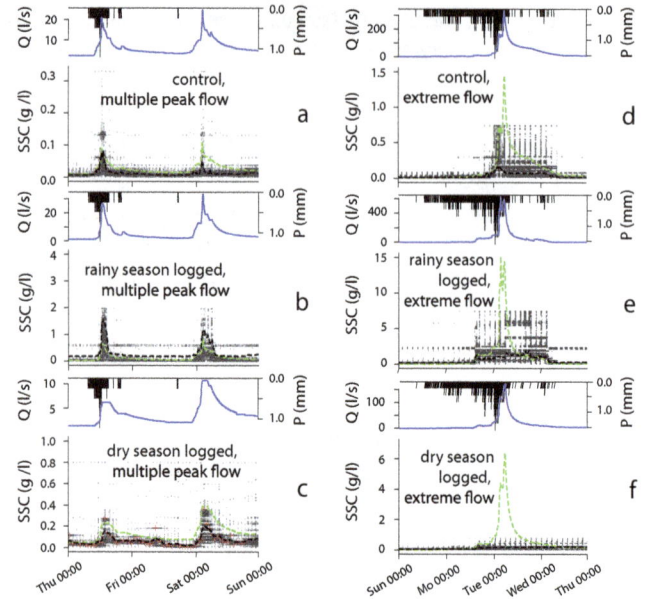

Figure 7. QRF model results of SSC dynamics during extreme peak flow for undisturbed and logged conditions during two rainfall events, i.e. 15–18 August 2010 (a–c) and 26–28 August 2010 (d–f). Density of SSC predictions of the QRF model for each time step encoded by grey histograms, black dashed lines are means of these predictions, and green dashed lines are SSC predictions based on SRCs. In (c), red crosses show measured SSC.

the time. Most of the sediment was transported during rare, large runoff events. The instantaneous flux rate ($g\,s^{-1}$) variability of these rare events spanned 4 orders of magnitude, and thus more than the variability of all other rates occurring over 95 % of the monitoring period (Fig. 5a). Our QRF data thus indicate that logging, regardless of its seasonal timing, coincided with a relatively increased contribution of moderate as opposed to extreme runoff events in terms of sediment transport. Thus, immediate post-logging effects on sediment transport involved shifting the geomorphic work towards less flashy and more moderate events. We found this effect to be more pronounced for dry-season than for rainy-season logging (Fig. 5b).

To rank the contributions of different environmental controls as predictors of sediment flux, we quantified their relative importance in terms of added total predictive accuracy (Fig. 6). We found that antecedent rainfall accumulated 28–81 min prior to a given SSC sample was most influential for unlogged conditions, whereas the timing of logging was not. In contrast, logged catchments did not respond to such short-term rainfall memory. Instead, rainfall accumulated over 244–729 min and 730–2181 min showed the highest importance for the catchments clear-cut during the rainy and dry seasons, respectively (Fig. 6). Near-instantaneous discharges cumulated over 1–3 and 4–9 min prior to SSC sampling were important for both unlogged and rainy-season logging catchments. Neither the day of year nor the timing of the 2010 earthquake showed any significant influence on the SSC predictions (Fig. 6).

5 Discussion

Our results show that high-frequency (3 min) time series of post-logging water and sediment fluxes are instructive with regard to understanding immediate hydro-geomorphic process response despite several unavoidable uncertainties. For one, we treat our bulk sediment flux measurements as minimum estimates given their low temporal resolution compared to the fast hydrological response. Under such restrictions, we assume that they do not fully capture potentially high but short-lived SSC during intense rainfall events. Thus, the total sediment yields based exclusively on bulk samples are a lower-bound estimate (Table 2). Furthermore, we find that conventional sediment rating curves (SRCs) are sensitive to outliers, resulting in implausibly high SSC (e.g. 10–15 g L^{-1}; Fig. 7e), but remain below our QRF predictions on average (Table 2, Supplement Table 5). Under the recorded low-flow regime (Huber et al., 2010), SRCs underestimate the hydro-geomorphic work of more frequent though lower sediment fluxes, while they overestimate the less frequent higher-magnitude events. This finding supports earlier work arguing that SRCs significantly underestimate sediment fluxes (e.g. Asselman, 2000).

Overall, the QRF predictions cast a much more detailed and consistent light on high-frequency post-logging sediment flux, particularly with regard to systematic shifts in the frequency–magnitude distribution of high-frequency sediment transport rates and total loads (Fig. 5). The choice of season for clear-cut logging is linked to distinct changes in the relative overall sediment transport efficiency with a general trend towards emphasizing moderate flows, while dampening the efficiency of more rare and extreme events, which we define here as those values above the 95th percentile of our data. Not only do our results significantly expand, down to the process timescale, the notion that extreme sediment transport events may perform the bulk of geomorphic work (Korup, 2012) but our findings also underscore the impact that logging may have on shifting the underlying frequency–magnitude distributions of water and sediment flux (Fig. 5). We interpret these as statistically robust changes, given that QRF avoids over-fitting by randomly selecting both data and predictor subsets, while providing objective measures of their relative importance (Fig. 6). Predictor importance also changes as a function of the logging season. Not only are the resulting predictions in line with the base flow-dominated discharge but they also maintain low uncertainty because of the averaging-out of low-precision predictions (Zimmermann et al., 2012). Nevertheless, QRF may have drawbacks for high-magnitude rainfall–runoff events (Fig. 7d–f) given the method's inability to extrapolate beyond the parameter space. Seasonal effects may not be fully represented in the

time series (Fig. 6) because our observation windows in time have significantly different rainfall patterns (Supplement Table 3). Our study area is dominated by frontal rainfall events instead of high-intensity convective storms, and we caution against extrapolating our results for rainfall–runoff events of higher magnitude. In essence, QRF is a robust and versatile method for hindcasting high-frequency time series of water and sediment discharge, but it is not designed for predicting future events.

Compared with similar studies on logging effects (e.g. Gomi et al., 2005), our QRF predictions indicate very low to even slightly decreasing post-logging sediment yields following dry-season logging when compared to unlogged conditions. Our observation of increased post-rainy-season logging sediment flux is consistent with previous work (e.g. Sidle et al., 2006). Yet the magnitude of this increase is small (e.g. Gomi et al., 2004), and our SSY estimates are within the range reported for natural, undisturbed forests (Zimmermann et al., 2012). The observed decreases of SSY following dry logging (Fig. 4b) may partly be due to prompt replanting of the logged slopes (Malmer and Grip, 1990). We exclude topographic controls on sediment flux since slope and stream gradient are highest in catchment 4 (Mohr et al., 2012), where the lowest sediment load had been observed. Sediment flux is expected to be even lower in this catchment assuming the appropriate adoption of best management practices such as maintenance of buffer strips (~ 7.5 m each) alongside the streams. Some maintenance works on a timber road in 2010 in the unlogged catchment are also likely to have contributed to elevating the local sediment supply given that unsealed timber roads may dominate sediment production per unit area in managed forests (e.g. Motha et al., 2003). Consequently, we expect that the thus elevated sediment flux in the unlogged catchment may have partly smothered the relative impact of logging in the other catchments.

We also exclude seasonal meteorological differences as drivers of the elevated SSYs in 2010, as rainfall was much higher in 2009 at comparable intensities (Table 2). Moreover, the timing of the 2010 earthquake did not notably distort any of the SSC predictions (Fig. 6) despite favourable conditions for post-seismic increases in sediment flux (Hovius et al., 2011). The limited earthquake response may be linked to a decisive lack of post-seismic rainfall in 2010, which may have otherwise triggered mass wasting. Plantations of *P. radiata* are prone to mass wasting because of their low root strength compared with other species (Watson et al., 1999). When logged, pine roots rapidly decay in root strength (Sidle, 1991). Shallow landslides are also promoted by slow root-strength recovery rates and cumulative effects of preceding rotations (Sidle et al., 2006). After logging, root decay leads to progressive weakening of slopes over time, whereas regrowth strengthens slopes. The opposing trends lead to maximum mass wasting rate at ~ 2–3 years after logging (Watson et al., 1999). Hence, our immediate post-logging predictions of SSC (Fig. 6) are consistent with mechanistic slope-stability models (Sidle et al., 2006). During the monitoring period of this study, however, we regard the contribution of mass wasting processes as minor, and hitherto insufficient, for explaining the hydro-geomorphic post-logging regimes.

The ORF-derived variable importance plot supports the notion of a predominant overland flow mechanism of recent harvest areas (Fig. 6). Infiltration capacity may increase on recently logged areas, thus impeding infiltration-excess overland flow generation under the low rainfall intensities observed (Mohr et al., 2013). Alternatively, high-duration rainfall is required to elevate groundwater levels, which in turn initiate erosive saturation-excess overland flow (Dunne and Black, 1970), and connect sediment sources to the drainage network. The relevance of several antecedent rainfall characteristics for predicting post-logging sediment fluxes reflects the local rainfall regime, where low-intensity and long-duration rainfall events successively saturate the soil layers over time, thus permitting erosive overland flow (Huber et al., 2010). Compared with rainy-season logging, the measured SSC following dry-season logging also responded to significantly longer time lags. The lower cohesion of wet soil elevates sediment supply for erosive overland flow during timber machinery action in the rainy season. Hence, the susceptibility to soil erosion is higher compared to logging in dry soil conditions. The distinct micro-topography left by heavy machinery persisted following dry-season logging until the subsequent rainy season, thus impeding overland-flow connectivity and requiring larger volumes of water to re-establish connectivity (Mohr et al., 2013). In contrast, SSC under unlogged conditions appears to be modified by more short-term antecedent rainfall characteristics (Fig. 6). Such flashiness may indicate effects of hydrophobic plantation forest cover (e.g. Miyata et al., 2009).

6 Conclusions

Our study provides novel insights into the immediate hydro-geomorphic process response to different seasonal timber harvest operations. We find that quantile regression forests (QRF) outperform sediment rating curves (SRC) in terms of accurately predicting post-logging sediment yields at the process scale. Using empirical sediment rating curves may lead to grave underestimates of sediment fluxes from managed forests. Our unprecedented high-frequency data on post-logging water and sediment fluxes from three Chilean headwater basins corroborates the widely held view that most sediment transport is accomplished within a few rare high-discharge events, particularly at the timescale of immediate hydro-geomorphic process response. Moreover, QRF-based hindcasting underlines that it is the seasonal timing of clear cutting that dictates the amount of shift in the frequency–magnitude relationship of sediment transport, eventually redistributing geomorphic work from rare, extreme events to

more moderate ones. Dry-season logging led to a much greater dampening of extreme events, whereas rainy-season logging accentuated the contrasts in instantaneous transport rates. Post-logging increases in sediment flux, most likely driven by saturation-excess overland flow, were an order of magnitude higher following rainy-season clear cutting.

With these findings in mind, we conclude that the task of predicting post-logging sediment yields remains a trade-off between model simplicity, applicability, and uncertainty. Our work provides a firm basis for longer-term studies that will complement our results by recording the cumulative net effects of logging and the avenues of hydro-geomorphic recovery. Still, given that data scarcity and variability are common for post-logging disturbances, we find that quantile regression forests are a robust and promising tool for quantifying in detail high-frequency time series of water and sediment fluxes following clear-cut operations.

Acknowledgements. We thank Andreas Bauer, Johannes Brenner, Franziska Faul, Rodrigo Bravo, and Cristian Frêne Conget for helping in the field and during data analysis. We are grateful to Forestal Mininco for providing access to our experimental catchments, and acknowledge Anton Huber for support and advice. We thank Simon Mudd and two reviewers for constructive comments on an earlier draft of this manuscript. This study is co-funded by the International Bureau of the German Federal Ministry of Education and Research and the Chilean government (CONICYT/BMBF 2009-092 and 2010-243).

Edited by: S. Mudd

References

Asselman, N. E. M.: Fitting and interpretation of sediment rating curves, J. Hydrol, 234, 228–248, 2000.

Breiman, L.: Random forests, Machine Learning, 45, 5–32, 2001.

Dunne, T. and Black, R. D.: An Experimental investigation of runoff production in permeable soils, Water Resour. Res., 6, 478–490, 1970.

FAO: Global Forest Resources Asssessment 2010, Main Report, Food and Agriculture Organization of the United Nations, Rome, 2010.

Francke, T., J.A. Lopez-Tarazon, and Schroder, B.: Estimation of suspended sediment concentration and yield using linear models, random forests and quantile regression forests, Hydrol. Processes, 22, 4892–4904, 2008a.

Francke, T., Lopez-Tarazon, J. A., Vericat, D., Bronstert, A., and Batalla, R. J.: Flood-based analysis of high-magnitude sediment transport using a non-parametric method, Earth Surf. Processes Landforms, 33, 2064–2077, 2008b.

Gomi, T., Moore, R. D., and Hassan, M. A.: Suspended sediment dynamics in small forest streams of the Pacific Northwest, J. Am. Water Resour. Assoc., 41, 877–898, 2005.

Gomi, T., R. C. Sidle, and Swanston, D. N.: Hydrogeomorphic linkages of sediment transport in headwater streams, Maybeso Experimental Forest, southeast Alaska, Hydrol. Processes, 18, 667–683, 2004.

Hovius, N., Meunier, P., Lin, C.-W., Chen, H., Chen, Y.-G., Dadson, S., Horng, M.-J., and Lines, M.: Prolonged seismically induced erosion and the mass balance of a large earthquake, Earth Planet. Sci. Lett., 304, 347–355, 2011.

Huber, A., Iroumé, A., Mohr, C. H., and Frene, C.: Effect of Pinus radiata and Eucalyptus globulus plantations on water resource in the Coastal Range of Biobio region, Chile, Bosque, 31, 219–230, 2010.

Iroumé, A.: Precipitación, escorrentía y producción de sedimentos en suspensión en una cuenca cercana a Valdivia, Chile, Bosque, 13, 15–23, 1992.

Korup, O.: Earth's portfolio of extreme sediment transport events. Earth-Sci. Rev., 112, 115–125, 2012.

Liaw, A., and Wiener, M.: Classification and Regression by randomForest, R News, 2, 18–22, 2002.

Malmer, A. and Grip, H.: Soil disturbance and loss of infiltrability caused by mechanized and manual extraction of tropical rainforest in Sabah, Malaysia, Forest Ecol. Manag., 38, 1–12, 1990.

Meinshausen, N.: Quantile regression forests, J. Machine Learning Res., 7, 983–999, 2006.

Miyata, S., Kosugi, K., Gomi, T., and Mizuyama, T.: Effects of forest floor coverage on overland flow and soil erosion on hillslopes in Japanese cypress plantation forests, Water Resour. Res., 45, doi:10.1029/2008WR007270, 2009.

Mohr, C. H., Montgomery, D. R., Huber, A., Bronstert, A., and Iroumé, A.: Streamflow response in small upland catchments in the Chilean coastal range to the M-W 8.8 Maule earthquake on 27 February 2010, J. Geophys. Res.: Earth Surf., 117, F02032, doi:10.111029/2011JF002138, 2012.

Mohr, C. H., Coppus, R., Huber, A., Iroumé, A., and Bronstert, A.: Runoff Generation and soil erosion processes after clear cutting, J. Geophys. Res.: Earth Surf., 118, 814–831, doi:10.1002/jgrf.20047, 2013.

Montgomery, D. R., Schmidt, K. M., Greenberg, H. M., and Dietrich, W. E.: Forest clearing and regional landsliding, Geology, 28, 311–314, 2000.

Motha, J. A., Wallbrink, P. J., Hairsine, P. B., and Grayson, R. B.: Determining the sources of suspended sediment in a forested catchment in southeastern Australia, Water Resour. Res., 39, doi:10.1029/2001WR000794, 2003.

Nearing, M. A.: Why soil erosion models over-predict small soil losses and under-predict large soil losses, Catena, 32, 15–22, 1998.

Schuller, P., Walling, D. E., Iroumé, A., Quilodran, C., Castillo, A. and Navas, A.: Using ^{137}Cs and ^{210}Pb and other sediment source fingerprints to document suspended sediment sources in small forested catchments in south-central Chile, J. Environ. Radioact., 124, 147–159, 2013.

Sidle, R. C.: A conceptual-model of changes in root cohesion in response to vegetation management, J. Environ. Qual., 20, 43–52, 1991.

Sidle, R. C., Ziegler, A. D., Negishi, J. N., Nik, A. R., Siew, R., and Turkelboom, F.: Erosion processes in steep terrain – Truths, myths, and uncertainties related to forest management in Southeast Asia, For. Ecol. Manage., 224, 199–225, 2006.

Strobl, C., Boulesteix, A. L., Kneib, T., Augustin, T., and Zeileis, A.: Conditional variable importance for random forests, BMC Bioinf., 9, doi:10.1186/1471-2105-9-307, 2008.

Walsh, R. P. D, Bidin, K., Blake, W. H., Chappell, N. A., Clarke, M. A., Douglas, I., Ghazali, R., Sayer, A. M., Suhaimi, J., Tych, W., and Annammala, K. V.: Long-term responses of rainforest erosional system at different spatial scales to selective logging and climatic change, Philos. Trans. R. Soc., B., 366, 3340–3353, 2011.

Watson, A., Phillips, C., and Marden, M.: Root strength, growth, and rates of decay: Root reinforcement changes of two tree species and their contribution to slope stability, Plant Soil, 217, 39–47, 1999.

Webb, A., Dragovich, D., and Jamshidi, R.: Temporary increases in suspended sediment yields following selective eucalypt forest harvesting, For. Ecol. Manage., 283, 96–105, 2012.

Zimmermann, A., Francke, T., and Elsenbeer, H.: Forests and erosion: Insights from a study of suspended-sediment dynamics in an overland flow-prone rainforest catchment, J. Hydrol, 428, 170–181, 2012.

10

Linking mineralisation process and sedimentary product in terrestrial carbonates using a solution thermodynamic approach

M. Rogerson, H. M. Pedley, A. Kelham, and J. D Wadhawan

Department of Geography, Environment and Earth Sciences, University of Hull, Cottingham Road, Hull, HU6 7RX, UK

Correspondence to: M. Rogerson (m.rogerson@hull.ac.uk)

Abstract. Determining the processes which generate terrestrial carbonate deposits (tufas, travertines and to a lesser extent associated chemical sediments such as calcretes and speleothems) is a long-standing problem. Precipitation of mineral products from solution reflects a complex combination of biological, equilibrium and kinetic processes, and the different morphologies of carbonate sediment produced by different processes have yet to be clearly demarked. Building on the groundbreaking work of previous authors, we propose that the underlying control on the processes leading to the deposition of these products can be most parsimoniously understood from the thermodynamic properties of their source solutions. Here, we report initial observations of the differences in product generated from spring and lake systems spanning a range of temperature–supersaturation space. We find that at high supersaturation, biological influences are masked by high rates of physico-chemical precipitation, and sedimentary products from these settings infrequently exhibit classic "biomediated" fabrics such as clotted micrite. Likewise, at high temperature ($> 40\,^\circ$C) exclusion of vascular plants and complex/diverse biofilms can significantly inhibit the magnitude of biomediated precipitation, again impeding the likelihood of encountering the "bio-type" fabrics.

Conversely, despite the clear division in product between extensive tufa facies associations and less spatially extensive deposits such as oncoid beds, no clear division can be identified between these systems in temperature–supersaturation space. We reiterate the conclusion of previous authors, which demonstrate that this division cannot be made on the basis of physico-chemical characteristics of the solution alone. We further provide a new case study of this division from two adjacent systems in the UK, where tufa-like deposition continuous on a metre scale is happening at a site with lower supersaturation than other sites exhibiting only discontinuous (oncoidal) deposition. However, a strong microbiological division is demonstrated between these sites on the basis of suspended bacterial cell distribution, which reach a prominent maximum where tufa-like deposits are forming.

We conclude that at high supersaturation, the thermodynamic properties of solutions provide a highly satisfactory means of linking process and product, raising the opportunity of identifying water characteristics from sedimentological/petrological characteristics of ancient deposits. At low supersaturation, we recommend that future research focuses on geomicrobiological processes rather than the more traditional, inorganic solution chemistry approach dominant in the past.

1 Introduction

Terrestrial carbonate deposits have outstanding potential to act as an archive of climatic change (Andrews, 2006) and/or changes in the behaviour of crustal fluids (Minissale, 2004). However, despite their high potential and the considerable research effort that has already been expended on them, there remains considerable debate and confusion over how mineral-forming processes can be linked to the sedimentology and petrology of tufas, travertines and other terrestrial chemical deposits. General recognition of the importance of terrestrial carbonate systems and deposits has never been higher. There is pressing need for more and better archives of past climatic and environmental change from the Holocene (Hori et al., 2008) to the Palaeoproterozoic (Brasier, 2014). Combined with this, the increasing focus on geochemical methods to provide proxies for reconstruction of the past means that having certainty that particular deposit records meteoric (or geothermal) processes is of fundamental importance. The discovery of terrestrial carbonate reservoirs hosting significant oil reserves in the South Atlantic (Wright, 2012) and recent suggestions that the carbon sink represented by these systems may be globally important (Liu et al., 2010) have yet further emphasised that our understanding of terrestrial carbonate production must be improved. In particular, it is a basic requirement that meteoric and geothermal processes can be disentangled before a fossil deposit can be understood, and we currently lack the tools and knowledge by which to make this distinction (Wright, 2012).

Early efforts to classify terrestrial carbonate deposits were based on straightforward applications of conventional carbonate petrographic schemes. The standard scheme of Dunham (1962) modified by Embury and Kloven (1972) works admirably for most freshwater limestones (for example, most laminated, microbial components are recognisably "boundstones"), but either has very little precision or is excessively overcomplicated (depending on how the basic scheme is modified) on account of the numerous biological elements present. These classifications also make no attempt to link product to process. Considerable advances from these early efforts have been made for geothermal deposits by numerous authors (Chafetz and Folk, 1984; Folk and Chafetz, 1983; Carrara et al., 1998; Carrara, 1994; Folk, 1994; Folk et al., 1985) and a useful overview of geothermal systems has been provided by Guo and Riding (1998). Much of the current petrographic terminology applied to ambient temperature tufas has also developed upon the Folk and Dunham schemes (see Pentecost, 2005). However, to date no single petrographic scheme has proven sufficiently effective to become standard for use in ambient temperature carbonate systems.

In contrast, a diverse group of field classification schemes have independently been developed specifically for freshwater carbonates (Stirn, 1964; Irion and Müller, 1968; Schneider et al., 1983; Pentecost and Lord, 1988) emphasising the role of botanical elements (Pentecost and Lord, 1988), their distinct geomorphological expression within the landscape (Symoens et al., 1951; Golubić, 1969; Nicod, 1981; Szulc, 1983), or combinations of biological and geomorphological aspects (Buccino et al., 1978; Ferreri, 1985; D'Argenio and Ferreri, 1988; Golubić et al., 1993). Others (e.g Julia, 1983; Golubić, 1969) have discussed a process-based approach frequently involving cyclic progression of the facies, and an important observation to arise from more recent classifications was the role played by the phytoherm (freshwater reef) in tufa accumulation (see discussion in Pedley (1992, 1990). This work was drawn together by Pedley (1990) and Ford and Pedley (1997), who offered an integrated, rationalised scheme for classifying freshwater carbonates which embraced both laboratory and field scales. Modifications of these schemes are now available for tropical tufas (e.g. (Carthew et al., 2006) and new controls on the processes controlling individual facies are being provided by continuing field (Drysdale et al., 2003b; Brasier et al., 2011) and laboratory-based research (Pedley and Rogerson, 2010b). Alternative field classifications continue to evolve (e.g. Glover and Robertson, 2003, and Peña et al., 2000, for ambient temperature systems, and Veysey II et al., 2008, for geothermal systems), and increasingly recognise the importance of the processes that control the distribution of specific products (Fouke et al., 2000, 2003). However, these schemes are rather site-specific and are yet to be updated to reflect recent developments in understanding biofilm processes (Decho, 2010).

Despite considerable progress on ambient and geothermal systems in isolation, little progress has been made towards a single scheme that adequately discriminates both geothermal and meteoric-type deposits simultaneously. Attempts have been made to bridge and update existing classification schemes to also encompass geothermal deposits (e.g. Jones and Renaut, 2010), but these remain focussed on products at the expense of knowledge about processes. In this regard, it is surprising that there has not yet been an attempt to classify these deposits on the basis of whether they are dominantly biomediated or physico-chemical deposits. In the light of new insights into these processes arising from experimental work (Pedley and Rogerson, 2010a; Pedley et al., 2009; Bissett et al., 2008a; Shiraishi et al., 2008b), this is a major shortcoming.

Ultimately, we are yet to achieve the goal of a universally effective classification for terrestrial deposits that adequately reflects their unique modes of origin. Understanding how processes can be read from products is a key step to achieve this goal. Major stumbling blocks remain to achieving this goal, however. Most seriously, the independent development of terminology between ambient temperature deposits and thermal deposits maintains the division of these two research communities. Even the basic terminology distinguishing terrestrial carbonates is not without conflict as some workers (e.g. Pentecost, 2005; Pentecost and Viles, 1994) recommend that all freshwater deposits should be called travertines, with the caveat that the prefixes "meteogene" or "thermogene" be applied for deposits in which the original precipitating source waters are either cool, shallow circulation-meteoric or hot, deep circulation (thermogene, with temperatures typically 20–60 °C at the point of emergence). Others (including the authors of this manuscript) recommend that travertine is reserved only for deposits in geothermal settings. The pressing need to see all terrestrial carbonate systems as related and not independent of each other is emphasised by the case of all "thermogene" or "travertine" systems, where vegetation colonisation occurs as waters cool towards ambient temperatures, resulting in deposits which are often indistinguishable from "meteogene" "tufas". While we argue that there is an advantage in retaining the terms "tufa" and "travertine" to distinguish between systems essentially driven by geothermal processes from meteoric processes, we simultaneously recognise that the downside of a simple division in terms of temperature is that alkaline lakes (e.g. Mono Lake, Connell and Dreiss, 1995) may generate deposits significantly different to normal "tufa" regardless of being ambient in terms of temperature.

Here, we attempt the first steps to overcome these stumbling blocks, building on the huge progress made in understanding the relationships between various types of deposits by previous workers outlined above, and propose a new direction in summarising precipitating system solution chemistry using simple thermodynamic concepts. Our thermodynamic approach is based on the assumption that physico-chemical forcing of precipitation will ultimately be a product of strongly negative Gibbs free energy of calcite precipitation, as first proposed by Dandurand (Dandurand et al., 1982). Gibbs free energy, as we use the concept here, therefore reflects the chemical potential of the system to undergo a change (Langmuir, 1997). The change we are interested in is the conversion of calcium and carbonate ions into solution to a combined mineralogical phase. The Gibbs energy therefore represents the potential of the chemical systems to change spontaneously. This ability to change must arise from another chemical process, often related to changes in the abundance of carbonate ions due to exchange of CO_2 gas at the air–water interface. However, there is considerable underlying complexity in understanding how the Gibbs free energy arises in the whole range of solutions possible in terrestrial carbonate systems. The chemical potential promoting precipitation could be outgassing of CO_2 or indeed ingassing of the same gas, especially in anthropogenic sites (Andrews et al., 1997). The ionic strength of the solution, which is allied to the salinity of the solution, reduces the ability of ions to collide and so to react and form a mineral. Consequently, precipitation of non-calcium or carbonate bearing mineral phases could be key in some sites. Mixing of very strong solutions arising from deep crustal sources with more dilute solutions near the surface could also cause a rise in saturation, so long as the weaker solution has significant calcium and/or carbonate concentration. Rather than attempt to identify the origin of the chemical potential at every individual site, we recom-

mend that an assessment of the Gibbs free energy as a representation of chemical potential is sufficient to describe the magnitude of the driving force behind precipitation regardless of its origin.

1.1 Carbonate precipitation mechanisms in terrestrial, open-water settings

1.1.1 Hydrochemical controls

To drive rapid precipitation, high levels of excess dissolved calcium carbonate are necessary in the ambient water. Consequently, the primary partition between tufa-forming and non-tufa river systems should coincide with some threshold in calcite supersaturation. However, Pentecost (1992) showed that in the Yorkshire Dales (UK) no systematic relationship between precipitation and degree of calcium carbonate supersaturation was present. An additional problem of this view is that the activation energy for calcite is $-48.1\,\mathrm{kJ\,mol^{-1}}$ (Inskeep and Bloom, 1985) whereas the underlying physical process promoting this, the degassing of CO_2, provides comparatively little driving-force (e.g. $\sim 5\,\mathrm{kJ\,mol^{-1}}$; Dandurand et al., 1982), leaving a wide energetic gap. It is therefore already abundantly clear that thermodynamic considerations are not sufficient to explain site-to-site differences at low supersaturation, that kinetic and "surface" conditions are responsible for driving mineral precipitation. Explanations of these additional conditions must be sought.

At higher supersaturation, physico-chemical precipitation (i.e. that regulated by classic equilibrium considerations) occurs either due to ingassing of atmospheric $CO_{2(g)}$ under hyperalkaline conditions due to the presence of excess $OH^-_{(aq)}$ (Clark and Fontes, 1990; Andrews et al., 1997) or, more normally, due to outgassing of CO_2 to atmosphere due to the presence of excess $HCO^-_{3(aq)}$ (Emeis et al., 1987). Rapid loss of $CO_{2(aq)}$ primarily occurs when fresh spring waters come into contact with air depleted in CO_2 relative to aquifer air (Chafetz and Folk, 1984). Degassing-driven precipitation occurs according to the le Chatellier principle via

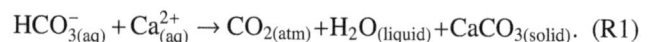

$$HCO^-_{3(aq)} + Ca^{2+}_{(aq)} \rightarrow CO_{2(atm)} + H_2O_{(liquid)} + CaCO_{3(solid)}. \quad (R1)$$

A consequence of this is that $HCO^-_{3((aq)}$ rapidly decreases downstream from point sources of spring water, driving rapid precipitation close to the spring and decreasing rate of precipitation in a downstream direction (Chen et al., 2004; Lorah and Herman, 1988; Merz-Preiss and Riding, 1999). A complicating factor is that calcite does not precipitate directly from bicarbonate, making the pH-related conversion of $HCO^-_{3(aq)}$ to $CO^{2-}_{3(aq)}$ potentially limiting of precipitation rate (Dreybrodt et al., 1997). This is demonstrated by the observation that precipitation can be slow immediately adjacent to the spring, but rapid a few 10s or 100s of metres downstream (Dandurand et al., 1982). Turbulent flow and active agitation increases the area of the air–water interface and therefore promotes gas exchange. Consequently, precipitation is considered to be enhanced at waterfalls and rapids by the same mechanism as occurs at springs (Chen et al., 2004), although physical calculations suggest this change is actually rather small (Hammer et al., 2010).

Carbon dioxide solubility is also strongly affected by temperature and evaporation will further promote increased supersaturation. Water at $0\,°C$ is capable of dissolving about three times as much CO_2 as at $30\,°C$ (Dramis et al., 1999) and as cold karst waters emerge its temperature increases, resulting in CO_2 outgassing and enhancing supersaturation of the water (Lorah and Herman, 1988). Mixing and high surface area : volume ratios promote warming, and consequently may potentially be enhanced at rapids and waterfalls, potentially overcoming the apparently minor changes in gas exchange expected from physical calculations. In temperate climates, slower downstream warming may result in reduced precipitation at proximal sites while warmer climates encourage more tufa formation closer to the spring (Drysdale et al., 2003a). High ambient air temperatures and low relative humidities will enhance levels of saturation, which is an important driver of precipitation in tropical tufa systems (Carthew et al., 2006; Drysdale and Gale, 1997). The impact of temperature and humidity can be large enough to significantly alter the large-scale morphology of major tufa systems, with high rates of precipitation reported from tufa systems in Spain relative to otherwise similar systems in the UK (Pedley et al., 1996).

In addition to these equilibrium processes acting in favour of precipitation, kinetic influences arising from degassing may also have significant impact, and may be particularly critical to initiating nucleation (Dandurand et al., 1982). If kinetics are dominant, different reaches and even different layers of water at a single reach of a freshwater system may have very different saturation characteristics. Surface layer kinetic influences (e.g. gas exchange) will ultimately be bound up with water flow structure, which regulates vertical ion transport rates. These authors are not aware of any specific study of flow structure in these settings. In contrast, bottom layer kinetic influences are well understood.

1.1.2   Bottom layer kinetic influences; the diffusive boundary layer

At sites with flowing water, flow separation occurs at the water–carbonate interface, forming a boundary layer across which ions are exchanged primarily by diffusion (Zaihua et al., 1995; Liu and Dreybrodt, 1997; Dreybrodt and Buhmann, 1991). Consequently, precipitation-inhibiting H^+ ions tend to accumulate at the water-carbonate interface via precipitation originating from bicarbonate:

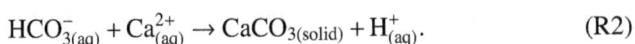

$$HCO_{3(aq)}^- + Ca_{(aq)}^{2+} \rightarrow CaCO_{3(solid)} + H_{(aq)}^+. \qquad (R2)$$

This drives down pH, causing conversion of $CO_{3(aq)}^{2-}$ to $HCO_{3(aq)}^-$ and reducing saturation. Investigation of tufa depositing rivers in China (Zaihua et al., 1995) showed that the

consequence of this behaviour was that deposition rates in fast flowing water are higher than still water by a factor of four. This mechanism has also been investigated via numerical modelling, which suggests that it may be a fundamental control on the patterns of development of tufa barrages (Hammer et al., 2007, 2008; Hammer, 2008; Veysey and Goldenfeld, 2008). Indeed, a degree of consensus is emerging from this work that "equilibrium" degassing at sites of enhanced flow has very little impact on the rate of precipitation, which is essentially driven by enhanced vertical ion flux, particularly where flows become unstable (Hammer et al., 2010). This consensus in the physical literature is yet to be fully recognised in the sedimentological literature.

1.2   Biological factors

The mechanisms and state of knowledge of microbially induced and influenced calcite precipitation has recently been thoroughly reviewed (Dupraz et al., 2009). Consequently, we here aim to only summarise the most relevant parts of the very extensive knowledge concerning microbe–calcite interaction.

1.2.1   Photosynthesis and respiration

Consumption of CO_2 by photosynthesis results in raised pH, and consequently in increased proportion of dissolved inorganic carbon present as $CO_{3(aq)}^{2-}$, close to the bodies of micro- and macrophytes as well as phototrophic microbes such as cyanobacteria. Consequently, when photosynthesis occurs in solutions at or near saturation with respect to $CaCO_3$, calcite may precipitate as sheaths around algal filaments and coatings on plants (Pentecost, 1978, 1987). Respiration operates in the inverse sense, tending to increase dissolved CO_2 levels, reduce pH and inhibit precipitation. Consequently, tufa systems exist within a dynamic equilibrium where precipitation may be triggered or prevented by the balance of microbial metabolic processes (Visscher and Stolz, 2005; Decho, 2010). This balance will primarily reflect the state of health of the biofilm itself, as a growing community must, by definition, be consuming more carbon than it is releasing and vice versa. The impact of microbial metabolisms is well reflected in an investigation of chemical fluxes associated with calcite precipitating solutions, which indicates increased flux of Ca^{2+} and HCO_3^- to biofilm-encrusted surfaces relative to non-encrusted surfaces in high light conditions, but the reverse under dark conditions (Shiraishi et al., 2008a). Curiously, biofilms are capable of modifying their microenvironment so that the pH of interstitial water remains constant regardless of changes in the macroenvironment (Bissett et al., 2008b), so this effect may actually be amplified under low ambient pH conditions. The observation of almost invariable pH within the biofilm has considerable significance for earlier ideas concerning the spatial control on tufa deposition, and potentially is one of the most important single observations made in this subject. For example, Pentecost (1992) suggested that the absence of tufa in certain streams in Yorkshire may reflect sporadic influxes of low pH water that would not be evident in field data derived from spot sampling. As the biofilm would buffer the growing tufa surface from such a change, this suggestion is now very difficult to sustain.

The degree to which photosynthesis alters the precipitation process has been debated for several decades (Shiraishi et al., 2008a). Monitoring of hydrochemical conditions over four diurnal cycles at a single station in Davys Creek, Australia, revealed little impact. Atmospheric temperature was identified as the dominant control on the $CO_{2(aq)}$ budget, implying that metabolic processes were of minor importance (Drysdale et al., 2003a). Conversely, studies of precipitation over diurnal cycles in China (Liu et al., 2008), Turkey (Bayari and Kurttas, 1995) and the UK (Spiro and Pentecost, 1991) indicate that macrophytes, algae and cyanobacteria play a dominant role in regulating precipitation rate. Laboratory studies indicate that precipitation to carbonate surfaces on the flow bed does not occur in the absence of biofilm (Shiraishi et al., 2008a, b; Pedley et al., 2009), and that photosynthesis may be critical to overcoming the energetic barrier to precipitation (Shiraishi et al., 2008a). This concept is given strong support by repeated observation of calcification of *Charophyta* bodies within water masses otherwise incapable of precipitating calcite (Anadon et al., 2002; Eremin et al., 2007; Pentecost et al., 2006). However, these effects still await proper quantification.

1.2.2 Organic hydrogel templating

The state of knowledge of "templating" – i.e. precipitation of calcite onto organic hydrogels – has recently been thoroughly reviewed (Decho, 2010), and we recommend this paper to interested readers. To avoid duplication here, we provide only a summary of the salient points. Templating is a well known, and widely used, means of accelerating, controlling and initiating mineral precipitation (Decho, 2010). Although it has been demonstrated that the chelation of Ca^{2+} ions to the natural organic gels formed by extracellular polymeric saccharides (EPS) in biofilms can be an inhibiting factor in carbonate mineral precipitation in seawater (Kawaguchi and Decho, 2002), in vitro experiments have shown this may not be the case in a freshwater setting with low ionic strength (Pedley et al., 2009; Rogerson et al., 2008). This appears to be a result of the fact that cation binding to EPS in these settings is more likely to be partial, with some aqueous ligands remaining intact (Saunders et al., 2014). As EPS gels generally account for between 50 and 90 % of the total organic matter of a biofilm (Wingender et al., 1999), the possibility that organic binding can be a first-order control on precipitation means that metalorganic intermediaries may play a critical role in determining both rate and character of carbonate precipitate

in terrestrial settings (Rogerson et al., 2010; Saunders et al., 2014).

As hinted above, the key process in regulating whether EPS promotes or impedes mineralisation is the immobilisation of calcium ions by anionic functional groups on EPS molecules (Decho, 2010). Understanding this process is still in its infancy, and highly dependent on nano-scale variations in EPS composition, which regulates the behaviour of the electrostatic properties of "microdomains". Promotion of precipitation is most likely to be enhanced where divalent cation binding is unidentate (i.e. only one electron volt of charge in the cation is offset by the ligands binding it to the organic molecule). Unidentate binding will considerably reduce the activation energy barrier which usually limits spontaneous precipitation (Dittrich and Sibler, 2010), making a significant contribution to promoting mineralisation. "Biologically influenced" precipitation arising from biopolymer templating is expected to deviate from "normal" physicochemical precipitation in terms of location, rate, morphology and chemistry. Before inference of a past environmental change can be inferred, it is therefore critical to determine whether "bio-influence" has been significant, or not. However, these processes remain understood only at descriptive levels and a means of distinguishing bio-mediated, bio-influenced and non-biological precipitation remains a major target for the research community (Decho, 2010).

1.2.3 Extracellular enzyme activity

As of the date of submission, we are not aware of any study specifically investigating the behaviour of extracellular enzyme activity in tufa systems. However, their role in regulating Ca^{2+} dynamics in karst soils (Li et al., 2007, 2005) and associated with cyanobacterial activity in soda lakes (Kupriyanova et al., 2003) and marine stromatolites (Kupriyanova et al., 2007) is well established and they are anticipated to be material to the mechanisms of "bio-influence" outlined in the preceding section (Decho, 2010). It may be that synthesis of these enzymes are partly responsible for the ability of biofilms to regulate their internal pH (Bissett et al., 2008b), and they are therefore material to understanding why the lower limit of precipitation of calcite (in terms of saturation) is so difficult to define (see Sect. 1.1.1).

1.2.4 Biofilms as bioreactors

It should be kept in mind that even when ambient water is fully oxygenated, the deeper parts of biofilms can routinely provide a substrate in which bacteria with anaerobic metabolism may flourish, often inhabiting niches within the sediment itself (Shiraishi et al., 2008c; Visscher and Stolz, 2005). Even when oxygen is available, local water chemistry may permit a wide range of metabolisms to be altering carbonate chemistry at a specific site (Fouke et al., 2003). Complex biofilm ecologies combine to create an "alkalinity

Table 1. Summary of microbial metabolic biogeochemistries, with full bicarbonate buffering. Adapted from Visscher and Stolz (2005).

Process	Non-calcite reagents	Non-calcite products	$CaCO_3$(s) per mole carbon reacted	Microbial groups
Oxygenic photoautotrophy	$2HCO_3^-{}_{(aq)} + Ca^{2+}{}_{(aq)}$	$CH_2O_{(o)} + O_{2(aq)}$	1	Cyanobacteria
Anoxygenic photoautotrophy	$3HCO_3^-{}_{(aq)} + Ca^{2+}{}_{(aq)} + HS^-{}_{(aq)}$	$2CH_2O_{(o)} + SO_4^{2-}{}_{(aq)}$	0.5	Purple- and green-sulfur bacteria
Aerobic respiration	$CH_2O_{(o)} + O_{2(aq)}$	$2HCO_3^-{}_{(aq)} + Ca^{2+}_{(aq)}$	-1	Heterotrophic bacteria
Fermentation	$3CH_2O_{(o)} + H_2O_{(l)}$	$2HCO_3^-{}_{(aq)} + Ca^{2+}{}_{(aq)} + C_2H_6O_{(aq)}$	-0.2	Many organisms
Dissimilatory iron reduction	$CH_2O_{(o)} + 4FeO_2H_{(aq)} + 7Ca^{2+}_{(aq)} + 6HCO_3^-{}_{(aq)}$	$4Fe^{2+}_{(aq)} + 6H_2O_{(l)}$	7	Iron-reducing bacteria
Dissimilatory nitrate reduction	$5CH_2O_{(o)} + 4NO_3^-{}_{(aq)}$	$6HCO_3^-{}_{(aq)} + 2N_{2(g)} + 2H_2O_{(l)} + Ca^{2+}_{(aq)}$	-0.2	Nitrate-reducing bacteria
Sulfate reduction	$2CH_2O_{(o)} + SO_4^{2-}{}_{(aq)} + Ca^{2+}_{(aq)} + OH^-_{(aq)}$	$CO_{2(g)} + 2H_2O_{(l)} + HS^-_{(aq)}$	0.5	Sulfate reducing bacteria[1]
Lithoautotrophic sulfate reduction	$8H_{2(aq)} + SO_4^{2-}{}_{(aq)} + 4HCO_3^-{}_{(aq)} + 2Ca^{2+}$	$HS^-_{(aq)} + CH_3CO\text{-}SCoA + 3H_2O_{(l)}$	2 moles per mole of sulfur reduced	Lithoautotrophic sulfate-reducing bacteria
Aerobic sulfide oxidation	$HS^-_{(aq)} + O_{2(aq)}$	$SO_4^{2-}{}_{(aq)} + Ca^{2+}{}_{(aq)} + HCO_3^-{}_{(aq)}$	-0.5 moles per mole of sulfur oxidised	Sulfide-oxidising chemolithoautotrophic bacteria
Anaerobic sulfide oxidation	$5HS^-_{(aq)} + 8O_3^-{}_{(aq)} + 3HCO_3^-{}_{(aq)} + 3Ca^{2+}_{(aq)}$	$4N_{2(g)} + 4H_2O_{(l)} + 5SO_4^{2-}{}_{(aq)}$	0.6 moles per mole of sulfur oxidised	Sulfide-oxidising chemolithoautotrophic bacteria
Ammonium oxidation	$NH_4^+(aq) + 1.5O2(aq) +$	$NO_2^-{}_{(aq)} + 2HCO_3^-{}_{(aq)} + 2Ca^{2+}_{(aq)}$	-2^2	Ammonium-oxidising chemolithoautotrophic bacteria

[1] Can be up to 1.5 moles $CaCO_3$ per mole of carbon oxidised with certain "complete oxidation" metabolisms. [2] Balanced by precipitation of 1 mole per mole of carbon fixed from $CO_{2(aq)}$ via the Calvin cycle. NB: other metabolic mechanisms such as two-step sulfide oxidation could be relevant, but only via variable microdomain ecology.

engine", the regulation, operation and impact of which are only partially understood (Dupraz et al., 2009). In solutions buffered by bicarbonate, anaerobic metabolisms will either promote or impede precipitation, depending on the specific stoichiometry of the electron donation–absorption system (Visscher and Stolz, 2005), as summarised in Table 1. Given the range of chemistries possible in alkaline spring systems, any specific microbial metabolism may be occurring at some site within the biofilm, meaning that many processes are simultaneously driving changes in the alkalinity engine. Although few detailed studies of them have been undertaken, "bioreactor" processes are probably highly site specific. For example, anaerobic sulfide oxidation by *Aquificales* spp. is clearly of fundamental importance at Angel Terrace, Yellowstone, USA; (Fouke et al., 2003), but is unlikely to be significant in karstic settings where reduced sulfur in solution is not abundant. In karstic settings, the most likely anaerobic metabolism found is sulfate reduction, as small amounts of sulfate are common in these solutions. Regardless, there is an apparent tendency for bioreactor processes to favour mineralisation of calcite. This deserves further investigation.

2   Methods and source of case study data

If we are to discriminate systems on the basis of whether biologically induced/influenced calcite precipitation is significant, and whether this precipitation is related to simple (exclusively microbial) or more complex ecologies (involving macrophytes) we first need to assess the chemical potential of the solution to promote physico-chemical precipitation of calcite.

In this study, we recommend using a thermodynamic representation of the calcite precipitation system rather than the more traditional saturation index. This is primarily due to the temperature range inherent in our data set, which is $> 90\,°C$. Temperature is a first-order control on chemical reactions, and the key advantage of our recommended parameter, Gibbs free energy, is that it fully incorporates this influence. The Gibbs free energy of an ion pair in solution (e.g. $Ca^{2+}_{(aq)}$ and $CO_3^{2-}{}_{(aq)}$) relative to its equilibrium position can be straightforwardly estimated from the ion activity product of the solution (Langmuir, 1997) with respect to a reaction. Throughout this study, the reaction of interest is the formation of solid calcite from solution ($Ca^{2+}_{(aq)} + CO_3^{2-}{}_{(aq)}$ $CaCO_{3(aq)}$).

The calculation is performed via

$$\Delta G_r = RT\ln\left(\gamma_1\left[Ca_{(aq)}^{2+}\right]\gamma_2\left[CO_{3(aq)}^{2-}\right]/K_{sp}\right),\qquad(1)$$

where ΔG_r ($kJ\,mol^{-1}$) is the Gibbs free energy available in the solution relative to equilibrium, and therefore of the chemical potential that a spontaneous change in the system will occur. R is the ideal gas constant ($8.314472\times10^{-3}\,kJ\,K^{-1}\,mol^{-1}$), T is the temperature (in Kelvin), γ_1 and γ_2 are the activity coefficients for $Ca_{(aq)}^{2+}$ and $CO_{3(aq)}^{2-}$ respectively, squared brackets indicate concentrations (in moles). K_{SP} is the saturation product, given by Plummer and Busenberg (1982):

$$K_{sp} = 10^{\left\{-171.9065-(0.077993T)+\left(\frac{2839.31}{T}\right)+(71.595(\log_{10}T))\right\}}.\qquad(2)$$

Calculation of ΔG_r could be achieved using a chemical model, such as PHREEQC. However, the use of different models, or indeed different generations of the same model, may cause differences in derived parameters such as ΔG_r. Moreover, different published works present different quantities of data regarding minor ions, making an approach based on model-based convergence of the solution charge balance partially dependent on the nature of the source. To circumvent these potential issues, here we estimate ΔG_r values analytically. A key advantage of using this approach is that it makes the internal calculations underlying our approach explicit to any reader.

Activity coefficients γ_1 and γ_2 are calculated according to extensions of the empirical Davies model for ion activity (Langmuir, 1997):

$$\log_{10}\gamma = -Az_i^2\left(\frac{\sqrt{I}}{1+\sqrt{I}}-0.3I\right),\qquad(3)$$

where I is the ionic strength of the solution, A is a Debye–Hückel parameter and z_i is ion charge.

$$A = 1.824928\times10^6\rho_0^{1/2}(\in T)^{-3/2},\qquad(4)$$

where ρ_0 is the density of water and $\in$ is the dielectric constant of water. Although it is dependent on temperature and salinity, here we use a constant value for ρ_0 for simplicity ($0.9997\,kg\,dm^{-3}$). This is due to the uncertainty of calculating the density effects of the various potential solutions represented in our whole data set. This simplification is unlikely to cause any significant error in our analysis. The dielectric constant is calculated from Langmuir (1997):

$$\in = 2727.586 + 0.6224107T - 466.9151\ln T - 52000.87/T.\qquad(5)$$

An additional complication of using existing field data is that the typical pH range for terrestrial carbonate producing systems is from 8 to 9, meaning that most dissolved inorganic carbon (DIC) is present as bicarbonate. This means in most cases $[CO_3]_{(aq)}^{2-}$ must be estimated from the Henderson-Hasselbach-like control on the equilibrium constant of the

second ionisation of DIC (Ka_2) exerted by the pH, which varies with temperature and ionic strength (Patterson et al., 1984). An analytical means to determine these effects are available derived from the Debye–Hückel limiting law (see Appendix 1 for derivation):

$$[CO_3^{2-}]_{(aq)} = Ka_2[HCO_3^-]_{(aq)}10^{3A\sqrt{I}+pH}.\qquad(6)$$

Ultimately, we find that using this analytical approach, ΔG_r, can be calculated from field-derived measurements of temperature, pH, $[Ca_{(aq)}^{2+}]$, $[HCO_{3(aq)}^-]$ and ionic strength (i.e. the sum of all charge arising from dissolved ions),which is a data set that can be reasonably created for any new system. For systems with pH > 10, we use field-derived values for $[CO_{3(aq)}^{2-}]$ instead as they can be accurately measured in these systems.

A potentially critically important, first-order parameter controlling the biological components the presence of which will affect the water–precipitate system is temperature. At high temperature (typically over $40\,°C$), macrophytes are excluded leaving only microbial processes to interact with mineral precipitation. At temperatures above ~ $60\,°C$, biofilms will be dominated by thermophiles, diversity falls and the potential for biomediated precipitation will thus be altered.

We therefore propose that – to first-order – sites can be classified in a binary sense according to their ΔG_r and water temperature, and that systems can be grouped according to whether they are high or low ΔG_r versus high or low temperature. For the purposes of the classification proposed here we use temperature in degrees Celsius rather than degrees kelvin, which is used in forthcoming calculations. This is for the convenience of field scientists, who tend to measure using the former unit rather than the latter.

2.1 Compilation of field data

We do not pretend that the compilation of field data presented here is in any way comprehensive, and simultaneously recognise the advantage to future classification schemes if it were more comprehensive. We therefore link this paper to an online resource to which further data can be added, calculations are performed using a single set of equations and calculated parameters added to our database (https://hydra.hull.ac.uk/resources/hull:7520). The field data used in the subsequent analyses are summarised in Table 2, which shows the wide range of compositions of near neutral–alkaline spring waters occurring in nature. To simplify these disparate data, a number of sub-populations are classified in terms of region of occurrence and water temperature (Table 2). Variance within these sub-populations remains quite high, with calcium concentration in cold Turkish springs varying by nearly 300 % of the mean (Table 2). Ionic strength, which is only partially dependant on the position of a location relative to source, is the most constant; ionic strength in the water at Plitvice (Croatia) varies only by about 10 % of the mean despite this being a

Table 2. Summary of field data (SD, standard deviation).

Type of System	T (°C)		Ca (mMol L^{-1})		CO$_3$ (mMol L^{-1})		ΔG_r (kJ mol^{-1})		n
	Mean	SD	Mean	SD	Mean	SD	Mean	SD	
Alkaline lake (Mono)	8.5	2.8	0.05	0.02	218.3	233.3	16.37	6.71	29
Japanese spring	13.6	3.6	1.6	0.8	0.01	0.009	0.25	5.81	13
Semi-arid temperate tufa river (Ruidera)	13.0	9.9	1.6	0	0.005	0.002	1.53	0.26	2
Cool-humid temperate tufa river (Ddol)	10.1	1.6	2.5	0.3	0.01	0.009	4.73	0.48	3
Cool Italian spring (< 20 °C)	11.7	2.4	2.5	1.8	0.004	0.004	0.40	2.19	63
Warm Italian spring (20–30 °C)	24.3	2.9	10.1	3.4	0.006	0.005	2.48	1.61	22
Thermal Italian spring (30–45 °C)	36.5	4.4	14.5	5.8	0.008	0.01	3.52	2.14	21
Hot Italian spring (> 45 °C)	53.5	5.5	15.6	3.3	0.01	0.01	5.14	2.31	13
Cool Turkish spring (< 20 °C)	26.1	13.2	1.0	2.9	17.8	25.3	10.92	3.70	43
Warm Turkish spring (20–30 °C)	21.1	1.3	0.5	0.1	36.2	47.2	15.98	2.12	4
Thermal Turkish spring (> 30 °C)	42.0	3.0	2.1	5.8	15.5	12.4	17.55	8.59	6
Warm-humid temperature river (Plitvice)	12.0	3.5	1.2	0.2	0.02	0.01	2.98	1.75	65
Hot silica spring	88.3	20.8	0.6	1.2	0.9	1.4	6.19	3.04	4

relatively large data set ($n = 65$). Even with the simplification provided by regional grouping, the data remain markedly unstructured in terms of the relationships between key parameters such as calcium and carbonate concentrations (Fig. 1a) or temperature and ionic strength (1b), all of which vary independently.

Comparison of water temperature and ΔG_r (Fig. 2) does not result in more structure within the data, but it does result in better clustering of the data than individual ion concentrations, as sites can differ between very high carbonate ion but low calcium ion concentration (e.g. Mono Lake) and vice versa (many Italian springs). The thermodynamic approach therefore provides a suitable simplification of the data for further analysis.

3 Towards linking process to product

The first-order analysis of sites based on ΔG_r versus temperature proposed above is shown in Fig. 3 (Fig. 3). There is a natural clustering of data in the region of the origin, reflecting the relatively high abundance of data from "meteogene" sources within karstic regions where deep crustal fluids are reaching the surface. Nevertheless, our collection of field data does encompass the majority of the T–ΔG_r pace reasonable to expect from natural systems. We first report a case study from each quartile of the diagram, then highlight some key features of the transition zones between (e.g. between "hot" and "cold" systems) and on the edges (e.g. at the lower limit of significant carbonate production) of these quartiles.

We emphasise that we provide only basic descriptions of the sedimentary product for each class. Further reading is recommended for a case study in each case.

3.1 "Hot, non-biomediated" systems with high temperature and ΔG_r

We recommend that these systems are informally designated as "super travertine", and a full description of a representative system (Chemerkeu, Kenya) can be found in (Renaut and Jones, 1997). Water temperature at Chemerkeu is around 99 °C, and Gibbs free energy is around 8.76 kJ mol^{-1}. A further example system of this type is Terme San Giovanni in Italy, where water temperature is 38 °C and Gibbs free energy is 10.81 kJ mol^{-1}.

In these systems, Gibbs free energy is high, providing significant chemical potential for mineralisation. Biota are restricted to thermophile microbes by high water temperature. Precipitation will be rapid and without material control from biological activity.

In high temperature thermal sites, calcium carbonate saturation levels are too extreme for extensive macrophytes or thermophile microbial biofilm development (Fig. 4). Such extreme hydrothermal waters equilibrate with surface conditions rapidly immediately upon resurgence, by cooling (increasing DIC solubility and altering chemical behaviour), ingassing (raising DIC), degassing (raising pH) mixing with ambient waters (reducing ionic strength) or precipitating other minerals (reducing ionic strength). These processes can trigger very rapid precipitation of calcite which nucleates onto the substrate, which provides surfaces with reduced activation energy further promoting calcite precipitation (Fig. 4a, d, e). Such spontaneous crystalline precipitates are typically made up of palisade stands of sparite crystals with their C-axes oriented normal to substrate and with individuals from 1–60 mm long (Fig. 4c). These fabrics develop during rapid cooling and degassing associated with highly alkaline, flowing waters which are confined within small channels. The resulting botryoidal crystalline masses are composed of fanning crystal clusters, each crystal with it's own well formed

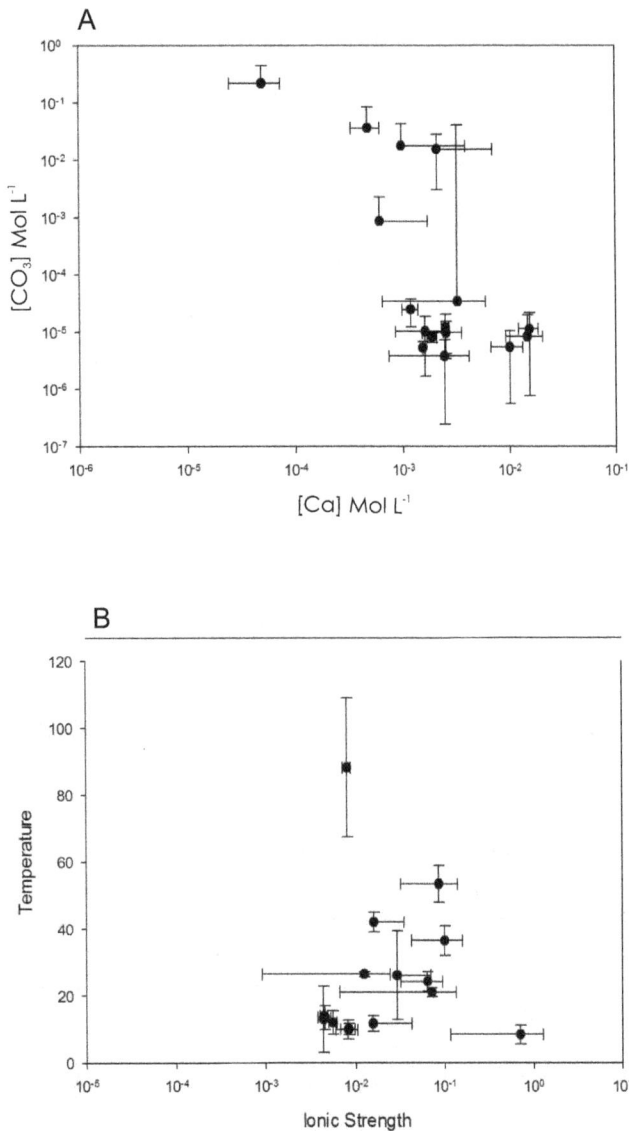

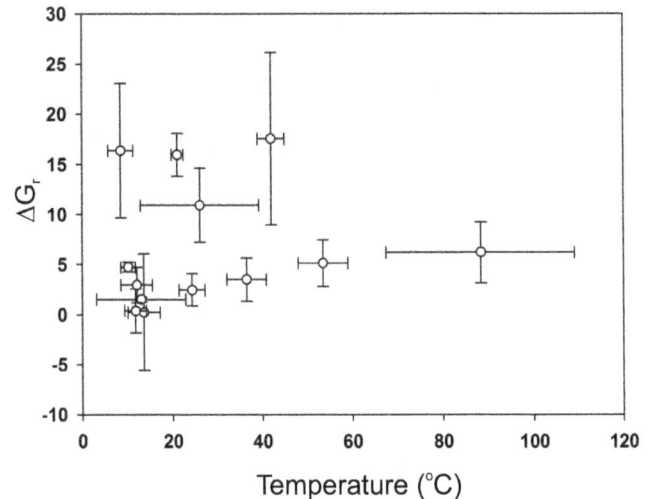

Figure 2. Summary of spontaneous nucleation index (dimensionless) and temperature (°C) from Table 2 (note logarithmic scales).

3.2 "Hot, biomediated" systems with high temperature and low ΔG_r

We recommend that these systems are informally described as "travertines", and the chemical, sedimentological and microbiological features of these systems are best described in Fouke et al., 2000 and 2003, and Veysey et al., 2008. At Angel Terrace, water temperature is around 73.3 °C and Gibbs free energy is around $6.84 \, kJ \, mol^{-1}$.

This is the largest category of precipitates associated with hydrothermal processes. In these systems, Gibbs free energy is too low to allow rapid spontaneous nucleation, so that rapid precipitation is dependent on some mechanism allowing the activation energy to be reduced. At Angel Terrace, microbiological activity provides this additional mechanism (Fouke et al., 2000). However, biota can be restricted by high water temperature meaning that despite low Gibbs free energy, slow precipitation as a result of physical/chemical processes may still dominate. The possibility of "bioreactor" processes (Sect. 1.2.4) sustaining high microbial activity may be critical in determining whether precipitation is fast and bio-influenced or slow and physico-chemical.

Deposits usually are confined to the immediate vicinity of the resurgence point and these may take the form of whaleback ridges (sublinear ridges often following fault lines), pinnacles and mounds. An apron of lower angle travertine sheets, with terracettes may also extend well beyond these constructions where flow rates are high (see details of Italian examples in Capezzuole and Gandin, 2005). Millimetre to centimetre thick laminae dominate all deposits and commonly are arranged into discrete bundles each separated by a truncation surface. These truncations may be karstified and buried by palaeosols indicating prolonged breaks in the depositional history. Individual laminae within bundles commonly show bubble, microlaminar and shrub fabrics (Fig. 5a,

Figure 1. Summary of field data shown in Table 2 (note logarithmic scales). (**A**) Mean and standard deviation of calcium and carbonate concentrations in molarity. (**B**) Mean and standard deviation of temperature (°C) and ionic strength (total electron volts dm⁻³).

crystal termination. These appear to be primary crystal fabrics, and are analogous to "columnar" fabrics described in speleothems growing in equilibrium with dripwater (Frisia et al., 2000). Clusters of calcite rosettes are also common. Crystal length appears to be controlled by water depth suggesting that standing pools best encourage this development. However, macroscopic rhombic calcite crystals also nucleate rapidly on the bottom of fast-flowing gutters to form complex fanning twins with overall botryoidal morphologies (Fig. 4b). When followed away from resurgence points these fabrics progressively give way to microbial dominated travertine and ultimately to ambient temperature paludal tufa laminites with dense macrophyte associations (Fig. 4d).

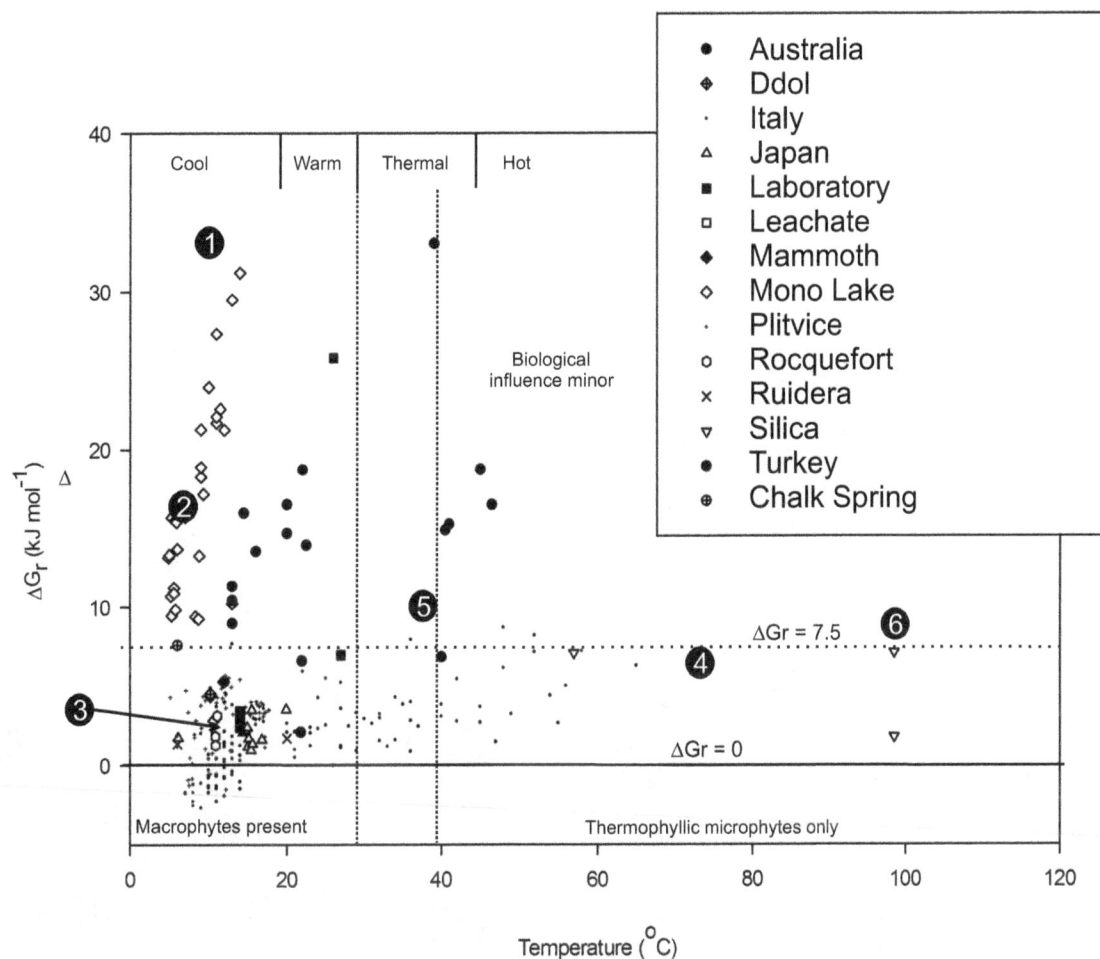

Figure 3. Summary of travertine and tufa spring water chemistries on thermodynamic grounds (see Sect. 3). 1: ,Brook Bottom (Sect. 3.3), 2: Mono Lake (Sect. 3.3), 3: Plitvice (Sect. 3.4), 4: Angel Terrace (Sect. 3.2), 5: Terme San Giovanni (Sect. 3.1), 6: Chemerkeu Spring (Sect. 3.1).

b, c). All macrophytes and many algae are excluded. Typically the deposits are extremely well cemented; the terracette laminae may show narrow rimstone pools and microgour topographies, whereas the faster flowing waters on steeper slopes (Fig. 5d) may create ripple developments.

3.3 "Cold, non-biomediated" systems with low temperature and high ΔG_r

We recommend that these systems are informally described as "super tufa". Naturally occurring examples include Mono Lake (USA), and more complete information about facies distributions and stratigraphy can be found in Newton, 1994. At Mono Lake, water temperature is around 8.5 °C and Gibbs free energy around 16.37 kJ mol^{-1}. Anthropogenic examples, arising from hyperalkaline leachates from various industrial process by-products, are also frequent in the developed world. An example of this type of system is Brook Bottom

in Derbyshire (UK), where water temperature is 10 °C and Gibbs free energy is in the region of 33 kJ mol^{-1}.

In these systems, the chemical potential represented by the high Gibbs free energy is sufficient for precipitation to happen readily. Biota may be abundant and diverse, but extreme composition of the water may also be limiting via toxicity. Regardless, abiotic mineral precipitation is dominant and sufficiently rapid for biological influence to be minor and therefore masked in resulting deposits. As well as natural systems, such as Mono Lake (USA), systems of this type occur abundantly where leachate from alkaline waste (from lime, steel, chromium, alumina and other industries) in landfill reaches the surface. These anthropogenic "super tufa" systems are potentially significant sinks for carbon, in addition to being important environmental hazards due to their extreme chemical composition (Mayes et al., 2008).

Extremely high alkalinity in these environments leads to continuous, pelagic lime mud precipitation which often results in a milky to pale turquoise colour to the water even

Figure 4. High temperature and ΔG_r. (**A**) Travertine build-up at the outflow point from the spar pool. Steep gradient microterracettes are typical of the deposit. Scale bar is 1 m long. Terme San Giovanni, Tuscany, Italy. (**B**) Crystalline calcium carbonate in botryoidal masses lining an outflow channel. Scale bar is 100 mm long. Terme San Giovanni, Tuscany, Italy. (**C**) Palisade calcite growing normal to substrate within a Holocene travertine site. The sequence is capped by a palaeosol. Scale bar is 300 mm long. Terme San Giovanni, Tuscany, Italy.

during cooler periods (Fig. 6). Aquatic organisms are generally excluded from such waters due to the extremely high precipitation rates which choke gill-breathing organisms and deeply encrust vegetation in precipitates after only a few hours.

The water itself may not be toxic, however, and extensive marginal vegetation may develop throughout the system (Pedley, 2000). Biofilms are thin and discontinuous and mainly located on the narrow spill-over points of rimstone pools in the paludal areas and in broader botryoidal areas of barrage growth across the outflowing watercourses (Fig. 6a). Precipitates are rapid with extensive development of rimstone pools each forming broad terracettes separated by narrow sinuous crested microdams. Pool depths in the paludal areas are generally a few centimetres deep and are infilled with lime muds (Fig. 6b, c). These muds are generated by precipitation within the pool areas and may be massive or finely laminated. Lithification generally is absent in the pool deposits and only weak within the rimstone dam areas. At

Figure 5. High temperature, low ΔG_r. (**A**) Laminated travertine with alternating bubble and laminate layering. Scale bar is 50 mm long. Holocene, Cava Oliviera, Tuscany, Italy. (**B**) Laminite-filled pools bounded by bacterial microherm dominated pool rims. Scale bar is 150 mm long. Early Quaternary Alcamo travertine, Sicily, Italy. (**C**) Typical laminated microbial travertine fabric. Scale bar is 150 mm long. Holocene Cava Oliviera, Tuscany, Italy. (**D**) Typical rim pool system from Saturnia, Tuscany, Italy. Note swimmers for scale, and steam rising from warm water (with thanks to Enrico Cappezuoli).

faster flowing sites, small transverse barrages of weakly lithified lime mud laminites develop into broad, vertical to overhanging botryoidal walls which impound small, deep pools floored with unlithified lime muds. These muds are typi-

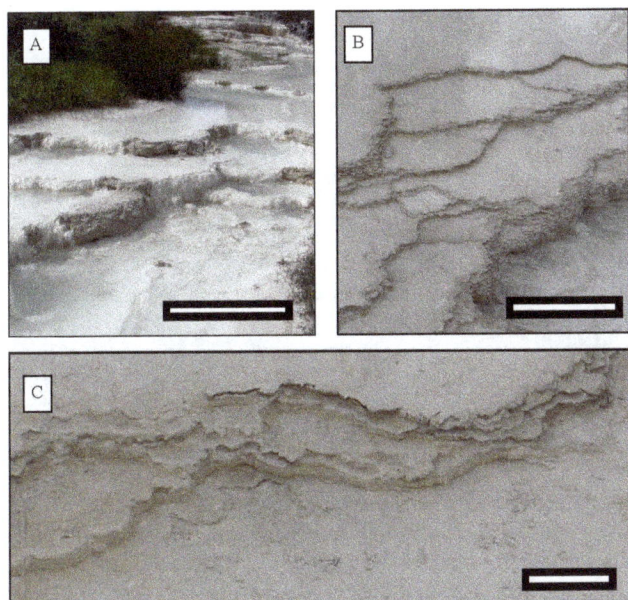

Figure 6. Low temperature, high ΔG_r. (**A**) Terrace and pool morphology in an active ambient temperature deposit. Note the large volume of lime mud within the pools. Scale bar is 1 m long. Brook Bottom, near Buxton, Derbyshire, UK. (**B**) Detail of the terracettes developed in gentle gradient sites. Note how the large upstream pools progressively overstep the smaller downstream features and progressively encourage the development of steps. Scale bar is 0.5 m long. Active, Brook Bottom, near Buxton, Derbyshire, UK. (**C**) Fine detail of the narrow pool rims and spillovers. Scale bar is 100 mm long. Brook Bottom, near Buxton, Derbyshire, UK.

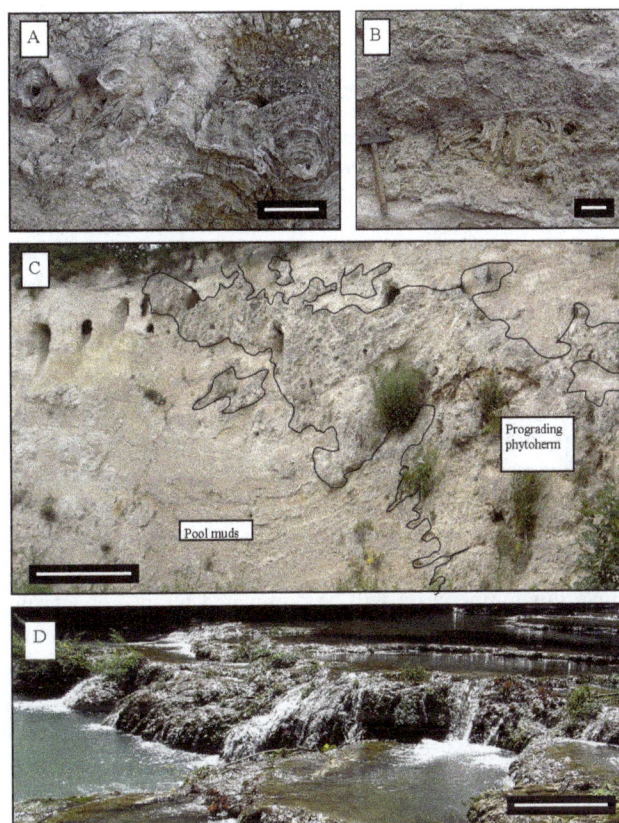

Figure 7. Low temperature, low ΔG_r. (**A**) Stromatolite microherm development in a fluvial tufa. Scale bar is 50 mm long. Early Quaternary Noto tufa, Sicily, Italy. (**B**) Burr reed Phytoherm cushion interlayered with detrital tufas above and lacustrine lime muds below. Scale bar is 150 mm long. Holocene Caerwys tufa, North Wales. (**C**) Profile view of a prograding phytoherm framestone sandwiched between adjacent lime mud pool deposits. Scale bar is 150 mm long. Holocene Caerwys tufa, North Wales. (**D**) An active pool and barrage system forming in Tuscany (cf. 2C). Scale bar is 1 m long. Colle val d'Elsa, Italy.

fied by bizarre, microcrystalline calcite growthforms with abundant imperfectly formed crystals, some with baroque faces, by hollow microspheres and by truncated skeletal crystal growthforms. Tufa towers may grow under similar conditions within alkaline lakes (e.g. Mono Lake, USA). These vertical structures also exhibit rapid growth fabrics with occasional microbial laminations, but are essentially inorganic carbonate precipitates.

3.4 "Cold, biomediated" systems with low temperature and ΔG_r

We recommend that these systems are informally described as "tufa", and the most widely used facies model for cold, biomediated terrestrial carbonate systems is Pedley (1990). An excellent example of this type of system is Plitvice (Croatia) where water temperature is on average 12.04 °C, and Gibbs free energy is 2.98 kJ mol^{-1}.

These are the classic, karstic tufa systems and are the most widespread form of terrestrial carbonate from spring, fluvial, lacustrine and paludal settings (Ford and Pedley, 1996). The Gibbs free energy is too low to allow rapid spontaneous mineral formation, so that precipitation is dependent on some mechanisms allowing the activation energy to be reduced.

Biota are abundant and diverse, permitting the full range of biomediation mechanisms to be active.

These deposits (Fig. 7) are extensively developed in association with phytoherm framestones in fluviatile perched springline, paludal and lacustrine settings (Pedley, 1990). Biofilms are developed on all available surfaces and all actively precipitate thick calcium carbonate laminae, often composed of alternations of spar and micrite cements and commonly developed into stromatolitic growths (Fig. 7a). Micritic precipitates with a characteristic laminar or peloidal texture may also dominate the system (Fig. 7b, c). A diverse range of invertebrates, vertebrates and vegetation are always associated with these sites. Precipitates rapidly lithify and progressively entomb earlier biomediated areas. Typically, they agrade and prograde into lithofacies-scale deposits (Fig. 7c) which bury macrophytes and are capable of infilling valleys. These impound bodies of water from pond to lake scale (Fig. 7d) within which further pelagic precipitation occurs and regularly laminated lake floor lime muds accumulate which may be associated with sapropels.

3.5 Interfacial systems with temperature in the region of 30–40 °C.

These deposits show overlapping similarities with travertine and tufa systems, and are likely to show interfingering of facies stratigraphically at individual sites. The most important impact of this in terms of product is that aquatic macrophytes are likely to be excluded by the high temperature, but may periodically establish when/where their physiological temperature threshold (ca. 30 °C) is reached. Microbiota may be abundant and diverse depending on local ecological factors, but the composition of biofilms will differ from systems developing at lower temperatures. Diatoms grow best at temperatures < 35 °C and photosynthetic elements of biofilms above this temperature will be dominated by cyanobacteria (Blanchard et al., 1996) meaning that the metabolic functioning of biofilms will alter through the 30–40 °C temperature range.

These deposits manifest in an abundance of thermophile and ambient temperature microbial colonisers associated with scattered, small cushion-shaped phytoherms (often single species of grass) throughout (Fig. 4a, b). There is also a marked tendency for laminite development, partially arising from the flow conditions of these sites which are typically found on the outermost gently inclined parts of travertine sheets and distal margins of thermal fissures (Fig. 4a). Here, the virtually ambient temperature of the outflowing former-hydrothermal waters is sufficiently low to permit scattered macrophyte colonisation. Figure 4b shows a profile view of the colonisers in the Cava Oliviera quarry. Note the non-preservation of roots and the thick accumulation of laminites around the vegetation. In some instances over 100 mm of laminate accumulation has occurred during the lifetime of individual plants, which attests to high nucleation rates in these cases.

3.6 Interfacial systems with moderate ΔG_r

In close analogy with temperature, systems which are marginal in terms of ΔG_r will present mixed petrologies and geomorphologies from the high and low ΔG_r type systems. Again in close analogy with high temperature systems, high ΔG_r systems will inevitably evolve into low ΔG_r systems in time and space. However, the ΔG_r range this transition will occur at is not well determined by our data, due to lack of specifically collected field petrographic evidence, beyond a general indication it will lie at $\Delta G_r \sim 7.5\,\mathrm{kJ\,mol}^{-1}$. This transition is an interesting target for future field research.

In terms of product, biomediated and physico-chemical type products will interfinger both in time and space in these interfacial systems, in the same manner that high and low

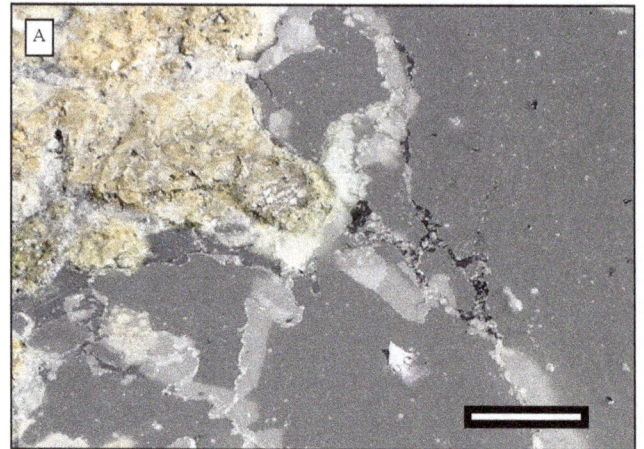

Figure 8. Marginally precipitating systems under high temperature. Living biofilm associated with superficial skins of calcite. Note the thin calcite rafts at the air–water interface. This is the only observable precipitate at the site. Scale bar is 100 mm long. Recent, Petreole, Tuscany.

temperature products interfinger in time and space in thermal interfacial systems. Again, there is a lack of field data concerning the diagnosis of these hydrochemical transition zones.

3.7 Defining the hydrochemical lower boundary of terrestrial carbonate precipitation

In these marginal systems where there is insufficient Gibbs free energy for calcite to nucleate spontaneously, precipitation will largely if not exclusively be biomediated. Consequently, at high temperatures where waters are too hot for macrophyte colonisation (although they may be extensively colonised by thermophilic microbial mats) and thus biomediation is limited, significant precipitates and petrographic data are sparse. Any carbonate precipitates will be thin and slow to develop. Typically, thin calcite rafts floating on the surface of resurgence pools are the only tangible precipitates (Fig. 8). Progressively as the precipitate slowly develops these rafts become too heavy to be held on the meniscus and they sink to produce chaotic pseudo-floe calcite sheet debris.

At low temperature, an extensive development of biofilm will coat available substrates (Fig. 9a, b). However, precipitates will tend to be restricted to thin, micritic laminae within the EPS discontinuous on the sub-metre scale. Generally, in cooler humid climatic situations these developments are localised to resurgence points, the resurgences sometimes being identified by a very restricted patch of thin tufas in its immediate vicinity (Fig. 9c). Commonly the carbonate precipitates are not visible in the field, but can be viewed by binocular microscope as whitish films or particulate micrite on and within the biofilm surface. A diverse range of biota is associated with these sites. A common type of development in the

Figure 9. Marginally precipitating systems under low temperature. (**A**) Sub-tufa oncoids in a flowing, shallow stream without any framework tufa associations. These typically build climbing ripple bedforms. Scale bar is 150 mm long. Welton Beck, Yorkshire, UK. (**B**) Close up of a random sample of living oncoids showing characteristic greyish green colour. Note the virtual absence of any detrital tufa or cements between the oncoids. Scale bar is 15 mm long. Welton Beck, Yorkshire, UK. (**C**) Oncoid in cross section from River Bradford, Derbyshire, UK. Note colonised surface (left and top) is not continuous, and laminae are apparent (e.g. light coloured middle section) but these are discontinuous on the scale of a few millimetres. Dark core material (centre) is a flake-shaped piece of carboniferous limestone. Scale bar is 50 mm long.

Yorkshire dales and in Belgium is associated with "Cron"-type sites (Pentecost, 2005) where discontinuous biofilms develop in the capillary zone on the surface of grass or moss cushions. The micrite precipitates may lithify into thin delicate sheets or may constantly wash or fall from these vertical surfaces and develop into peripheral, detrital lime mud deposits. In flowing waters there is a common tendency for marginal systems to be associated with thin calcite films precipitated in association with epiphyllous algae on the under surface of river weed such as *Potamogeton*.

Thinly encrusted oncoids in which a few thin skins of carbonate are wrapped around suitable sized clastic nuclei are another common association of sub-tufa environments (Fig. 9a, b). Continental carbonate oncoids are laminated, microbially precipitated, millimetre to centimetre diameter spheroids which grow under turbulent conditions at the sediment–water interface and have laminae constructed of sparitic, thrombolitic or micritic carbonate, or an alternation of more than one of these. Although oncoids were originally described from the marine realm (Peryt, 1983; Riding, 1991) they are also common in ambient (Pedley, 1990, 2009) and thermal freshwater (Chafetz and Guidry, 2003; Jones et al., 1998) situations and can routinely be found in fossil terrestrial carbonate deposits (Fig. 9b).

However, we do not find that it is possible to draw a horizontal line in Fig. 3 dividing sites precipitating significant and insignificant carbonate. This echos a previous assessment of surface water in the Yorkshire Dales (UK), where no spatial relationship between water chemistry and precipitate occurrence could be identified (Pentecost, 1992) and other efforts in which precipitation was simultaneously present and absent in hydrochemically similar sites (Hagele et al., 2006; Ledger et al., 2008). Thus, we do not find the failure of thermodynamics to determine this boundary a surprising result, as ΔG_r decreases and biomediation becomes more important to carbonate creation, small-scale variations in ΔG_r will inevitably become immaterial. Clearly another direction is needed to determine this boundary.

4   Towards a new paradigm for defining the lower limit on carbonate precipitation at low ΔG_r

As biofilms play a demonstrable role in promoting calcite precipitation (Bissett et al., 2008a; Shiraishi et al., 2008a, b; Pedley et al., 2009; Rogerson et al., 2008) and typically demonstrate a tendency to alter the macroenvironment rather than be altered by it (Rogerson et al., 2010; Bissett et al., 2008b) it is likely that a new paradigm for the process determining the lower limit of terrestrial carbonate precipitation will arise from better understanding of microbial processes. This will require a fundamental reassessment of how fieldwork in these systems is performed, as even rather basic microbiological information is very seldom collected for terrestrial carbonate systems. Using a case study from Lincolnshire (UK) (Fig. 10), we propose a potential way forward.

4.1   Lincolnshire case study; demonstrating the coincidence of high microbial activity and precipitation of significant carbonate in "borderline" systems

In this case study, we emphasise the failure of thermodynamic assessments to determine the lower boundary of significant terrestrial carbonate production, and simultaneously attempt to link that boundary to a relatively simple assessment of on-site microbial activity. Although probably not significant players in biomediation of precipitate, local maxima in the suspended bacterial count will reflect locally enhanced productivity. This close association of higher biomass and

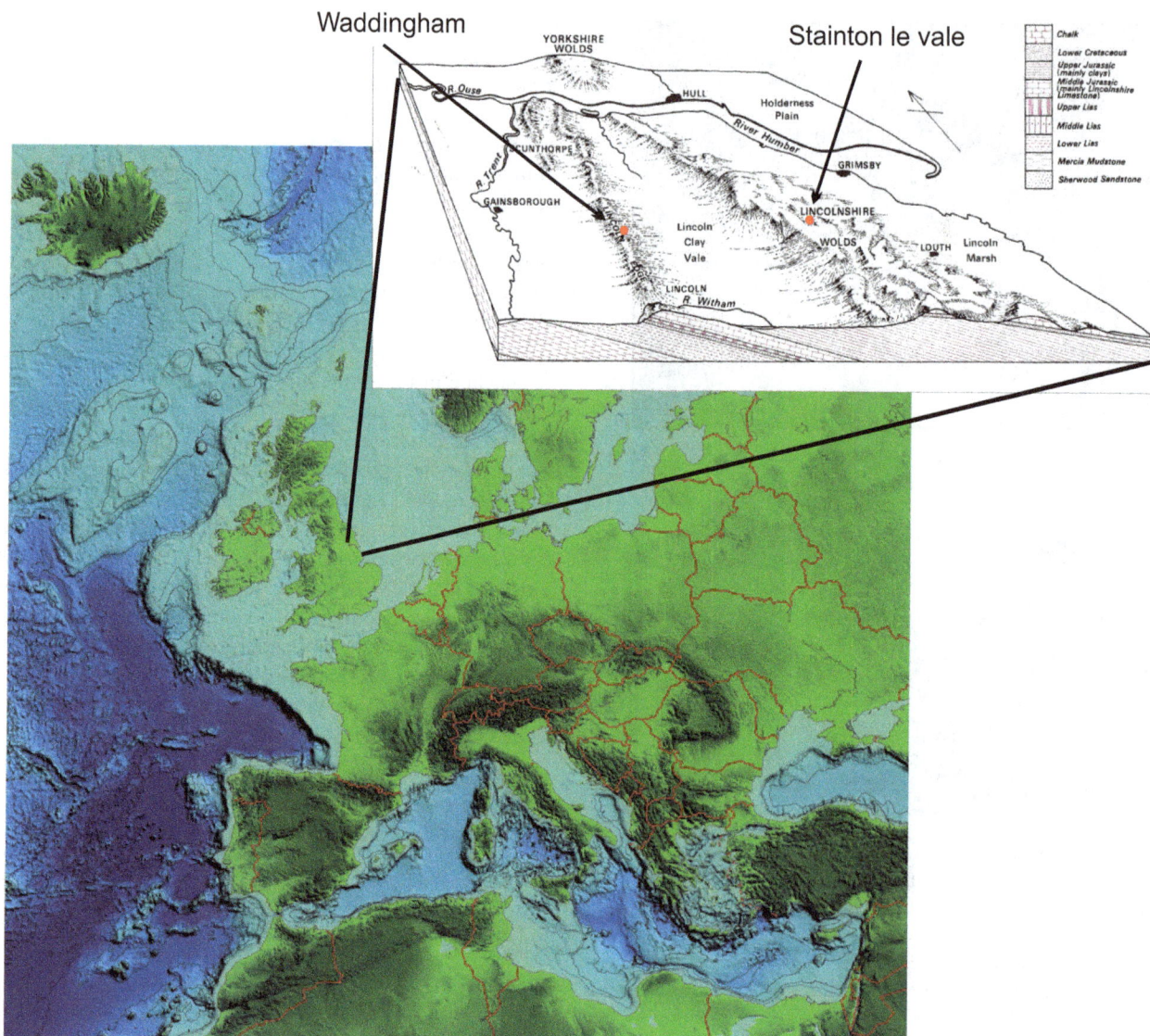

Figure 10. Location of tufa and sub-tufa sites in Lincolnshire, UK. Insert: geological map from Kent et al. (1980).

diversity at tufa sites relative to non-tufa sites is not our suggestion, but a known feature of these systems (Marks et al., 2006). To incorporate a simple microbiological measurement to the classification framework we propose here, we use suspended bacterial count as a proxy for the ecological energy flux (Marks et al., 2006) in the same way that water chemistry measurements traditionally act as a proxy for the gradient of thermodynamic forcing (see Sect. 2).

4.1.1 Background information on sites

The annual rainfall in Lincolnshire is 577 mm, and average summer and winter air temperatures are 18 and 5 °C respectively. The first location, Stainton le Vale (UK grid reference TF17622 BNG93804; Fig. 10), lies on the Cretaceous chalk Wold hills and aquifer water emerges from two springs po-

sitioned a few metres apart at the base of a quartzarenite bed that occurs within the chalk. Limited carbonate precipitation occurs along the stream banks and onto the biofilm-encrusted carbonate surfaces of a small (1.5 m high) tufa cascade (Fig. 11). Three sites have been studied from this valley: S1 is upstream near the springs where there is no geomorphologically significant carbonate forming, S2 is adjacent to the waterfall where crystalline patches of older carbonate outcrop at the water surface–stream bank interface but is barely producing a significant "tufa" product, S3 is downstream where the stream emerges from woods and flows through cultivated land, and no deposition is evident. The second location is at Waddingham, which lies on the eastern side of the Lincolnshire Edge, Jurassic limestone hills (UK grid Reference SK96457 BNG95311; Fig. 10). Again three sites were studied, one of which shows only oncoid development (W2)

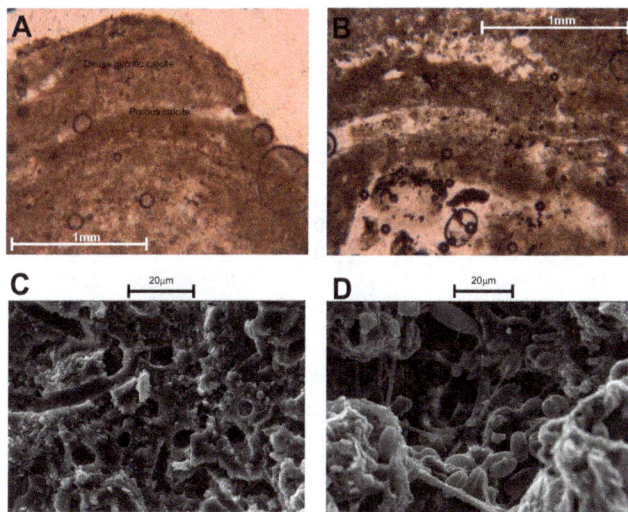

Figure 11. Photomicrographs of tufa deposit from Stainton, Lincolnshire, UK. (**A**) and (**B**) transmitted-light images showing alternative layers of micrite and porous calcite. (**C**) SEM image showing tubular holes in calcite left by loss of cyanobacterial filaments. (**D**) clotted calcite bodies with cyanobacterial filaments and diatoms in the centre of the view.

and two which are respectively immediately upstream and immediately downstream of the oncoid site (W1 and W3) and produce no significant precipitation.

The precipitate at site S2 is a friable, porous deposit adhering to the outer surface of siliciclastic stream bank muds. The precipitate is closely associated with organic matter, moss, fungi, heterotrophic bacteria and cyanobacteria from stream waters and surrounding vegetation. Microscopic grains cemented into the carbonate fabric enable small-scale aggradational seasonal laminae to be seen (Fig. 11). The sheet-like morphology of both precipitate and biofilm laminae consist of alternative clotted micrite and spar calcite. The degree of lithification, scarcity of detrital grains and association with organic components together identify this deposit as a tufa according to the terminology first proposed by Pedley (1994).

4.1.2 What makes the tufa site "Stainton 2" (S2) different from the other non-product producing sites (S1, S3 and W1–3)?

Counter-intuitively, calcium (mean $= 130$ mg L^{-1}; Table 3) and bicarbonate (mean $= 168.5$ mg L^{-1}; Table 3) concentrations are all higher at the Waddingham oncoid site than at Stainton (78.5 mg L^{-1} and up to 93.5 mg L^{-1}; Table 3), possibly reflecting the source water being derived from friable limestone rather than the less solution-prone, well-lithified chalk. Both systems have pH in the range 7.5–8.3, and magnesium concentrations are of the same order of magnitude (Table 3) . It is therefore completely unsustainable to assume the tufa–sub-tufa boundary is controlled by the thermody-

namic forcing of source waters; the simple water chemistry approach we otherwise advocate in this paper would indicate that Waddingham was more likely to be above the tufa to sub-tufa boundary than is Stainton. There must, therefore, be another mechanism dominating these systems which is not reflected within the water analyses.

The difference between Stainton and Waddingham must therefore arise from microbiological effects. Suspended bacterial abundance (Table 3; established via a simple cell-counting procedure (Hobbie et al., 1977) where cells in 50 ml of formaldehyde-fixed water were stained with Acridine orange, drawn through a filter membrane with 0.22 μm pore size and counted under an epifluorescent microscope) strongly differs between the S2 site where tufa is forming (22 617 cells mL^{-1}) and both other sites at Stainton (3490 and 3134 cells mL^{-1} for S1 and S3, respectively) and all locations at Waddingham (1325, 1228 and 1131 cells mL^{-1} respectively for stations W1–W3).

Although we do not know if these suspended cells themselves are inducing precipitation, it is likely that they at least reflect a more diverse and abundant benthic ecology at the S2 site than at any of the other sites studied in this case study. The difference in microbiological activity at sites precipitating significant calcite and those not doing so has previously been reported (Hagele et al., 2006; Marks et al., 2006; Ledger et al., 2008), but requires significant, further field-based research before it is understood at the fundamental biogeochemical level. A fundamental reassessment of well-known systems is therefore required, and testing the concept that suspended cells can be used as a proxy for enhanced biologically induced (Lowenstam, 1981) precipitation may be a fruitful initial avenue to explore.

5 Conclusions

Binary analysis of ΔG_r and water temperature of terrestrial carbonate-producing systems appears to be a fruitful avenue to pursue with the goal of a single, unified process-oriented classification of these systems in mind. Both parameters are material to regulating whether biological influences are key in the precipitation process. As understanding past precipitation environments are critically dependent on establishing whether precipitation was via metalorganic intermediaries or direct from solution, determining which quartile of temperature–ΔG_r space a precipitate reflects will underpin all subsequent investigation of that material and its environmental significance.

We do not currently have the information available to determine the ΔG_r level at which non-biological precipitation begins to dominate, other than that this level appears to be in the region of 7.5. This is likely to be a fruitful avenue of research to pursue. Once this level has been determined, it will be possible to identify the dominant precipitation mechanics occurring in whichever specific system is being studied

Table 3. Mean field data for sites at Stainton and Waddingham, Lincolnshire, UK.

Location	pH	water temp. (°C)	air temp. (°C)	humidity %	flow rate m s^{-1}	Ca mg L^{-1}	Mg mg L^{-1}	HCO$_3^-$ mg L^{-1}	CO$_3^{2-}$ mg L^{-1}	NO$_3^-$ mg L^{-1}	PO$_4^{3-}$ mg L^{-1}	DOC mg L^{-1}	Cell count per mL
Waddingham 1	8.1	10.3	10.7	76.5	0.2	140.4	1.4	172.8	12.5	< 10	1.7	5.5	1325
Waddingham 2	8.1	10.3	11.9	73.5	0.3	130.0	1.8	168.5	17.7	< 10	0.7	5.6	1228
Waddingham 3	7.8	10.4	12.3	73.9	0.4	117.5	1.2	170.3	7.8	< 0	2.0	5.9	1131
Stainton 1	7.7	9.5	11.6	72.3	0.4	80.6	5.7	91.5	0	< 0	1.8	7.3	3490
Stainton 2	8.0	9.5	11.5	76.6	0.3	78.5	3.6	93.5	7.7	< 0	1.0	7.3	22 617
Stainton 3	8.1	9.5	11.7	76.6	0.3	73.0	3.3	93.5	10	< 0	1.9	6.8	3134

allowing much more sophisticated analysis of process. For fossil systems, this will also allow us to determine whether a specific product indicates conditions above or below this critical value.

We also do not have the information available to understand the lower limit of ΔG_r at which significant precipitation takes place. Indeed, we find it is unlikely any such division can be determined, but that precipitation dynamics at this level is regulated by biology rather than purely physical processes. We emphasise this problem with a new case study. We also propose a simple field-based means of further investigation of the role of biology in regulating this system, and show that at sites of precipitation suspended cell count is up to an order of magnitude higher than adjacent reaches of the same river. This is also likely to be a fruitful avenue for future research.

Acknowledgements. We sincerely thank the many field workers, without whose published data this contribution would have been impossible. We also thank John Adams for his help with suspended cell count measurements. Associate editor Edward Tipper is warmly thanked for his careful and tolerant editing of this submission, and the reviewers Alex Brasier, Adrian Immenhauser and Nick Tosca for their very significant contributions, which immeasurably improved our work.

Edited by: E. Tipper

References

Anadon, P., Utrilla, R., and Vazquez, A.: Mineralogy and Sr-Mg geochemistry of charophyte carbonates: a new tool for paleolimnological research, Earth Planet. Sci. Lett., 197, 205–214, 2002.

Andrews, J. E.: Palaeoclimatic records from stable isotopes in riverine tufas: Synthesis and review, Earth-Science Reviews, 75, 85–104, 2006.

Andrews, J. E., Gare, S. G., and Dennis, P. F.: Unusual isotopic phenomena in Welsh quarry water and carbonate crusts, Terr. Nova, 9, 67–70, 1997.

Bayari, C. S. and Kurttas, T.: Algae: An important agent in deposition of karstic travertines: Observations on natural-bridge Yerkopru travertines, Aladaglar, Eastern Taurids, Turkey, 5th International Symposium and Field Seminar on Karst Waters and Environmental Impacts, Antalya, Turkey, ISI:A1997BH23Q00034, 269–280, 1995.

Bissett, A., DeBeer, D., Schoon, R., Shiraishi, F., Reimer, A., and Arp, G.: Microbial mediation of tufa formation in karst-water creeks., Limnol. Oceanogr., 53, 1159–1168, 2008a.

Bissett, A., Reimer, A., de Beer, D., Shiraishi, F., and Arp, G.: Metabolic Microenvironmental Control by Photosynthetic Biofilms under Changing Macroenvironmental Temperature and pH Conditions, Appl. Environ. Microbiol., 74, 6306–6312, 2008b.

Blanchard, G. F., Guarini, J. M., Richard, P., Gros, P., and Mornet, F.: Quantifying the short-term temperature effect on light-saturated photosynthesis of intertidal microphytobenthos, Mar. Ecol.-Prog. Ser., 134, 309–313, 1996.

Brasier, A. T.: Archaean Soils, Lakes and Springs: Looking for Signs of Life, in: Evolution of Archean Crust and Early Life, edited by: Dilek, Y. and Furnes, H., Modern Approaches in Solid Earth Sciences, Springer Netherlands, the Netherlands, 367–384, 2014.

Brasier, A. T., Andrews, J. E., and Kendall, A. C.: Diagenesis or dire genesis? The origin of columnar spar in tufa stromatolites of central Greece and the role of chironomid larvae, Sedimentology, 58, 1283–1302, 2011.

Buccino, G., D'Argenio, B., Ferreri, V., Brancaccio, L., Ferreri, M., Panachi, C., and Stazione, D.: I travertini della basse valle del Tanagro (Campania), Studio geomorphologico, sedimentologico e geochimico, Boll. Soc. Geol. Ital., 97, 617–646, 1978.

Carrara, C.: I travertini di Canino (Viterbo, Italia Centrale): elementi di cronolitostratigrafia, di geochimica isotopica e loro significato ambientale e climatico, Quaternario, 7, 73–90, 1994.

Carrara, C., Ciufarella, L., and Paganin, G.: Inquadramento geomorfologico e climatico-ambientale dei travertini di Rapolano Terme (SI), Quaternario, 11, 3119–3329, 1998.

Carthew, K. D., Taylor, M. P., and Drysdale, R. N.: An environmental model of fluvial tufas in the monsoonal tropics, Barkly karst, northern Australia, Geomorphology, 73, 78–100, 2006.

Chafetz, H. S. and Folk, R. L.: Travertines - Depositional morphology and the bacterially controlled constituents, J. Sediment. Petrol., 54, 289–316, 1984.

Chafetz, H. S. and Guidry, S. A.: Deposition and diagenesis of Mammoth Hot Springs travertine, Yellowstone National Park, Wyoming, USA, Can. J. Earth Sci., 40, 1515–1529, 2003.

Chen, J. A., Zhang, D. D., Wang, S. J., Xiao, T. F., and Huang, R. G.: Factors controlling tufa deposition in natural waters at waterfall sites, Sediment. Geol., 166, 353–366, 2004.

Clark, I. D. and Fontes, J. C.: Paleoclimatic reconstruction in Northern Oman based on carbonates from hyperalklaine groundwaters, Quat. Res., 33, 320–336, 1990.

D'Argenio, B. and Ferreri, V.: Ambiente di deposizione e litofacies dei travertino quaternari dell' Italia centrale-meridionale, Mem Soc. Geol. It., 41, 861–868, 1988.

Dandurand, J. L., Gout, R., Hoefs, J., Menschel, G., Schott, J., and Usdowski, E.: Kinetically controlled variations of major components and carbon and oxygen isotopes in a calcite-precipitating spring, Chem. Geol., 36, 299–315, 1982.

Decho, A. W.: Overview of biopolymer-induced mineralization: What goes on in biofilms?, Ecol. Engin., 36, 137–144, 2010.

Dittrich, M. and Sibler, S.: Calcium carbonate precipitation by cyanobacterial polysaccharides, in: Speleothems and Tufas: Unravelling Physical and Biological controls, edited by: Pedley, H. M. and Rogerson, M., Geological Society of London Special Publication, Geological Society of London, London, 2010.

Dramis, F., Materazzi, M., and Cilla, G.: Influence of climate change on freshwater travertine deposition: a new hypothesis, Physical, Chem. Earth, 24, 893–897, 1999.

Dreybrodt, W. and Buhmann, D.: A mass-transfer model for dissolution and precipitation of calcite from solutions in turbulent motion, Chem. Geol., 90, 107–122, 1991.

Dreybrodt, W., Eisenlohr, L., Madry, B., and Ringer, S.: Precipitation kinetics of calcite in the system $CaCO_3$-H_2O-CO_2: The conversion to CO_2 by the slow process $H+HCO_3-> CO_2+H_2O$ as a rate limiting step, Geochimica Et Cosmochimica Acta, 61, 3897–3904, 1997.

Drysdale, R., Lucas, S., and Carthew, K.: The influence of diurnal temperatures on the hydrochemistry of a tufa-depositing stream, Hydrol. Process., 17, 3421–3441, 2003a.

Drysdale, R. N. and Gale, S. J.: The Indarri Falls travertine dam, Lawn Hill Creek, northwest Queensland, Australia, Earth Surface Processes and Landforms, 22, 413–418, 1997.

Drysdale, R. N., Carthew, K. D., and Taylor, M. P.: Larval caddis-fly nets and retreats: a unique biosedimentary paleocurrent indicator for fossil tufa deposits, Sediment. Geol., 161, 207–215, 2003.

Dunham, R. J.: Classification of carbonate rocks according to depositional texture., in: Classification of Carbonate Rocks, edited by: Ham, W. E., Am. Assoc. Petrol. Geol., , Tulsa, Oklahoma, 108–121, 1962.

Dupraz, C., Reid, R. P., Braissant, O., Decho, A. W., Norman, R. S., and Visscher, P. T.: Processes of carbonate precipitation in modern microbial mats, Earth-Science Reviews, 96, 141–162, 2009.

Embrey, A. F. and Kloven, E. J.: Absolute water depth limits of Late Devonian paleoecological zones, Geol. Rundsch., 61, 1972.

Emeis, K. C., Richnow, H. H., and Kempe, S.: Travertine Formation in Plitvice-National-Park, Yugoslavia – Chemical Versus Biological-Control, Sedimentology, 34, 595–609, 1987.

Eremin, A., Bulychev, A., Krupenina, N. A., Mair, T., Hauser, M. J. B., Stannarius, R., Muller, S. C., and Rubin, A. B.: Excitation-induced dynamics of external pH pattern in Chara corallina

cells and its dependence on external calcium concentration, Photochem. Photobio. Sci., 6, 103–109, 2007.

Ferreri, V.: Criteri di analisi facies e classificazione dei travertini pleistocenici dell'Italia meridionale., Rend. Acc. Sci. Fis. Mat. Serie, 4, 1–47, 1985.

Folk, R. L.: Practical petrographic classification for limestones., Bull. Amer. Assoc. Petroleum Geologists., 43, 1–38, 1959.

Folk, R. L.: Interaction between bacteria, nannobacteria and mineral precipitation in hot springs of Central Italy., Geogr. Phys. Quatern., 48, 233–246, 1994.

Folk, R.L. and Chafetz, H. S.: Pisoliths (pisoids) in Quaternary travertines of Tivoli, Italy, in: Coated Grains, edited by: Peryt, T. M., Springer-Verlag, Berlin, 474–487, 1983.

Folk, R.L., Chafetz, H. S., and Tiezzi, P. A.: Bizarre forms of depositional and diagenetic calcite in hot spring travertines, central Italy, in: Carbonate Cements Soc., edited by: N.B.Schneidmann, and Harris, P., Econ. Paleont. Miner. Spec. Publ., 349–369, 1985.

Ford, T. D. and Pedley, H. M.: A review of tufa and travertine deposits of the world, Earth-Sci. Rev., 41, 117–175, 1996.

Ford, T. D. and Pedley, H. M.: Tufa and travertine deposits of the Grand Canyon, Cave and Karst Science, 24, 107–116, 1997.

Fouke, B. W., Farmer, J. D., Marais, D. J. D., Pratt, L., Sturchio, N. C., Burns, P. C., and Discipulo, M. K.: Depositional Facies and Aqueous-Solid Geochemistry of Travertine-Depositing Hot Springs (Angel Terrace, Mammoth Hot Springs, Yellowstone National Park, USA), J. Sediment. Res., 70, 565–585, 2000.

Fouke, B. W., Bonheyo, G. T., Sanzenbacher, B., and Frias-Lopez, J.: Partitioning of bacterial communities between travertine depositional facies at Mammoth Hot Springs, Yellowstone National Park, USA, Can. J. Earth Sci., 40, 1531–1548, 2003.

Frisia, S., Borsato, A., Fairchild, I. J., and McDermott, F.: Calcite Fabrics, Growth Mechanisms, and Environments of Formation in Speleothems from the Italian Alps and Southwestern Ireland, J. Sediment. Res., 70, 1183–1196, 2000.

Glover, C. and Robertson, A. F. H.: Origin of tufa (cool water carbonate) and related terraces in the Antalya area, edited by: S. W. Turkey, Geol. J., 38, 329–358, 2003.

Golubić, S.: Cyclic and non-cyclic mechanisms in the formation of travertine., Verh. Int. Ver.Theor. Ang. Limnol., 17, 956–961, 1969.

Golubić, S., Violante, C., Ferreri, V., and D'Argenio, B.: Algal control and early diagenesis in Quaternary travertine formation (Rocchetta a Volturno, Central Apennines., in: Studies on Fossil Benthic Algae, edited by: Barattolo, F., Boll. Soc. Paleontol. Ital., Spec. vol., 231–247, 1993.

Guo, L. and Riding, R.: Hot spring travertine facies and sequences, Late Pleistocene Rapolano Terme, Italy, Sedimentology, 45, 163–180, 1998.

Hagele, D., Leinfelder, R., Grau, J., Burmeister, E. G., and Struck, U.: Oncolds from the river Alz (southern Germany): Tiny ecosystems in a phosphorus-limited environment, Palaeogeogr. Palaeoclimat. Palaeoecol., 237, 378–395, 2006.

Hammer, O.: Watch your step, Nature Physics, 4, 265–266, 2008.

Hammer, O., Dysthe, D. K., and Jamtveit, B.: The dynamics of travertine dams, Earth Planet. Sci. Lett., 256, 258–263, 2007.

Hammer, O., Dysthe, D. K., Lelu, B., Lund, H., Meakin, P., and Jamtveit, B.: Calcite precipitation instability under laminar, open-channel flow, Geochimica Et Cosmochimica Acta, 72, 5009–5021, 2008.

Hammer, Ø., Dysthe, D. K., and Jamtveit, B.: Travertine terracing: patterns and mechanisms, Geological Society, London, Special Publications, 336, 345–355, 2010.

Hobbie, J. E., Daley, R. J., and Jasper, S.: Use of Nuclepore filters for counting bacteria by fluorescence microscopy, Appl. Environ. Microbiol., 33, 1225–1228, 1977.

Hori, M., Hoshino, K., Okumura, K., and Kano, A.: Seasonal patterns of carbon chemistry and isotopes in tufa depositing groundwaters of southwestern Japan, Geochimica Et Cosmochimica Acta, 72, 480–492, 2008.

Inskeep, W. P. and Bloom, P. R.: An evaluation of rate equations for calcite precipitation kinetics at pCO2 less than 0.01 atm and pH greater than 8 Geochimica et Cosmochimica Acta, 49, 2165–2180, 1985.

Irion, and Müller, G.: Mineralogy, petrology and chemical composition of some calcareous tufa from Swabische Alb, Germany, in: Recent developments in carbonate sedimentology in Central Europe, edited by: Müller, G., and Freidman, G. M., Springer-Verlag, 156–171, 1968.

Jones, B. and Renaut, R. W.: Calcareous Spring Deposits in Continental Settings, Sedimentology, 61, 177–224, 2010.

Jones, B., Renaut, R. W., and Rosen, M. R.: Microbial biofacies in hot-spring sinters: a model based on Ohaaki Pool, North island, New Zealand, J. Sediment. Res., 68, 413–434, 1998.

Julia, R.: Travertines, in: Carbonate Depositional Environments, edited by: Schole, P. A., Bebout, D. G., and Moore, C., AAPG Mem., 64–72, 1983.

Kawaguchi, T. and Decho, A. W.: Isolation and biochemical characterization of extracellular polymeric secretions (EPS) from modern soft marine stromatolites (Bahamas) and its inhibitory effect on $CaCO_3$ precipitation, Preparat. Biochem. Biotechnol., 32, 51–63, 2002.

Kent, P., Gaunt, G. D., and Wood, C. J.: British Regional Geology, Eastern England from the Tees to the Wash. 2nd edition., Institute of Geological Sciences, Natural Environ. Res. Council, HMSO, London, 1980.

Kupriyanova, E., Villarejo, A., Markelova, A., Gerasimenko, L., Zavarzin, G., Samuelsson, G., Los, D. A., and Pronina, N.: Extracellular carbonic anhydrases of the stromatolite-forming cyanobacterium Microcoleus chthonoplastes, Microbiology, 153, 1149–1156, 2007.

Kupriyanova, E. V., Lebedeva, N. V., Dudoladova, M. V., Gerasimenko, L. M., Alekseeva, S. G., Pronina, N. A., and Zavarzin, G. A.: Carbonic anhydrase activity of alkalophilic cyanobacteria from soda lakes, Russ. J. Plant Physiol., 50, 532–539, 2003.

Langmuir, D.: Aqueous Environmental Geochemistry, Prentice-Hall Inc., Upper Saddle River, New Jersey, 600 pp., 1997.

Ledger, M. E., Harris, R. M. L., Armitage, P. D., and Milner, A. M.: Disturbance frequency influences patch dynamics in stream benthic algal communities, Oecologia, 155, 809–819, 2008.

Li, W., Yu, L. J., Yuan, D. X., Wu, Y., and Zeng, X. D.: A study of the activity and ecological significance of carbonic anhydrase from soil and its microbes from different karst ecosystems of Southwest China, Plant Soil, 272, 133–141, 2005.

Li, W., Yu, L. J., Wu, Y., Jia, L. P., and Yuan, D. X.: Enhancement of Ca2+ release from limestone by microbial extracellular carbonic anhydrase, Bioresource Technol., 98, 950–953, 2007.

Liu, Z., Dreybrodt, W., and Wang, H.: A new direction in effective accounting for the atmospheric CO_2 budget: Considering the combined action of carbonate dissolution, the global water cycle and photosynthetic uptake of DIC by aquatic organisms, Earth-Sci. Rev., 99, 162–172, 2010.

Liu, Z. H. and Dreybrodt, W.: Dissolution kinetics of calcium carbonate minerals in H_2O-CO_2 solutions in turbulent flow: The role of the diffusion boundary layer and the slow reaction $H_2O + CO_2$ reversible arrow H++HCO3, Geochimica Et Cosmochimica Acta, 61, 2879–2889, 1997.

Liu, Z. H., Liu, X. L., and Liao, C. J.: Daytime deposition and nighttime dissolution of calcium carbonate controlled by submerged plants in a karst spring-fed pool: insights from high time-resolution monitoring of physico-chemistry of water, Environ. Geol., 55, 1159–1168, 2008.

Lorah, M. M. and Herman, J. S.: The chemical evolution of a Travertine-depositing stream: geochemical processes and mass transfer reactions, Water Resour. Res., 24, 1541–1552, 1988.

Lowenstam, H. A.: Minerals formed by organisms, Science, 211, 1126–1131, 1981.

Marks, J. C., Parnell, R., Carter, C., Dinger, E. C., and Haden, G. A.: Interactions between geomorphology and ecosystem processes in travertine streams: Implications for decommissioning a dam on Fossil Creek, Arizona, Geomorphology, 77, 299–307, 2006.

Mayes, W. M., Younger, P. L., and Aumonier, J.: Hydrogeochemistry of alkaline steel slag leachates in the UK, Water Air Soil Pollut., 195, 35–50, 2008.

Merz-Preiss, M. and Riding, R.: Cyanobacterial tufa calcification in two freshwater streams: ambient environment, chemical thresholds and biological processes, Sedimentary Geol., 126, 103–124, 1999.

Minissale, A.: Origin, transport and discharge of CO_2 in central Italy, Earth-Sci. Rev., 66, 89–141, 2004.

Newton, M. S.: Holocene fluctuations of Mono Lake, California: the sedimentary record, in: Sedimentology and Geochemistry of Modern and Ancient Saline Lakes, edited by: Renaut, R. W. and Last., W. M., SEPM Special Publications, SEPM, 143–157, 1994.

Nicod, J.: Repartition, classification, relation avec les milieux karstiques et karstification, Bull. Assoc. Geogr. Fr, 479/480, 181–187, 1981.

Patterson, C. S., Busey, R. H., and Mesmer, R. E.: Second ionization of carbonic acid in NaCl media to 250 °C, J. Solut. Chem., 13, 647–661, 1984.

Pedley, H. M.: Classification and environmental models of cool freshwater tufas, Sediment. Geol., 68, 143–154, 1990.

Pedley, H. M.: Prokaryote microphyte biofilms: a sedimentological perspective, Kaupia, Darmstader Betr. Naturgesh., 4, 45–60, 1994.

Pedley, H. M.: Tufas and travertines of the Mediterranean region: a testing ground for freshwater carbonate concepts and developments, Sedimentology, 56, 221–246, 2009.

Pedley, H. M. and Rogerson, M.: Introduction to tufas and speleothems, in: Speleothems and Tufas: Unravelling Physical and Biological controls, edited by: Pedley, H. M., and Rogerson, M., Geologicasl Society Special Publication, Geological Society, London, 2010a.

Pedley, H. M. and Rogerson, M.: In vitro investigations of the impact of different temperature and flow velocity conditions on tufa microfabric. , in: Speleothems and Tufas: Unravelling Physical and Biological controls, edited by: Pedley, H. M., and Rogerson,

M., Geological Society Special Publication, Geological Society of London, London, 193–210, 2010b.

Pedley, H. M., Rogerson, M., and Middleton, R.: The growth and morphology of freshwater calcite precipitates from in Vitro Mesocosm flume experiments; the case for biomediation, Sedimentology, 56, 511–527, 2009.

Pedley, M.: Fresh-Water (Phytoherm) Reefs – the Role of Biofilms and Their Bearing on Marine Reef Cementation, Sediment. Geol., 79, 255–274, 1992.

Pedley, M., Andrews, J., Ordonez, S., delCura, M. A. G., Martin, J. A. G., and Taylor, D.: Does climate control the morphological fabric of freshwater carbonates?, A comparative study of Holocene barrage tufas from Spain and Britain, Palaeogeogr. Palaeoclimat. Palaeoecol., 121, 239–257, 1996.

Peña, J.L., Sancho, C., and Lozano, M. V.: Climatic and tectonic significance of late Pleistocene and Holocene tufa deposits in the Mijares River canyon, eastern Iberian range, northeastern Spain., Earth Surf. Proc. Land., 25, 1403–1417, 2000.

Pentecost, A.: Blue-green-algae and freshwater carbonate deposits, Proceedings of the Royal Society of London Series B-Biological Sciences, 200, 43–61, 1978.

Pentecost, A.: Growth and calcification of the fresh-water cyanobacterium Rivularia haematites, Proceedings of the Royal Society of London Series B-Biological Sciences, 232, 125–136, 1987.

Pentecost, A.: Carbonate chemistry of surface waters in a temperature karst region – the southern Yorkshire Dales, UK., J. Hydrol., 139, 211–232, 1992.

Pentecost, A.: Travertine, Springer, Berlin, 2005.

Pentecost, A. and Lord, T. C.: Postglacial tufas and travertines from the Craven district Yorkshire., Cave Science,, 15, 15–19, 1988.

Pentecost, A. and Viles, H. A.: A review and reassessment of travertine classification., Géogr. Phys. Quaternaire, 48, 305–314, 1994.

Pentecost, A., Andrews, J. E., Dennis, P. F., Marca-Bell, A., and Dennis, S.: Charophyte growth in small temperate water bodies: Extreme isotopic disequilibrium and implications for the palaeoecology of shallow marl lakes, Palaeogeogr.Palaeoclimat. Palaeoecol., 240, 389–404, 2006.

Peryt, T. M.: Oncoids, Comments on recent developments, in: Coated Grains, edited by: Peryt, T. M., Springer-Verlag, Berlin, Heidelburg, 273–275, 1983.

Plummer, L. N. and Busenberg, E.: The solubilities of calcite, aragonite and vaterite in CO_2-H_2O solutions between 0 and 90 °C, and an evaluation of the aqueous model for the system $CaCO_3$-CO_2-H_2O, Geochimica Et Cosmochimica Acta, 46, 1011–1040, 1982.

Renaut, R. W. and Jones, B.: Controls on aragonite and calcite precipitation in hot spring travertines at Chemurkeu, Lake Bogoria, Kenya, Can. J. Earth Sci., 34, 801–818, 1997.

Riding, R.: Chapter 2, Classification of microbial carbonates, in: Calcareous Algae and Stromatolites, edited by: Riding, R., Springer-Verlag, Berlin, Heidelburg, 21–51, 1991.

Rogerson, M., Pedley, H. M., Wadhawan, J. D., and Middleton, R.: New Insights into Biological Influence on the Geochemistry of Freshwater Carbonate Deposits, Geochimica et Cosmochimica Acta, 72, 4976–4987, 2008.

Rogerson, M., Pedley, H. M., and Middleton, R.: Microbial Influence on Macroenvironment Chemical Conditions in Alka-line (Tufa) Streams; Perspectives from In Vitro Experiments, in: Speleothems and Tufas: Unravelling Physical and Biological controls, edited by: Pedley, H. M., and Rogerson, M., Geological Society Special Publication, Geological Society of London, London, 65–81, 2010.

Saunders, P. V., Rogerson, M., Pedley, H. M., Wadhawan, J., and Greenway, G.: Mg/Ca ratios in freshwater microbial carbonates: Thermodynamic , Kinetic and Vital Effects. , Geochimica Et Cosmochimica Acta, in review, 2014.

Schneider, J., Schneider, H. G., Campion, S.L., and Alsumard, T.: Algal micro-reefs-coated grains from freshwater environments, in: Coated Grains, edited by: T.M. Peryt, Springer, Berlin, 284–298, 1983.

Shiraishi, F., Bissett, A., de Beer, D., Reimer, A., and Arp, G.: Photosynthesis, respiration and exopolymer calcium-binding in biofilm calcification (Westerhfer and deinschwanger creek, germany), Geomicrobiol. J., 25, 83–94, 2008a.

Shiraishi, F., Reimer, A., Bissett, A., de Beer, D., and Arp, G.: Microbial effects on biofilm calcification, ambient water chemistry and stable isotope records in a highly supersaturated setting (Westerhofer Bach, Germany), Palaeogeogr. Palaeoclimat. Palaeoecol., 262, 91–106, 2008b.

Shiraishi, F., Zippel, B., Neu, T. R., and Arp, G.: In situ detection of bacteria in calcified biofilms using FISH and CARD-FISH, J. Microbiol. Methods, 75, 103–108, 2008c.

Spiro, B. and Pentecost, A.: One Day in the Life of a Stream – a Diurnal Inorganic Carbon Mass Balance for a Travertine-Depositing Stream (Waterfall Beck, Yorkshire), Geomicrobiol. J., 9, 1–11, 1991.

Stirn, A.: Kalktuffvorkommen und Kalktufftypen der Schwäbischen Alb, Abh. Karst Höhlenknd, 1–91, 1964.

Symoens, J. J., Duvigneaud, P., and van den Bergen, C.: Aperçu sur la végétation des tufs calcaires de la Belgique, Bull. Soc. R. Bot. Belg., 83, 329–352, 1951.

Szulc, J.: Genesis and classification of travertine deposits, Przegl. Geol., 31, 2311–2236, 1983.

Veysey, J. and Goldenfeld, N.: Watching rocks grow, Nature Physics, 4, 310–313, 2008.

Veysey II, J., Fouke, B. W., Kandianis, M. T., Schickel, T. J., Johnson, R. W., and Goldenfeld, N.: Reconstruction of Water Temperature, pH, and Flux of Ancient Hot Springs from Travertine Depositional Facies, J. Sediment. Res., 78, 69–76, 2008.

Visscher, P. T. and Stolz, J. F.: Microbial mats as bioreactors: populations, processes, and products, Palaeogeogr. Palaeoclimat. Palaeoecol., 219, 87–100, 2005.

Wingender, J., Neu, T., and Flemming, H.-C.: What are Bacterial Extracellular Polymeric Substances?, in: Microbial Extracellular Polymeric Substances, edited by: Wingender, J., Neu, T., and Flemming, H.-C., Springer Berlin Heidelberg, 1–19, 1999.

Wright, V. P.: Lacustrine carbonates in rift settings: the interaction of volcanic and microbial processes on carbonate deposition, Geological Society, London, Special Publications, 370, 39–47, 2012.

Zaihua, L., Svensson, U., Dreybrodt, W., Daoxian, Y., and Buhmann, D.: Hydrodynamic control of inorganic calcite precipitation in Huanglong Ravine, China: Field measurements and theoretical prediction of deposition rates Geochimica et Cosmochimica Acta, 59, 3087–3097, 1995.

11

Morphodynamics of river bed variation with variable bedload step length

A. Pelosi[1] and G. Parker[2]

[1]Department of Civil Engineering, Università degli Studi di Salerno (UNISA), Via Giovanni Paolo II – 84084 Fisciano (Salerno), Italy
[2]Department of Civil & Environmental Engineering and Department of Geology, Hydrosystems Laboratory, University of Illinois, 301 N. Mathews Ave., Urbana, IL 61801, USA

Correspondence to: A. Pelosi (apelosi@unisa.it)

Abstract. Here we consider the 1-D morphodynamics of an erodible bed subject to bedload transport. Fluvial bed elevation variation is typically modeled by the Exner equation, which, in its classical form, expresses mass conservation in terms of the divergence of the bedload sediment flux. An entrainment form of the Exner equation can be written as an alternative description of the same bedload processes, by introducing the notions of an entrainment rate into bedload and of a particle step length, and assuming a certain probability distribution for the step length. This entrainment form implies some degree of nonlocality, which is absent from the standard flux form, so that these two expressions, which are different ways to look at same conservation principle (i.e., sediment continuity), may no longer become equivalent in cases when channel complexity and flow conditions allow for long particle saltation steps (including, but not limited to the case where particle step length has a heavy tailed distribution) or when the domain of interest is not long compared to the step length (e.g., laboratory scales, or saltation over relatively smooth surfaces). We perform a systematic analysis of the effects of the nonlocality in the entrainment form of the Exner equation on transient aggradational/degradational bed profiles by using the flux form as a benchmark. As expected, the two forms converge to the same results as the step length converges to zero, in which case nonlocality is negligible. As step length increases relative to domain length, the mode of aggradation changes from an upward-concave form to a rotational, and then eventually a downward-concave form. Corresponding behavior is found for the case of degradation. These results may explain anomalously flat, aggradational, long profiles that have been observed in some short laboratory flume experiments.

1 Introduction

The Exner equation of sediment conservation, when combined with a hydrodynamic model and a sediment transport model, is a central tool to evaluate the bed evolution (e.g., aggradation and degradation) in the field of morphodynamics of Earth's surface.

The Exner equation, in its classical formulation, relates the bed evolution to the divergence of the bedload sediment flux (q), which is assumed to be a local function of the flow and the topography. However, certain sediment dynamics, such as (i) particle diffusion in river bedload (e.g., Nikora et al., 2002; Bradley et al., 2010; Ganti et al., 2010; Martin et al., 2012), (ii) bed sediment transport along bedrock channels (Stark et al., 2009) and (iii) particle displacements on hillslopes (Foufoula-Georgiou et al., 2010) may show nonlocal behavior that is not easily captured by the classical form of the Exner equation (the notation used throughout the manuscript is defined and listed after the conclusions).

The nonlocality of interest here is embedded in the step length r of a bedload particle, i.e., the distance that a particle, once entrained into motion, travels before being deposited. The existence of a finite step length r implies a nonlocal connection between point x (where a particle is deposited) and point $x - r$ (where it was entrained). The degree of nonlocality can be characterized in terms of the probability density (PDF) of step lengths $f_s(r)$. This PDF can be hypothesized to be thin-tailed (e.g., exponential) or heavy-tailed (e.g., power).

In recent years, considerable emphasis has been placed on asymptotic nonlocality associated with heavy-tailed PDFs for step length (e.g., Schumer et al., 2009; Bradley et al., 2010; Ganti et al., 2010). This is motivated by the desire to preserve nonlocality in the limit of long time, thus leading to fractional advective–diffusive equations (fADE) for pebble tracer dispersion corresponding to the now-classical fADE model (e.g., Schumer et al., 2009). Here we consider nonlocality in a more general sense, as outlined below.

Experiments conducted under the simplest possible conditions (including steady, uniform flow, single-sized sediment and the absence of bedforms) yield thin-tailed and, more specifically, exponential distributions for step length PDF (Nakagawa and Tsujimoto, 1980; Hill et al., 2010). Ganti et al. (2010), however, showed (a) the bed to consist of a range of sizes, (b) the PDF of size distribution to obey a gamma distribution and (c) the PDF of for step length of each grain size to be exponential, the resulting PDF for step length would be heavy-tailed. Hassan et al. (2013) analyzed 64 sets of field data on pebble tracer dispersion in mountain rivers (which by nature contain a range of sizes). They found that all but 5 cases either showed thin-tailed PDFs, or could be rescaled as thin-tailed PDFs. Their results, combined with those of Ganti et al. (2010), however, do suggest that the gradual incorporation of the many factors in nature that lead to complexity can also lead to nonlocal behavior mediated by heavy-tailed PDFs.

Here, however, we focus on the case of nonlocality mediated by thin-tailed (exponential) PDFs for step length. Regardless of the thin tail of the PDF, the degree of nonlocality nevertheless increases with increasing mean step length $\bar{r}$. This nonlocality may become dominant when $\bar{r}$ approaches the same order of magnitude as the domain length L_d under consideration. We show that patterns of bed aggradation and degradation are strongly dependent on the ratio $\bar{r}/L_d$, a parameter that may be surprisingly large in some small-scale experiments. Our results may explain anomalously flat, aggradational, long profiles that have been observed in some short laboratory flume experiments, without relying on either of the fractional partial differential equations or heavy-tailed distributions invoked or implied by Voller and Paola (2010). We use our framework to explore the consequences of heavy-tailed PDFs for step lengths as well.

2 Methods

2.1 Theoretical framework

1-D riverbed elevation variation is classically described by the 1-D Exner equation of sediment conservation in flux form (or equivalently in the 2-D case, divergence form):

$$\frac{\partial \eta(x,t)}{\partial t} = -\frac{\partial q(x,t)}{\partial x}, \tag{1}$$

where η (L) denotes the bed elevation, t (T) denotes the time, x (L) denotes the streamwise distance and q (L^{2}T^{-1}) is the volume bedload transport rate per unit width. (Here, the porosity of the bed sediment is set to 0 and bedload only is considered, both for the sake of simplicity.) There is, however, a completely equivalent entrainment form of sediment conservation (e.g., Tsujimoto, 1978):

$$\frac{\partial \eta(x,t)}{\partial t} = D(x,t) - E(x,t), \tag{2}$$

where E (L T^{-1}) denotes the volume rate of entrainment of bed particles into bedload per unit area per unit time and D (L T^{-1}) denotes the volume rate of deposition of bedload material onto the bed per unit area per unit time.

The deposition rate can be related to the entrainment rate by means of the probability density of the step length $f_s(r)$ (L^{-1}); that is, the probability density of the distance that an entrained particle moves before being re-deposited. Assuming that, once entrained, a particle undergoes a step with length r before depositing, and that this step length has the probability density $f_s(r)$ (PDF of step length), the volume deposition rate D can be specified as follows in terms of entrainment rate upstream and travel distance (e.g., Parker et al., 2000; Ganti et al. 2010),

$$D(x) = \int_0^\infty E(x-r)f_s(r)\mathrm{d}r, \tag{3}$$

so that the entrainment form of sediment mass conservation can be written as

$$\frac{\partial \eta}{\partial t} = -E(x) + \int_0^\infty E(x-r)f_s(r)\mathrm{d}r. \tag{4}$$

As has been shown by Tsujimoto (1978), the two Eqs. (1) and (4), are in principle equivalent in so far as the following equation precisely describes the bedload transport rate:

$$q(x) = \int_0^\infty E(x-r) \int_r^\infty f_s(r')\mathrm{d}r'\mathrm{d}r. \tag{5}$$

Yet in any given implementation, Eqs. (1) and (4) are rarely equivalent. More specifically, in most implementations of the flux Eq. (1), q is taken to be a local function of the flow (e.g., bed shear stress), whereas in most implementations of the entrainment Eq. (4), E is taken to be a local function of the flow (again, e.g., bed shear stress). The presence of the spatial convolution term in the entrainment Eqs. (3) and (4) ensures nonlocality in the entrainment form as compared to the flux form. This nonlocality is present regardless of whether the PDF of step length $f_s(r)$ is thin-tailed or heavy-tailed, and vanishes only when $f_s(r)$ becomes proportional to $\delta(r)$, where δ denotes the Dirac function. In the present implementation of Eq. (4), then, we take E to be a local function of flow conditions, so that q is nonlocal according to Eq. (5).

It should be pointed out that the formulation of Eq. (4) involves a purely kinematic description of particle step length, with the trajectory of the particle unmodified by intervening flow conditions. This is in line with the work of Einstein (1950), Nakagawa and Tsujimoto (1980) and Ganti et al. (2010). In a more detailed analysis, particle momentum balance, and in particular relaxation effects involving, e.g., particle inertia (Parker, 1975; Charru, 2006), should be included. In so far as a step length generally consists of many individual particle saltations (Nino et al., 1994), however, the present kinematic formulation may be sufficient for a first-order analysis. One other way to formulate the problem is in terms of mass and momentum conservation of two sediment phases; i.e., a static bed phase and a moving bedload phase above it, with exchange between the two (e.g., Charru, 2006). An Eulerian version of such a model would, however, preclude the analysis of nonlocality associated with varying step length, which is a purely Lagrangian parameter. A Lagrangian analysis that includes the dynamics of a particle as it saltates its way through one step length could lead to an improved formulation.

Before continuing, it is of value to specifically indicate what we mean by nonlocality. Equations (3) and (4) are nonlocal in so far as the deposition rate is not determined at a point, but is instead determined from a convolution involving the entrainment rate at every point upstream. The problem is thus nonlocal in the sense of Du et al. (2012). The problem becomes nonlocal in the asymptotic sense only when the PDF of step length $f_s(r)$ is heavy-tailed, such that moments beyond a specific value fail to exist (e.g., Schumer et al., 2009). Both cases are considered here; we specifically address the problem of asymptotic nonlocality in Sect. 4.

Here we explore the consequences of nonlocality by comparing the local and nonlocal Eqs. (1) and (4) for Exner over a range of conditions. To do this, we assume that the PDF $f_s(r)$ has a mean, and consider the dimensionless parameter ε:

$$\varepsilon = \frac{\bar{r}}{L_d}, \tag{6}$$

where $\bar{r}$ (L) denotes the mean particle step length and L_d (L) denotes the length of the domain of interest (e.g., flume length or length of river reach). The flux and entrainment

forms become strictly equivalent only under the constraint:

$$\varepsilon = \frac{\bar{r}}{L_d} \ll 1. \tag{7}$$

Here we demonstrate that this equivalence for $\varepsilon \ll 1$ breaks down with increasing ε. This is because a finite mean step length $\bar{r}$ in and of itself implies nonlocality, regardless of whether or not the probabilistic distribution of particle step length $f_s(r)$ is thin- or heavy-tailed. A further degree of nonlocality can be introduced by adopting a heavy-tailed distribution for $f_s(r)$.

The standard thin-tailed form for the particle step length probability density function is the exponential distribution (e.g., Nakagawa and Tsujimoto, 1980; Hill et al., 2010):

$$f_s(r) = \frac{1}{\bar{r}} \exp\left(-\frac{r}{\bar{r}}\right), \; \begin{cases} r > 0 \\ \bar{r} > 0 \end{cases}. \tag{8}$$

The heavy-tailed Pareto distribution with a shift, which ensures that the maximum value of the distribution is realized at $r = 0$, can be considered as an alternative:

$$f_s(r) = \frac{\alpha r_0^\alpha}{(r + r_0)^{\alpha+1}}, \; \begin{cases} r_0 > 0 \\ \alpha > 0 \end{cases}, \tag{9}$$

where α is the shape parameter and r_0 (L) is the scale parameter. The mean value $\bar{r}$ of the distribution of Eq. (9) can be written as

$$\bar{r} = \frac{\alpha r_0}{\alpha - 1} - r_0, \; \begin{cases} r_0 > 0 \\ \alpha > 0 \end{cases}. \tag{10}$$

2.2 Numerical model

Here we solve the flux and entrainment formulations under parallel conditions, the only exception being the formulation for step length. To simplify the problem and focus on this point, we approximate the flow as obeying the normal (steady, uniform) approximation. Momentum conservation then dictates that bed shear stress τ_b ($ML^{-1}T^{-2}$) can be represented as proportional to the product of depth H (L) and slope S (1):

$$\tau_b = \rho u_*^2 = \rho g H S, \tag{11a}$$

$$S = -\frac{\partial \eta}{\partial x}, \tag{11b}$$

where u_* (LT^{-1}) is the shear velocity.

The dimensionless Shields number governing particle mobility is defined as

$$\tau^* = \frac{\tau_b}{\rho R g D_c}, \tag{12}$$

where ρ (ML^{-3}) is water density, D_c (L) is characteristic bed grain size (here taken to be uniform for simplicity) and R denotes the submerged specific gravity of the sediment (~ 1.65 for quartz).

The flow can be computed by introducing the Manning–Strickler resistance relation:

$$\frac{U}{u_*} = \alpha_r \left(\frac{H}{k_c}\right)^{1/6},\tag{13}$$

where U (L T^{-1}) is the depth-averaged flow velocity, α_r is a dimensionless coefficient between 8 and 9 (Chaudhry, 1993), and k_c (L) denotes a composite roughness height. In absence of bedforms, k_c is equivalent to the roughness height k_s (L), which is proportional to grain size D_c by means of a dimensionless coefficient with typical values between 2 and 5 (Parker, 2004). Here, α_r is set equal to 8.1, as suggested by Parker (1991) for gravel-bed streams, while k_c, in absence of bedforms, is taken to be 2.5 times the grain size D_c (Parker, 2004).

The equation for water conservation for quasi-steady flow is

$$Q_w = UBH,\tag{14}$$

where Q_w (L^{3}T^{-1}) is the water discharge and B (L) denotes the channel width.

Combining Eqs. (11–14), we relate the dimensionless Shields number to the flow properties:

$$\tau^* = \left[\frac{(k_c)^{1/3} Q_w^2}{\alpha_r^2 g B^2}\right]^{3/10} \frac{S^{7/10}}{RD_c}.\tag{15}$$

The basis for our morphodynamic calculations is the form of Meyer-Peter and Müller (1948), as modified by Wong and Parker (2006). It takes the form

$$q = \gamma \sqrt{RgD_c} D_c (\tau^* - \tau_c^*)^{3/2},\tag{16}$$

where g (L T^{-2}) denotes the gravitational acceleration. The parameter τ_c denotes the threshold Shields number and γ is a coefficient of proportionality; these parameters take the respective values 0.0495 and 3.97 (as specified by Wong and Parker, 2006).

The volume bedload transport rate per unit width q at equilibrium can also be written as

$$q = E \cdot \bar{r},\tag{17}$$

(Einstein, 1950) so that the entrainment rate takes the form

$$E = \frac{\gamma}{\beta} \sqrt{RgD_c} (\tau^* - \tau_c^*)^{3/2},\tag{18a}$$

$$\beta = \frac{\bar{r}}{D_c}.\tag{18b}$$

Here β is a dimensionless parameter. Einstein (1950), suggested, based on a simple flume-like configuration, that $\bar{r}/D_c$ takes a value on the order of 100–1000, so that a step length is about 100–1000 grain sizes. This order of magnitude has been confirmed by the experiments of Nakagawa and Tsujimoto (1980), Wong et al. (2007) and Hill et al. (2010).

In systems with higher degrees of complexity, however, β is likely to vary over a wide range. Combinations of multiple grain sizes, bedforms, scour and fill and partially exposed bedrock are likely to give rise to connected pathways along which particles may travel for an extended distance, so giving rise to larger values of $\bar{r}$ (e.g., Parker, 2008). In order to capture this effect in a simplified 1D model, we allow the ratio $\bar{r}$, and thus $\beta = \bar{r}/D_c$ to vary freely, so that the ratio $\bar{r}/L_d$ of step length to domain length can vary from 0 (in which case the flux and entrainment formulations become equivalent) to unity (in which case a particle starting at the upstream end of the domain reaches the downstream end in a single step).

Linking Eq. (18a), the following relation arises at equilibrium conditions:

$$\frac{q}{\sqrt{RgD_c}D_c} = \beta \frac{E}{\sqrt{RgD_c}}.\tag{19}$$

Our formulation is such that increased step length is adjusted against reduced entrainment, so that the equilibrium bedload transport rate is the same whether the flux or entrainment formulation is used. A difference, however, arises under disequilibrium conditions, in which case Eq. (16) is solved in conjunction with Eq. (1) in the flux case, and Eq. (18a) is solved in conjunction with Eq. (4) in the entrainment case. This allows us to capture the difference between the two formulations in a comparable way.

The flux formulation, Eq. (1), corresponds to a nonlinear diffusion equation, i.e.,

$$\frac{\partial \eta(x,t)}{\partial t} = \frac{\partial}{\partial x}\left(\nu \frac{\partial \eta}{\partial x}\right),\tag{20}$$

where according to Eqs. (11), (15) and (16), the kinematic diffusivity ν (L^{2}T^{-1}) is a function of bed slope $S = -\partial \eta / \partial x$:

$$\nu = \frac{\sqrt{RgD_c}D_c}{S} \gamma \left\{\left[\frac{(k_c)^{1/3} Q_w^2}{\alpha_r^2 g B^2}\right]^{3/10} \frac{S^{7/10}}{RD_c} - \tau_c^*\right\}^{3/2}.\tag{21}$$

The governing equation is second order in x, and thus requires two boundary conditions. Here we require that the bed elevation at the downstream end is zero, and that the sediment transport rate at the upstream end is given as a constant, specified feed rate:

$$\eta|_{x=L_d} = 0,\tag{22a}$$

$$q|_{x=0} = q_f.\tag{22b}$$

The entrainment formulation of Eq. (4), however, is only first order in x, in so far as the entrainment rate E is a specified function of bed slope $S = -\partial \eta / \partial x$ according to Eqs. (4) and (18a). Thus there can be only one boundary condition in x; here we use Eq. (22a) for this, so that both the flux and entrainment formulations satisfy the condition of vanishing bed elevation (corresponding to set base level) at the downstream end.

Although no boundary condition can be set at the upstream end for the entrainment formulation, it is still possible to choose conditions so that the sediment transport rate at the upstream equals the feed value under equilibrium conditions.

To do this, we assume that the entrainment rate everywhere upstream of $x = 0$ equals a specified value E_f, specified as follows:

$$E_f = \frac{q_f}{\bar{r}}. \tag{23}$$

The deposition rate $D(x)$ of Eq. (3) can then be rewritten in terms of the sum of particles that originate within the domain $(x - r \geq 0)$ and those that originate upstream of the domain $(x - r < 0)$:

$$
\begin{aligned}
D(x) &= \int_0^\infty E(x-r) f_s(r) dr \\
&= \int_0^x E(x-r) f_s(r) dr + \int_x^\infty E(x-r) f_s(r) dr \\
&= \int_0^x E(x-r) f_s(r) dr + E_f f_{ls}(x),
\end{aligned} \tag{24}
$$

where

$$f_{ls}(x) = \int_x^\infty f_s(r) dr \tag{25}$$

is the probability (L^{-1}), that a particle travels at least a distance x.

The entrainment form of sediment mass conservation thus takes the ultimate form

$$\frac{\partial \eta}{\partial t} = -E(x) + \int_0^x E(x-r) f_s(r) dr + E_f f_{ls}(x). \tag{26}$$

For the numerical computation, we nondimensionalize Eqs. (1) and (26). We assume that the computation begins from some equilibrium initial condition with spatially constant slope S_{in}, bedload transport rate and entrainment rate $q_{in} = \bar{r} E_{in}$. At $t = 0$, however, the supply of sediment is impulsively altered, causing subsequent bed aggradation or degradation, but with an altered sediment feed rate for $t > 0$. We normalize against initial equilibrium conditions using the following definitions:

$$\hat{\eta} = \frac{\eta}{L_d \cdot S_{in}}, \tag{27a}$$

$$\hat{x} = \frac{x}{L_d}, \tag{27b}$$

$$\hat{r} = \frac{r}{L_d}, \tag{27c}$$

$$\hat{t} = \frac{E_{in} \cdot \varepsilon}{L_d \cdot S_{in}} t, \tag{27d}$$

$$\hat{s} = \frac{S}{S_{in}}. \tag{27e}$$

In addition, we nondimensionalize the entrainment rate (for the entrainment formulation) and the bedload transport

Upstream conditions   *Downstream condition*

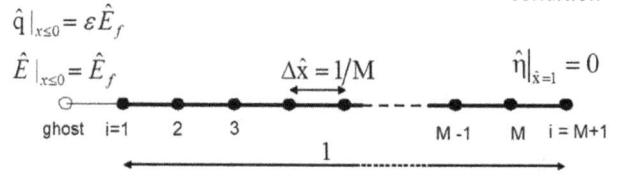

$$\hat{q}|_{x \leq 0} = \varepsilon \hat{E}_f$$

$$\hat{E}|_{x \leq 0} = \hat{E}_f \qquad \Delta \hat{x} = 1/M \qquad \hat{\eta}|_{\hat{x}=1} = 0$$

ghost   i=1   2   3   M-1   M   i = M+1

Figure 1. Discretization of the domain.

rate (for the flux formulation) as

$$\hat{E} = \frac{E}{E_{in}}, \tag{27f}$$

$$\hat{q} = \varepsilon \cdot \hat{E}. \tag{27g}$$

Then, the nondimensional flux and entrainment forms of the sediment mass conservation, Eqs. (1) and (26) take the respective forms

$$\frac{\partial \hat{\eta}}{\partial \hat{t}} = -\frac{1}{\varepsilon} \frac{\partial \hat{q}}{\partial \hat{x}} = -\frac{\partial \hat{E}}{\partial \hat{x}}, \tag{28}$$

$$\frac{\partial \hat{\eta}}{\partial \hat{t}} = -\frac{1}{\varepsilon} \hat{E}(\hat{x}) + \frac{1}{\varepsilon} \int_0^{\hat{x}} \hat{E}(\hat{x} - \hat{r}) \tilde{f}_s\left(\frac{\hat{r}}{\varepsilon}\right) d\hat{r} + \frac{1}{\varepsilon} \int_{\hat{x}}^\infty \tilde{f}_s\left(\frac{\hat{r}}{\varepsilon}\right) d\hat{r}, \tag{29}$$

where

$$\tilde{f}_s\left(\frac{\hat{r}}{\varepsilon}\right) = \frac{1}{\varepsilon} \exp\left(\frac{\hat{r}}{\varepsilon}\right) \tag{30}$$

is the dimensionless step length PDF for the exponential distribution, and

$$\tilde{f}_s\left(\frac{\hat{r}}{\varepsilon}\right) = \frac{\alpha \hat{r}_0^\alpha}{(\hat{r} + \hat{r}_0)^{\alpha+1}} \tag{31}$$

is the corresponding form for the Pareto distribution, where $\hat{r}_0$ is the dimensionless scale parameter equal to r_0/L_d.

These are the upstream conditions, for the entrainment formulation

$$\hat{E}(x,t)\big|_{\hat{x} \leq 0} = \hat{E}_f \tag{32}$$

and for the flux formulation

$$\hat{q}(x,t)\big|_{\hat{x} \leq 0} = \varepsilon \hat{E}_f. \tag{33}$$

The downstream boundary condition is the same for both

$$\hat{\eta}(x,t)\big|_{\hat{x}=1} = 0. \tag{34}$$

Here $\hat{E}_f$ is an imposed upstream entrainment rate, and $\varepsilon \hat{E}_f$ is an imposed upstream bedload feed rate, chosen to be different from the initial equilibrium values so

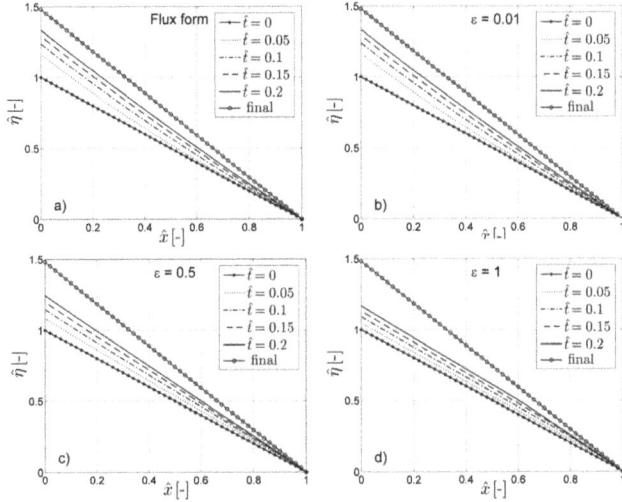

Figure 2. Bed profile evolution for the case $\hat{E}_f = 2$: (**a**) flux form; (**b**) entrainment form for $\varepsilon = \bar{r}/L_d = 0.01$, (**c**) entrainment form for $\varepsilon = 0.5$ and (**d**) entrainment form for $\varepsilon = 1$, using the thin-tailed exponential step length function of Eq. (8). As ε increases, it is clearly seen that the differences between the results for the two formulations increases. More specifically, as ε increases, nonlocality effects mediate a transition from upward concave transient profiles to downward concave transient profiles.

that the bed is forced to aggrade (or degrade) toward a new equilibrium state.

Manipulating the relations of Eqs. (15) and (18a), with the definitions of Eqs. (27), $\hat{E}$ can be at any given time as

$$\hat{E} = \left(\frac{\tau_{in}^* \hat{s}^{7/10} - \tau_c^*}{\tau_{in}^* - \tau_c^*} \right)^{3/2}, \tag{35}$$

where τ_{in}^* is the dimensionless Shields number, calculated from Eq. (15) with the initial flow and bed conditions and $\hat{s}$ is the local dimensionless slope.

The key parameter of interest here in describing the difference between the entrainment and flux formulations is ε. In the case $\varepsilon \ll 1$, both formulations become identical. We show below, however, that as ε increases, the response to change in sediment supply differs between the two cases.

We discretize the relation between dimensionless slope and dimensionless bed elevation as follows:

$$\hat{s} = \begin{cases} \frac{\hat{\eta}_1 - \hat{\eta}_2}{\Delta \hat{x}}, & i = 1 \\ \frac{\hat{\eta}_{i-1} - \hat{\eta}_{i+1}}{2\Delta \hat{x}}, & i = 2 \dots M \\ \frac{\hat{\eta}_M - \hat{\eta}_{M+1}}{\Delta \hat{x}}, & i = M+1 \end{cases} \tag{36}$$

The discretization of the domain is schematized in Fig. 1: a central finite-difference scheme is used to solve Eqs. (28) and (29).

3 Results

Here we compare the results for aggradation and degradation for the entrainment formulation with varying values of

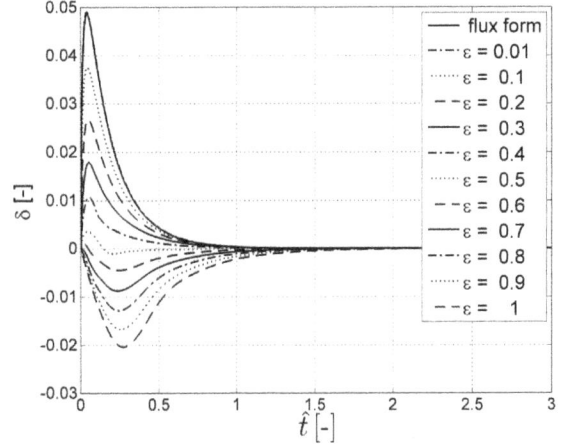

Figure 3. Aggradation case: variation in time of the dimensionless concavity parameter δ in the case of the flux formulation, and in the cases of the entrainment formulation for different values of ε ranging from 0.01 to 1. The result for the flux form overlaps with the result for the entrainment form with $\varepsilon = 0.01$. Note the reversal in behavior as ε increases beyond about 0.5.

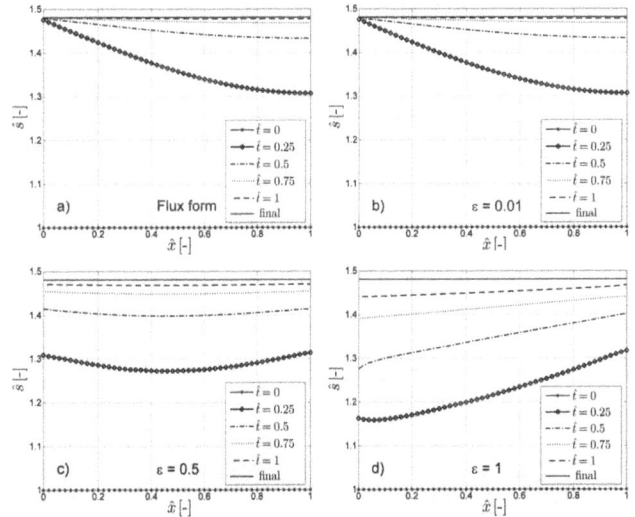

Figure 4. Slope profile evolution for the case $\hat{E}_f = 2$: (**a**) flux form; (**b**) entrainment form for $\varepsilon = \bar{r}/L_d = 0.01$; (**c**) entrainment form for $\varepsilon = 0.5$; and (**d**) entrainment form for $\varepsilon = 1$, using the thin-tailed exponential step length function of Eq. (8). In the case of the flux form and the entrainment form with $\varepsilon = 0.01$, slope increase is first realized upstream and then propagates downstream in time. For the case $\varepsilon = 0.5$, slope more or less increases simultaneously everywhere, corresponding to the rotational evolution in Fig. 2c. In the case $\varepsilon = 1$, slope first increases downstream, the effect then gradually propagating upstream in time.

ε against those for the flux formulation. In Fig. 2, bed elevation profiles are shown, having set as an upstream boundary condition $\hat{E}_f = 2$, so forcing the bed to aggrade. Case (a) is the solution for the flux form of Eq. (28), while cases (b), (c)

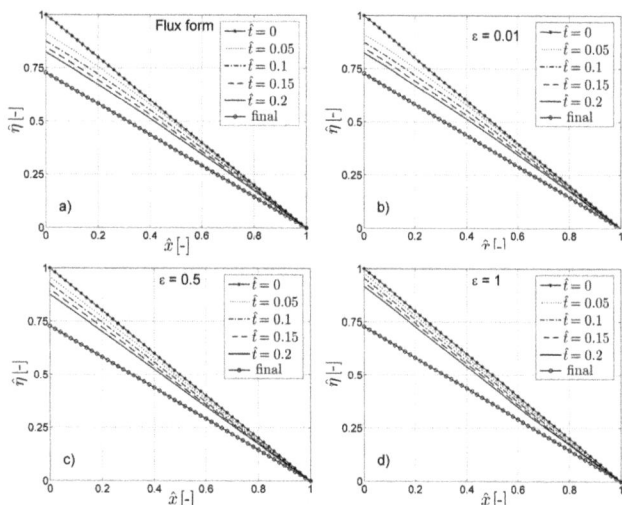

Figure 5. Bed profile evolution for the case $\hat{E}_f = 1/2$: (a) flux form; (b) entrainment form for $\varepsilon = \bar{r}/L_d = 0.01$; (c) entrainment form for $\varepsilon = 0.5$; and (d) entrainment form for $\varepsilon = 1$, using the thin-tailed exponential step length function of Eq. (8). As ε increases, the differences between the results from the two forms increase because of the nonlocality of particle movement, with an evolution from downward-concave transient profiles to upward-concave ones.

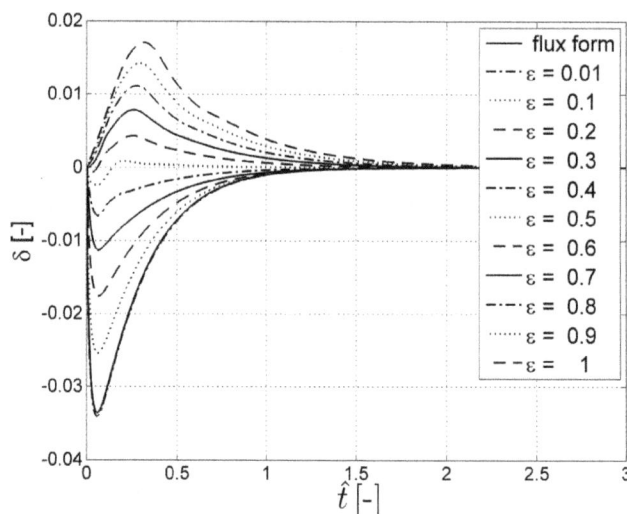

Figure 6. Degradation case: variation in time of the dimensionless concavity parameter δ in the case of the flux formulation and in the cases of the entrainment formulation for different values of ε ranging from 0.01 to 1. The result for the flux form overlaps with the result for the entrainment form with $\varepsilon = 0.01$. Note the reversal in transient behavior as ε increases beyond about 0.5.

and (d) are the solutions for the entrainment form of Eq. (29) solved, respectively, for $\varepsilon = 0.01, 0.5$, and 1.

As expected, the solutions of Eqs. (28) and (29) collapse to nearly the same results in the case $\varepsilon = 0.01$; i.e., when the mean particle step length is short compared to the length of the domain. Under this condition the local (flux) form, essentially coincides with the entrainment form. For higher values of ε, however, the differences between the results increase because the entrainment form is able to capture the nonlocal feature of the particle movement. For the flux form and the case $\varepsilon = 0.01$, the aggradational profile is strongly upward concave, with bed slope declining downstream. The transient aggradational bed profiles tend to assume a nearly linear profile, and thus the bed rotates upward for values of ε close to 0.5. For higher values a downward-concave form profile is realized.

To highlight and quantify this change in shape, we introduce a concavity parameter δ, which measures the deviation, in the center of the profile at $\hat{x} = 0.5$ relative to the constant initial slope:

$$\delta = \frac{0.5\,\hat{\eta}|_{\hat{x}=0} - \hat{\eta}|_{\hat{x}=0.5}}{\hat{\eta}|_{\hat{x}=0}}, \tag{37}$$

where $\hat{\eta}|_{\hat{x}=0}$ denotes the dimensionless bed elevation at $\hat{x} = 0$ and $\hat{\eta}|_{\hat{x}=0.5}$ denotes the same quantity in the center of the profile ($\hat{x} = 0.5$). Positive δ indicates upward concavity, while negative δ indicates downward concavity. In Fig. 3, the variation in time of δ is shown for the flux case, and different values of ε for the entrainment case. It is seen that δ is positive for smaller ε and but becomes negative for ε

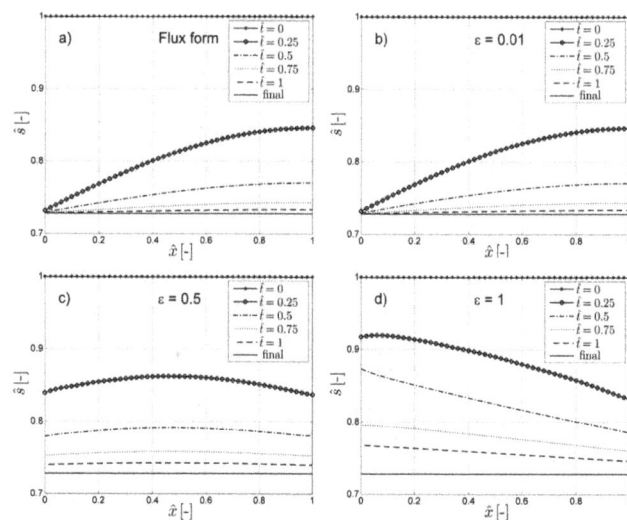

Figure 7. Slope profile evolution for the case $\hat{E}_f = 1/2$: (a) flux form; (b) entrainment form for $\varepsilon = \bar{r}/L_d = 0.01$; (c) entrainment form for $\varepsilon = 0.5$; and (d) entrainment form for $\varepsilon = 1$, using the thin-tailed exponential step length function of Eq. (8). The observed behavior corresponds to that of Fig. 4. In the case of the flux form and the entrainment form with $\varepsilon = 0.01$, slope decrease is first realized upstream, and then propagates downstream in time. For the case $\varepsilon = 0.5$, slope more or less decreases simultaneously everywhere, corresponding to the rotational evolution in Fig. 5c. In the case $\varepsilon = 1$, slope first decreases downstream, the effect then gradually propagating upstream in time.

greater than 0.5. The results for the flux form overlap with the form for $\varepsilon = 0.01$.

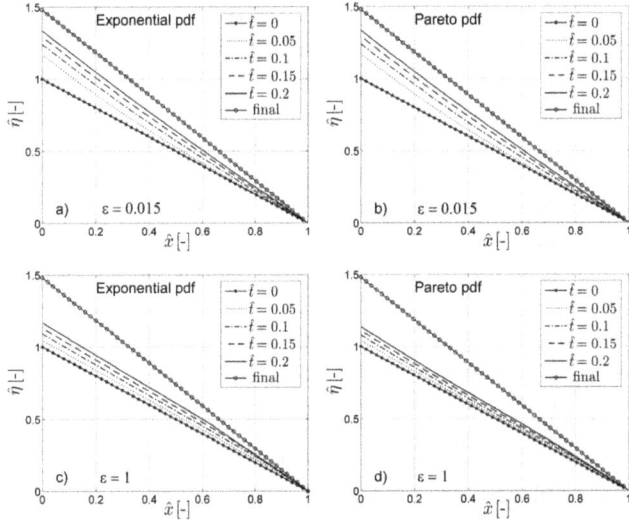

Figure 8. Bed profile evolution for the case $\hat{E}_f = 2$. (i) $\varepsilon = 0.015$: (a) thin-tailed exponential step length PDF; (b) heavy-tailed Pareto step length PDF ($\alpha = 1.5$, $r_0 = 1.5$ m). (ii) $\varepsilon = 1$ (c) thin-tailed exponential step length PDF; (d) heavy-tailed Pareto step length PDF ($\alpha = 1.5$, $r_0 = 100$ m). The shape of the tail of the step length PDF does not significantly change the results for $\varepsilon = 0.015$, but does result in some change compared to the thin-tailed case $\varepsilon = 1$. It should be realized that the numerical calculation has been carried out under the constraints $0 < \varepsilon = \bar{r}/L_d \leq 1$ and $0 \leq \hat{x} = x/L_d \leq 1$, constraints that preclude the evolution of asymptotic behavior.

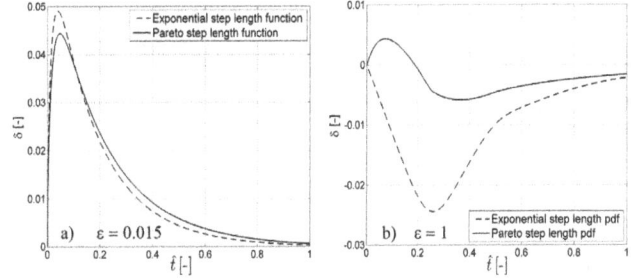

Figure 9. Variation in time of the concavity parameter δ for the case of the thin-tailed exponential distribution for step length, and the case of heavy-tailed Pareto distribution for step length. The parameter $\varepsilon = \bar{r}/L_d$ takes the value 0.015 in (a) and 1.0 in (b).

In Fig. 4, the slope evolution is plotted: the typical upward concave shape for the flux case and $\varepsilon = 0.01$ is due to the preferential proximal deposition of sediment, which causes the sediment load and thus the Shields number τ^* to decrease downstream (Parker, 2004). Thus, according to Eq. (15), a downstream decreasing slope is realized (Fig. 4a, b). However, a downward concave shape for $\varepsilon = 1$ is characterized by an increasing slope downstream (Fig. 4d). This corresponds to bedload particles that can jump from the upstream end of the domain to the downstream end in one step.

For completeness, the case of degradation, due to an imposed entrainment and feed rate upstream $\hat{E}_f = 1/2$, is described by Figs. 5, 6 and 7. The results show a congruent behavior with the aggradation case. In Fig. 5, for $\varepsilon = 0.01$ and $\hat{E}_f = 1/2$, it is seen that the two profiles more or less agree. In Fig. 6, the concavity parameters δ also more or less agree for this case. When ε increases to 1, the concavity of the transient degradational profiles changes from downward to upward. In Fig. 7, slope changes from increasing downstream to decreasing upstream. When $\varepsilon = 0.5$, it is shown in Fig. 7 that the transient profiles tend to keep a straight shape, and the evolution of the bed is essentially rotational about the downstream end.

Summarizing, (i) the flux model and the entrainment model yield essentially the same results for $\varepsilon = 0.01$; (ii) for $\varepsilon = 0.5$, nearly rotational aggradation and degradation are obtained; and (iii) for $\varepsilon = 1$, the pattern of concavity is reversed compared to the flux case.

Then, a Pareto distribution with a shift, i.e., Eq. (9) for particle step length distribution, is considered as well so as to compare the case of heavy tail of the PDF of step length with the thin-tail exponential form. In the calculations for the entrainment rate with $\hat{E}_f = 2$, two cases are evaluated: (a) $\varepsilon = 0.015$ and (b) $\varepsilon = 1$. It is seen that the two profiles more or less agree for case (a). A more substantial difference is seen for case (b), but the concavity is quite small for both the cases of thin-tailed and heavy-tailed PDFs for step length. Assuming $L = 200$ m, with a thin-tailed PDF the value $\varepsilon = 0.015$ corresponds to a mean step length equal to 3 m, and the value $\varepsilon = 1$ corresponds to 200 m. We have set the shape parameter α in the Pareto PDF equal to 1.5, and the scale parameter r_0 equal to 1.5 m for case (a), and t at 100 m for case (b). This yields values of $\bar{r}$ from Eq. (10) that are respectively equal to 3 and 200 m; i.e., the same values as for the thin-tailed case.

The analysis shows that the shape of the tail of the step length PDF does not significantly change the results for $\varepsilon = 0.015$ but does result in some change compared to the thin-tailed case $\varepsilon = 1$. Figure 8 shows the long profiles resulting from both the thin-tailed and heavy-tailed cases, and Fig. 9 shows the corresponding evolution of concavity. As seen in Fig. 9c and d corresponding to the case of aggradation with $\varepsilon = 1$, the profiles are downward-concave for the thin-tailed PDF of step length, and upward-concave for the heavy-tailed case. The concavity in both cases, however, is so small that the same rotational behavior for profile adjustment is seen, as documented in Fig. 8c and d.

In interpreting the results regarding the thin-tailed and heavy-tailed cases, it should be recalled that the problem is solved numerically only over the domain $0 < \hat{x} \leq 1$, with the further constraint $0 < \varepsilon = \bar{r}/L_d \leq 1$. This constraint prevents attainment of an asymptotic nonlocal state. An example of an asymptotical form is given below.

4   Sample asymptotic nonlocal relation for entrainment form of mass conservation

Taking the spatial Fourier transform of Eq. (4) results in the form

$$\frac{\partial \hat{\eta}}{\partial t} = -\hat{E} + \hat{E}\hat{f}_s, \tag{38}$$

where the Fourier transform of any parameter $Z(x)$ is given as

$$\hat{Z}(k) = \int_{-\infty}^{\infty} Z(x)e^{-ikx}\mathrm{d}x. \tag{39}$$

Following the analysis of e.g., Ganti et al. (2010), we assume that the PDF $f_s(r)$ has a mean $\bar{r}$ but no standard deviation, so that $\hat{f}_s(k)$ can be expanded in asymptotic form

$$\hat{f}_s(k) \cong 1 - ik\bar{r} + c_\alpha, (ik)^\alpha, \tag{40}$$

where $1 < \alpha < 2$. The implication of this is that $f_s(r)$ has a power-law tail. Substituting Eq. (40) into Eq. (38), inverse-transforming back to real space and reducing with Eq. (17), it is found that Eq. (4) reduces to

$$\frac{\partial \eta}{\partial t} \cong -\frac{\partial q}{\partial x} + \frac{c_\alpha}{\bar{r}}\frac{\partial^\alpha q}{\partial x^\alpha}. \tag{41}$$

In so far as q is specified by Eq. (16), Eq. (41) takes the form of a nonlinear fractional PDE (partial differential equation). While the asymptotic form is of interest from a theoretical point of view, numerical solutions of specific problems are more easily carried out in terms of the original convolution form of Eq. (4).

5   Discussion and conclusions

The main goal of the work is to show how the entrainment form of the Exner equation of sediment continuity diverges from the flux form of the Exner equation when nonlocal behavior in particle motion arises: (i) as the mean particle step length $\bar{r}$ increases from 0 to the order of magnitude of the domain length L_d for a thin-tailed step length PDF and (ii) as a heavy-tailed PDF for particle step length is used.

The dimensionless parameter ε is defined as the ratio between the mean step length $\bar{r}$ and the length of the domain of interest L_d. We analyzed the effect of variation of ε on bed aggradational/degradational profiles by solving the entrainment form of the Exner equation, with the assumption of a thin-tailed PDF for particle step length. As expected, the two forms collapse in the case $\varepsilon \ll 1$.

For high values of ε, however, the differences between the results from the two forms increase because of the nonlocality of particle movement, which is not captured by the classical flux form of the Exner equation: the transient aggradational (degradational) bed profiles tend to assume, for ε

greater than 0.5, a downward (upward) concave shape, rather than the upward (downward) concave shape of the flux form. When the value of ε is close to 0.5, an interesting behavior for both cases of aggradation and degradation has been found: the transient profiles tend to rotate around the downstream point, keeping almost a straight shape. For a value of ε in the range [0,0.5], the concavity of the bed profiles is still upward for aggradation and downward, for degradation, but by increasing ε to 0.5, the concavity is nearly vanishing. These results may serve as an explanation for relatively flat aggradational bed profiles, which have been achieved in some short laboratory experiments (e.g., Muto, 2001; and Voller and Paola, 2010), where the value of the ratio between mean particle step length and length of the domain of interest may not be negligible. At the laboratory scale, the mean step length becomes comparable to domain length so that the inclusion of nonlocal effects in the PDF of step length, which this circumstance entails, should clearly be evaluated in order to properly model the bed evolution.

The analysis also investigates the effect of the heavy tailedness in the PDF of step length on the bed profile. For the case studied, we show that the variation of the shape of the step length distribution from thin- to heavy-tailed does not significantly influence the results when step length is small. This is probably due to the "short" domain length compared to the tail of the power law distribution. There is a somewhat larger difference in the case when step length equals domain length, but the bed elevation profiles are nearly linear for both thin-tailed and heavy-tailed PDFs. Voller and Paola (2010) introduced heavy-tailed behavior to explain profiles that evolve with concavity that is small compared to the standard flux case of Eq. (1). Here we find that a heavy-tailed behavior is not necessary to obtain this result.

Recently Falcini et al. (2013) have presented a nonlocal formulation for sediment transport and bed evolution that bears comparison to the present work. They assume a locally-determined "reference [sediment transport rate] q^L whose physical interpretation requires some care", and then integrate this with a nonlocal weighting function to determine the actual sediment transport rate q at a section. Their analysis can yield an upward concave nonlinear final equilibrium state in the absence of subsidence (which is not included in this analysis as well), whereas the present analysis predicts only equilibrium states with constant slope. That is, in our analysis, profile concavity or convexity is a transient phenomenon. We suggest that our analysis has a somewhat clearer basis than that of Falcini et al. (2013), who determine their weighting function from heuristic considerations.

Long step lengths of bedload particles in the field may result from any bed pattern that induces preferential paths for transport, including grain size mixtures (Ganti et al., 2010), bedforms, scour and fill, and intermittent bedrock exposure (Stark et al, 2009). Thus our results may be applicable to these cases. The case of sediment suspension can also be represented in entrainment form (e.g., Parker,

2004). This case is generally associated with much longer mean path lengths than the case of bedload. As a result, the suspension-dominated case may show much more non-local behavior than the bedload case. This case deserves further investigation.

Notation

B	channel width (L)
$D(x)$	deposition rate (L T^{-1})
D_c	characteristic bed grain size (L)
$E(x)$	entrainment rate (L T^{-1})
$\hat{E}$	normalized E against initial conditions
E_f	entrainment rate upstream of $x=0$ (L T^{-1})
$\hat{E}_f$	normalized E_f against initial conditions
$f_{ls}(x)$	probability (L^{-1}) that a particle travels at least a distance x
$f_s(r)$	probability density (PDF) of step lengths (L^{-1})
$\tilde{f}_s\left(\frac{r}{\varepsilon}\right)$	dimensionless step length PDF
g	gravitational acceleration (L T^{-2})
H	water depth (L)
k_c	composite roughness height (L)
k_s	roughness height (L)
L_d	domain length (L)
L	lenght unit
M	mass unit
q	volume bedload transport rate per unit width (L^{2}T^{-1})
$\hat{q}$	normalized q against initial conditions
q_f	sediment transport rate at the upstream end (L^{2}T^{-1})
Q_w	water discharge (L^{3}T^{-1})
r	step length r of a bedload particle (L)
$\bar{r}$	mean particle step length (L)
$\hat{r}$	dimensionless particle step length
r_0	scale parameter for the Pareto step length PDF (L)
R	submerged specific gravity of the sediment
S	bed slope
$\hat{s}$	normalized slope against initial conditions
T	time unit
t	time (T)
$\hat{t}$	dimensionless time
u_*	shear velocity (L T^{-1})
U	depth-averaged flow velocity (L T^{-1})
x	streamwise distance (L)
$\hat{x}$	dimensionless streamwise distance
α	shape parameter for the Pareto step length PDF
α_r	dimensionless coefficient in the Manning–Strickler resistance relation
β	ratio between particle mean step length and grain size
γ	coefficient of proportionality in the bedload transport formulae
δ	concavity parameter
ε	ratio between particle mean step length and domain length
η	bed elevation (L)
$\hat{\eta}$	dimensionless bed elevation
ν	kinematic diffusivity (L^{2}T^{-1})
ρ	water density (M L^{-3})
τ_b	bed shear stress (M L^{-1}T^{-2})
τ^*	Shields number
τ_c^*	threshold Shields number

Acknowledgements. Anna Pelosi was supported by the Ph.D in Civil and Environmental Engineering program of the University of Salerno and hosted by the CEE Department of the University of Illinois at Urbana-Champaign. This research was motivated in part by a comment from D. Jerolmack.

Edited by: F. Metivier

References

Bradley, D. N., Tucker, G. E., and Benson, D. A.: Fractional dispersion in a sand bed river, J. Geophys. Res., 115, F00A09, doi:10.1029/2009JF001268, 2010.

Charru, F.: Selection of the ripple length on a granular bed sheared by a liquid flow, Phys. Fluids, 18, 121508, 2006.

Chaudhry, M. H.: Open-Channel Flow, Prentice-Hall, Englewood Cliffs, p. 483, 1993.

Du, Q., Gunzburger, M., Lehoucq, R. B., and Zhou, K.: Analysis and approximation of nonlocal diffusion problems with volume constraints, SIAM Review, 54, 667–696, 2012.

Einstein, H. A.: The Bed-load Function for Sediment Transportation in Open Channel Flows, Technical Bulletin 1026, US Dept. of the Army, Soil Conservation Service, Washington, DC, 1950.

Exner, F. M.: Zur physik der dunen, Akad. Wiss. Wien Math. Naturwiss. Klasse, 129, 929– 952, 1920.

Exner, F. M.: U ber die wechselwirkung zwischen wasser und geschiebe in flussen, Akad. Wiss. Wien Math. Naturwiss. Klasse, 134, 165– 204, 1925.

Falcini, F., Foufoula-Georgiou, E., Ganti, V., Paola, C., and Voller, V. R.: A combined nonlinear and nonlocal model for topographic evolution in channelized depositional systems, J. Geophys. Res., 118, 1617–1627, 2013.

Foufoula-Georgiou, E., Ganti, V., and Dietrich, W.: A nonlocal theory of sediment transport on hillslopes, J. Geophys. Res., 115, F00A16, doi:10.1029/2009JF001280, 2010.

Ganti, V., Meerschaert, M. M., Foufoula-Georgiou, E., Viparelli, E., and Parker, G.: Normal and anomalous diffusion of gravel tracer particles in rivers, J. Geophys. Res., 115, F00A12, doi:10.1029/2008JF001222, 2010.

Hassan, M. A., Voepel, H., Schumer, R., Parker, G., and Fraccarollo, L.: Displacement characteristics of coarse fluvial bed sediment, J. Geophys. Res. Earth Surf., 118, 155–165, doi:10.1029/2012JF002374, 2013.

Hill, K. M., DellAngelo, L., and Meerschaert, M. M.: Heavy-tailed travel distance in gravel bed transport: An exploratory enquiry, J. Geophys. Res., 115, F00A14, doi:10.1029/2009JF001276, 2010.

Martin, R. L., Jerolmack, D. J., and Schumer, R.: The physical basis for anomalous diffusion in bed load transport, J. Geophys. Res., 117, F01018, doi:10.1029/2011JF002075, 2012.

Meyer-Peter, E. and Müller, R.: Formulas for Bed-Load Transport, Proceedings, 2nd Congress, International Association of Hydraulic Research, Stockholm, 39–64, 1948.

Muto, T: Shoreline autoretreat substantiated in flume experiment. J. Sed. Res., 71, 246–254, 2001.

Nakagawa, H. and Tsujimoto, T.: A Stochastic Model for Bed Load Transport and its Applications to Alluvial Phenomena, in: Application of Stochastic Processes in Sediment Transport, Water Resources Publications, edited by: Shen, H. W. and Kikkawa, H., Colorado, USA, 1-1–11-54, 1980.

Nikora, V., Habersack, H., Huber, T., and McEwan, I.: On bed particle diffusion in gravel bed flows under weak bed load transport, Water Resour. Res., 38, 1081, doi:10.1029/2001WR000513, 2002.

Nino, Y., Garcıa, M. and Ayala, L.: Gravel saltation: 1. Experiments, Water Resour. Res., 30, 1907–1914, 1994.

Paola, C. and Voller, V. R.: A generalized Exner equation for sediment mass balance, J. Geophys. Res., 110, F04014, doi:10.1029/2004JF000274, 2005.

Paola, C., Heller, P. L., and Angevine, C. L.: The largescale dynamics of grain-size variation in alluvial basins, 1: Theory, Basin Res., 4, 73–90, 1992.

Parker, G.: Sediment inertia as a cause of river antidunes, J. Hydraul. Eng., 101, 211–221, 1975.

Parker, G.: Selective sorting and abrasion of river gravel, II: Applications, J. of Hydraul. Eng., 117, 150–171, 1991.

Parker, G.: 1D Sediment Transport Morphodynamics with Applications to Rivers and Turbidity Currents, Copyrighted e-book, available at: http://hydrolab.illinois.edu/people/parkerg//morphodynamics_e-book.htm, 2004.

Parker, G.: Transport of gravel and sediment mixtures, Sedimentation engineering processes, measurements, modeling and practice, ASCE Manual and Reports on Engineering Practice, 110, edited by: Garcia, M., Am. Soc. of Civ. Eng., New York, 3, 165–252, 2008.

Parker, G., Paola, C., Whipple, K., and Mohrig, D.: Alluvial fans formed by channelized fluvial and sheet flow. I: Theory, J. Hydraul. Eng., 124, 985–995, 1998.

Parker, G., Paola, C., and Leclair, S.: Probabilistic Exner sediment continuity equation for mixtures with no active layer, J. Hydrual. Eng., 126, 818–826, 2000.

Postma, G., Kleinhans, M. G., Meijer, P. T., and Eggenhuisen, J. T.: Sediment transport in analogue flume models compared with real world sedimentary systems: A new look at scaling sedimentary systems evolution in a flume, Sedimentology, 55, 1541–1557, doi:10.1111/j.1365-3091.2008.00956.x., 2008.

Sayre, W. and Hubbell, D.: Transport and dispersion of labeled bed material, North Loup River, Nebraska, US Geol. Surv. Prof. Pap., 433-C, 48 pp., 1965.

Schumer, R., Benson, D. A., Meerschaert, M. M., and Baeumer B.: Fractal mobile/immobile solute transport, Water Resour. Res., 39, 1296, doi:10.1029/2003WR002141, 2003.

Schumer, R., Meerschaert, M. M., and Baeumer, B.: Fractional advection-dispersion equations for modeling transport at the Earth surface, J. Geophys. Res., 114, F00A07, doi:10.1029/2008JF001246, 2009.

Stark, C. P., Foufoula-Georgiou, E., and Ganti, V.: A nonlocal theory of sediment buffering and bedrock channel evolution, J. Geophys. Res., 114, F01029, doi:10.1029/2008JF000981, 2009.

Tsujimoto, T.: Probabilistic model of the process of bed load transport and its application to mobile-bed problems, Ph.D. thesis, Kyoto Univ., Kyoto, Japan, 174 pp., 1978.

Voller, V. R. and Paola, C.: Can anomalous diffusion describe depositional fluvial profiles?, J. Geophys. Res., 115, F00A13, doi:10.1029/2009JF001278, 2010.

Voller, V. R., Ganti, V., Paola, C., and Foufoula-Georgiou, E.: Does the flow of information in a landscape have direction?, Geophys. Res. Lett., 39, L01403, doi:10.1029/2011GL050265, 2012.

Whipple, K. X., Parker, G., Paola, C., and Mohrig, D.: Channel Dynamics, Sediment Transport, and the Slope of Alluvial Fans: Experimental Study, J. Geol., 106, 677–693, 1998.

Wong, M. and Parker, G.: Reanalysis and Correction of Bed-Load Relation of Meyer-Peter and Müller Using Their Own Database, J. Hydraul. Eng., 132, 1159–1168, 2006.

Wong, M., Parker, G., DeVries, P., Brown, T. M., and Burges, S. J.: Experiments on dispersion of tracer stones under lower-regime plane-bed equilibrium bed load transport, Water Resour. Res., 43, W03440, doi:10.1029/2006WR005172, 2007.

12

Development of a meandering channel caused by the planform shape of the river bank

T. Nagata[1], Y. Watanabe[2], H. Yasuda[3], and A. Ito[1]

[1]Civil Engineering Research Institute for Cold Region, Public Works Research Institute, Japan
[2]Department of Civil and Environmental Engineering, Kitami Institute of Technology, Japan
[3]Research Center for Natural Hazard & Disaster Recovery, Niigata University, Japan

Correspondence to: T. Nagata (03213@ceri.go.jp)

Abstract. Due to a typhoon and a stationary rain front, record amounts of rain fell in September 2011, and the largest class of discharge in recorded history was observed in the Otofuke River of eastern Hokkaido in Japan, and extensive bank erosion occurred in various parts of the river channel. Damages were especially serious in the middle reaches, where part of a dike was washed out. The results of a post-flood survey suggested that the direct cause of the dike breach was lateral advance of the bank erosion associated with the development of meandering channels. As the related development mechanism and predominant factors have not yet been clarified, this remains a priority from the viewpoint of disaster prevention. A past study on the development of meandering channels was reported by Shimizu et al. (1996). In this study, the meandering channel development process was reproduced using a slope failure model that linked bank erosion with bed changes. The study attempted to clarify the meandering development mechanism in the disaster and its predominant factors by using this model. The analysis properly reproduced the characteristics of the post-flood meandering waveforms. Therefore, it is suggested that the development of meandering during the flood attributed to the propagation of meandering downstream, which is triggered by the meandering flow from the meandering channel in the upstream, which also suggests that this propagated meandering then caused a gradual increase of meandering amplitude accompanied by bank erosion in the recession period of the flood.

1 Introduction

Due to a typhoon and a stationary rain front, a record amount of rain fell in September 2011. Discharge of the largest class was observed in the Otofuke River of the Tokachi River basin, and extensive bank erosion occurred in various parts of the river channel (Fig. 1). In the area near the left bank of KP18.2 at the middle reaches, where the erosion was the most severe, part of the river dike was almost entirely washed away (Fig. 2). Post-flood surveys revealed the direct cause of the dike breach to be the bank erosion that progressed during the development of meandering flow in the low-water channel. However, since the development of meandering flow on such a great scale was never observed for this river, the mechanism and dominant factors of this phenomenon are not fully understood. Clarifying the mechanism and dominant factors

of this phenomenon is an urgent issue toward developing and implementing appropriate and effective preventive measures.

A distinguishing feature of this flood is that the extreme discharge continued for a long time, filling the low-flow channel almost to the crest (40 h at the average maximum yearly discharge of 155 m^3 s^{-1}), which shows that channel migration during the flood was dominated by the action of the water flow running through the low-water channel and suggests that channel migration was associated with the mechanism of sandbar development. Additionally, the flow channel geometry left after the flood had the pattern of a single low watercourse that extensively meandered between the dikes due to the erosion of the low-water channel, which suggests that the channel migration was associated with the mechanism of meandering channel development. That is to

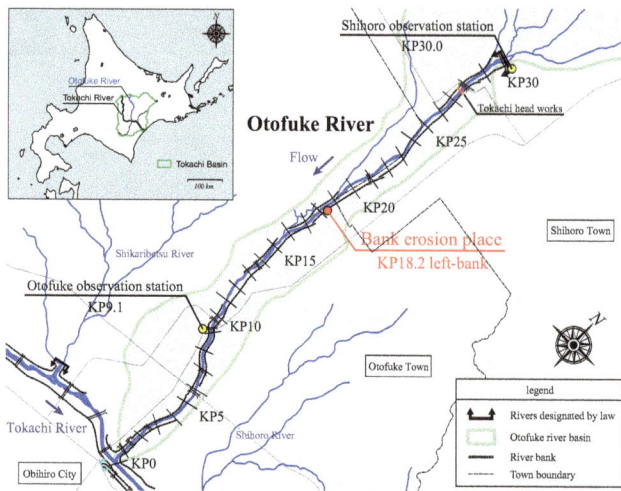

Figure 1. Bank disaster area (Otofuke River, Hokkaido, Japan).

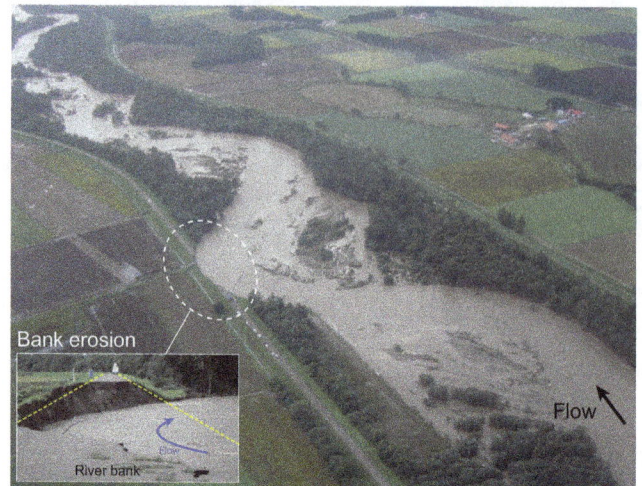

Figure 2. Situation at the time of bank disaster (7 Sep. 2011).

say, there is a possibility that the development of meandering flows in the flood occurred under the interrelated influence between the mechanism of sandbar development and the mechanism of meandering channel development operating in the low-water channel.

Previous studies conducted by the authors (e.g., Nagata et al., 2013) addressed sandbar topography as a factor in the development of meandering flow. Experiments and analyses confirmed that the topography of alternating bars can be a factor in the development of meandering channels. However, the phenomenon that occurred in the area around the damaged bank of the Otofuke River was so dynamic that even the wave number of meandering channels decreased; therefore, it is difficult to fully explain some aspects of this phenomenon only by the development of gravel-bar-derived meandering flow.

The most dominant factor after sandbar topography is the planar configuration of the riverbank. As confirmed in a field survey, the riverbank of a low-water channel sometimes forms with an extremely developed sandbar; therefore, there is no substantial difference between them as topography-derived factors. A sandbar whose wave height has increased to the height of the low-water riverbank is assumed to behave like a low-water riverbank, in the sense that the sandbar redirects the flood flow that runs in the low-water channel.

In light of the above, this study addresses both the sandbar topography and the planar configuration of the low-water riverbank and conducts various examinations using numerical analysis toward identifying major factors in damage to this river dike.

2   River channel evolution process

2.1   Major external forces and the river channel formation process

To estimate the factors that brought about the bank disaster, the river channel formation process was investigated in the section extending from KP17.0 to KP21.0, which includes the damaged location. The aerial photos in Fig. 3 indicate the typical changes that took place in the river channel during the roughly 30 year period from the late 1970s to the post-flood time. The chronological table (on the left) shows river improvement work, which is an unnatural external force; major floods, which are a nature-derived external force; and an image of the meandering channel that developed as a result of those forces. In this section, large-scale river work was performed in the 1970s to straighten the low-water channel, and this period is marked as a starting point of the river channel formation that has been continuing up to the present. After that, the largest recorded flood took place, in 1981, and it triggered the further development of meandering channels.

Figure 4 shows the flow regime of the period when discharge decreased during the 1981 flood. The red line represents the riverbank of the low-water channel along the normal line of river channel, and the blue line represents the main streamline. At this point, three large meandering parts had already formed in the upstream section (KP18.4–KP21.0), and as shown in Fig. 3, these flows (hereinafter: M-1, M-2 and M-3 from downstream to upstream) gradually increased the degree of meandering over the period of 30 years.

In addition, due to the state of the river channel in 2005, river work was again performed to straighten the low-water channel immediately downstream of KP18.6, where the turning point of meandering curvature had come close to the river dike. The upstream side of KP18.6 is surrounded by

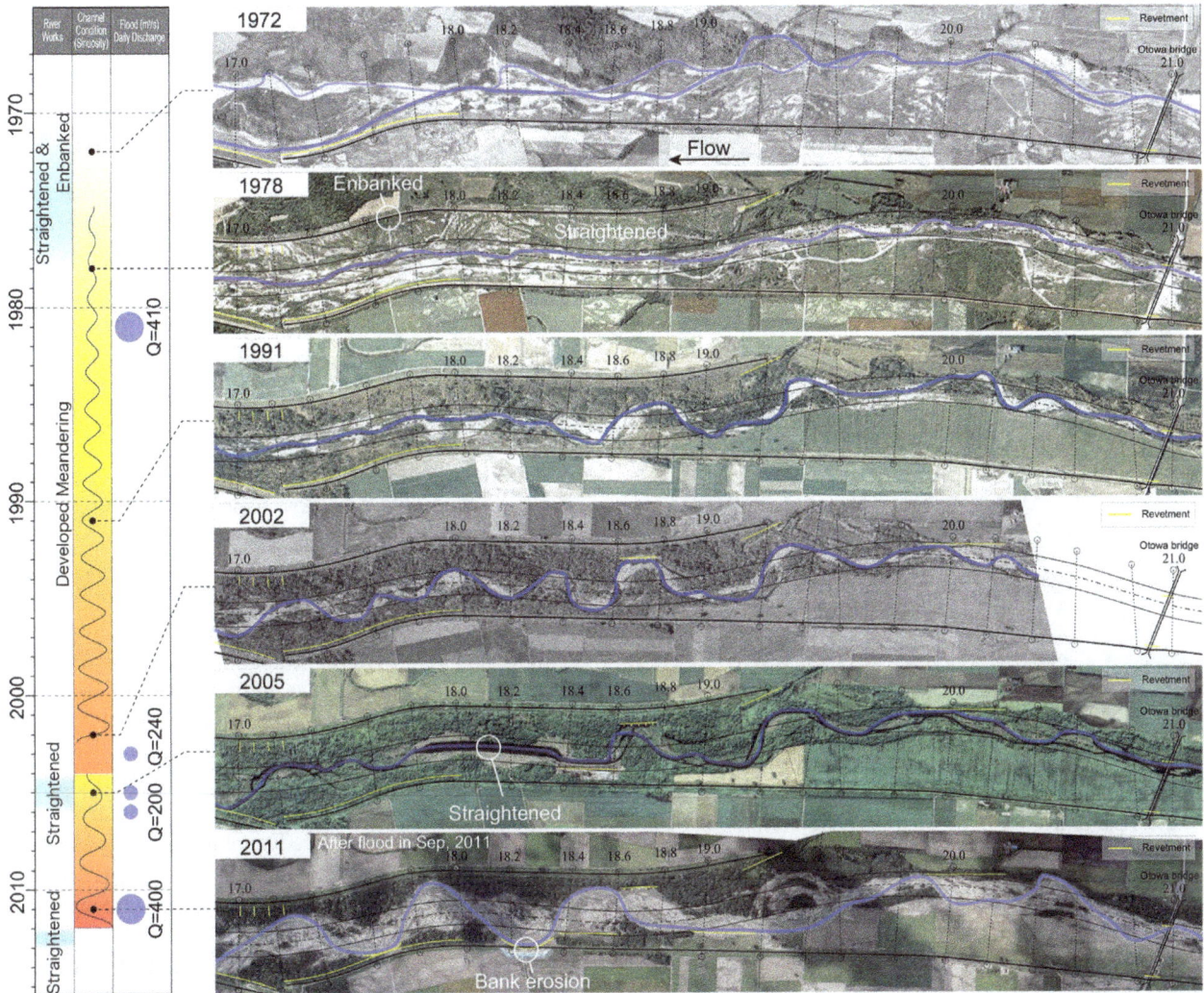

Figure 3. Migration history of the river channel (Otofuke River, KP17.0–KP21.0).

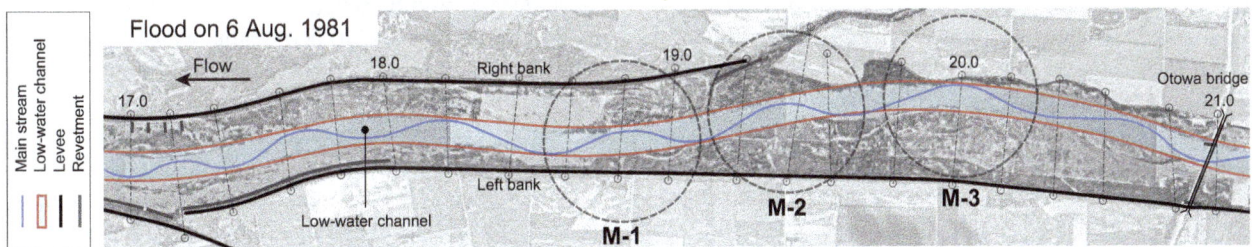

Figure 4. Main streamline and front bank line of the low-water channel during the 1981 flood (recession period).

a floodplain and terraces that have been serving as an embankment; thus, safety has been ensured. Therefore, large-scale river improvement work had never been performed there. Hence, it is reasonable to assume that there was a major difference in the state of the river channel between the upstream side of the KP18.6 area and downstream side of the KP18.6 area. In light of this on-site situation, analysis was performed separately for the upstream section versus for the downstream section. KP18.4–KP21.0, which is on the upstream side, is referred to as Section-1; KP17.0–KP19.0, which is on the downstream side, is referred to as Section-2. In addition, there is an overlap between two sections, because the meandering part in M-1 plays a key role in this analysis, which will be discussed later.

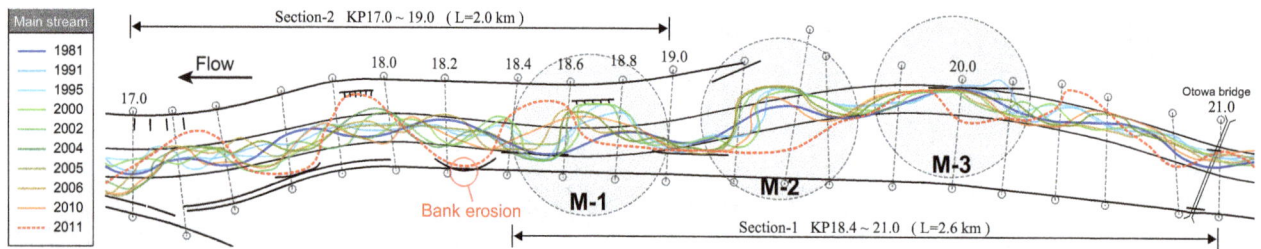

Figure 5. Changes in the configuration of meandering channels (1981–2011, KP17.0–KP21.0).

Figure 6. Planform shape of the low-water channel in 1978.

2.2   Watercourse change

Figure 5 diagrams the horizontal curve of the main stream-line from the aerial photos taken in 10 different years over the past 30 years. The changes from 1981 to 2010 show that the watercourse shifted repeatedly and irregularly toward the left and right banks immediately downstream of KP18.6. In contrast, M-1, M-2 and M-3, which are on the upstream side of KP18.6, shifted their phase slightly downstreamward, while shifting their waveforms forward. Also, M-1, M-2 and M-3 increased the degree of meandering in a single direction; however, the sandbar remained at an almost fixed position. Thus, the upstream side shows changes that are obviously different from those of the downstream side.

In general, in a straight river channel, alternating bars tend to move along the direction of flow; in a meandering channel, however, they have the property of remaining roughly stationary. Kinoshita found that sandbar movement and lack of movement are determined by the meandering wavelength, the channel width and the meandering angle of the channel, and that the meandering angle has a certain limiting gradient at which sandbars stop moving (e.g., Kinoshita et al., 1974). As clearly shown in Fig. 5, the normal line in the low-water channel forms a large curvature whose vertex is near KP20.0. That is to say, the above finding suggests the possibility that

the planform shape of the riverbank of the low-water channel induced the development of point bars in M-1, M-2 and M-3.

3   Relation between bend of the normal line in low-water channels and development of point bars

Given the above background, an analysis was made of Section-1 (KP18.4–21.0) at first with the aim of evaluating the relation between the bend of the normal line in low-water channels and the development of point bars in M-1, M-2 and M-3. In identifying dominant factors in phenomena that occur in the field, it is more appropriate to conduct model tests under simplified conditions than to analyze complex field data. Therefore, the authors had already conducted a movable-bed hydraulic model test under various conditions before conducting this study. The reproducibility of the sediment transport formula used in this study was tested on the basis of measurement values obtained in this model test.

In the model test, in order to maintain the similarity of the hydraulic and sediment transport phenomena between the site and the model, various experimental conditions in the model test were set such that the values of the dimensionless quantity (Fr, τ_*), which have a dominant impact on both phenomena, would be consistent. This resulted in maintaining the similarity of width/depth ratio, which greatly affects the formation and configuration characteristics of alternating bars, at the same time. However, perfect similarity cannot be maintained between the actual river and the model (similarity of the Reynolds number of the particles is not fulfilled). Therefore, it is unclear how much applicability the sediment transport formula, whose reproducibility was confirmed in the model-scale experiment, will have in a full-scale experiment. In addition, it is difficult to measure sediment transport during a flood, which means that it is also difficult to perform comparative verification using measured values.

Given the above, it was determined to be appropriate to perform the analysis at a model scale that was confirmed to be able to reproduce hydraulic and sediment transport phenomena, and then to convert the obtained results to those at actual scale and discuss the phenomena that occurred at the actual place. Note that, hereinafter, in order to achieve consistency with previous studies (e.g., Nagata et al., 2013), the analysis was conducted at 1 : 100 scale; however, in order to

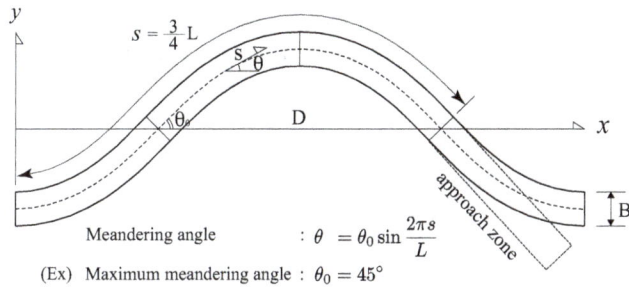

Figure 7. Definition of the sine-generated curve.

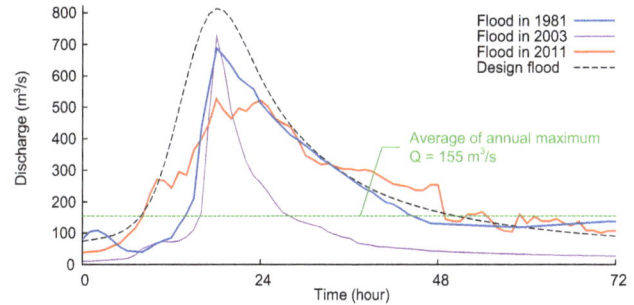

Figure 8. Discharge hydrographs of the major floods.

facilitate the comparison with the actual site, in this paper, the numerical values obtained in the calculation results are converted into actual measurement values.

3.1 Calculation condition (Section-1)

As the initial condition of the river channel in Section-1, the channel configuration after the river was channelized into a straighter channel with only low-amplitude curvature (Fig. 6: before the 1981 flood) and was simplified as follows. In the planform shape of the low-water channel, the normal line in the low-water channel was approximated by the sine-generated-curve shown in Fig. 7, and the low-water channel represented by the red line in Fig. 5 was designed to be a meandering channel with a meandering angle θ_0 of 13° and river width of 100 m. On the basis of previous survey data, the cross-section profile of the low-water channel was designed to have a 2 m-high bank with a slope gradient of 2 : 1, and the riverbed surface was designed to give particle-sized disturbance to the flat bed. The left and right sides of the low-water channel were provided with a 100 m-wide floodplain that allows bank erosion, and the entire calculation area was a movable riverbed. A longitudinal slope of 0.00610 was used, which was the average value for Section-1.

Incidentally, since a bridge (Otowa Bridge) was built near KP21.0 at the actual site in Otofuke, this point was determined to be the upstream end of the analysis section, and a straight river channel that did not include curvature was set on the farther upstream side as an approach zone and was connected to the analysis section.

Figure 8 compares discharge hydrographs of previous major floods. The results of the analysis for Section-1 make it possible to understand the characteristics of bed morphology formed by the 1981 flood, in addition to how the riverbed responded to the steady flow. The figure shows that the discharge of the 1981 flood had a scale comparable to the design-flood discharge. In addition, the 2011 flood fell below the 1981 flood in terms of discharge during peak discharge; however, the 2011 flood exceeded the 1981 flood in terms of the duration of the average annual maximum discharge.

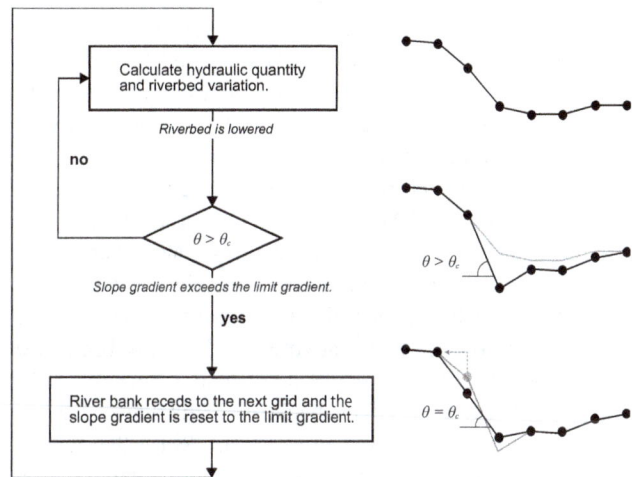

Figure 9. Slope failure model (refer to Nays2D solver manual).

3.2 Calculation model

The analysis performed in this study used the iRIC river analysis software package and its solver Nays2D v4.0 developed by Shimizu (e.g., Shimizu, 2003). Governing equations used in the model are a two-dimensional plane, shallow-flow equation for the unsteady flow and a continuous flow formula; and the amount of riverbed evolution is calculated by a sediment transport formula and the continuous formula of sediment transport. Details are omitted here. Please refer to the website of iRIC (http://i-ric.org/en/) for more information. In addition, for calculating the sediment transport, Eq. (1) was used, which was developed on the basis of the Ashida–Michiue formula by adjusting the coefficients of the formula in light of previous experimental results.

As Wong and Parker showed that the coefficient of Meyer-Peter and Müller equation is excessive in the past study (e.g., Wong and Parker, 2006), and Lajeunesse and others organized the coefficient of sediment transport equation that was used in the past (e.g., Lajeunesse et al., 2010), various values are proposed for the coefficient of sediment transport equation. Equation (1) shows a near value by the Meyer-Peter and Müller equation that Wong and Parker showed in comparison to an original Ashida–Michiue formula (coefficient: 17).

$$q_{\mathrm{b}} = 13\,\tau_*^{1.5}\left(1 - \frac{\tau_{*c}}{\tau_*}\right)\left(1 - \sqrt{\frac{\tau_{*c}}{\tau_*}}\right)\sqrt{sgd^3} \tag{1}$$

The details are as follows: q_{b}, bedload transport rate per unit width (m^2 s^{-1}); τ_*, dimensionless tractive force; τ_{*c}, dimensionless critical bed shear stress (Iwagaki formula); s, specific gravity of sand grains; g, gravitational acceleration (m^2 s^{-1}); and d, sand grain size (m).

$$n = \frac{d^{1/6}}{6.8\,\sqrt{g}} \tag{2}$$

The grain size of bed material was determined to be $d_{60} = 50$ mm from the survey results of 2011, and the Manning–Strickler formula shown in Eq. (2) was used to obtain the roughness coefficient. In setting a condition for sediment transport it was determined to use only bedload sediment, which was regarded as having the same grain size.

Furthermore, in the present study, a slope failure model was used to reproduce the bank erosion phenomenon (Fig. 9). This model is designed such that the low-water riverbank is simulated to collapse naturally when the slope gradient exceeds a certain limit; thus, the bank erosion phenomenon is reproduced indirectly by moving the riverbank backward to maintain the limiting gradient; at which time the sediment budget is balanced by backfilling the lower part of the riverbed with the collapsed sediment. Since the present model is not intended to physically solve for bank erosion phenomena, there still remains the challenge that the results of the analysis depend on the choice of computational grid; however, previous studies have proved that the development of meandering flow is able to be reproduced to a certain degree. In this analysis, the limiting gradient of the slope is set as $\theta_c = 25°$. Additionally, in all the simulations in this study, uniform flow depth was given as a boundary condition at the downstream edge of the analysis section and sediment budget at the upstream edge was assumed to maintain a state of equilibrium.

3.3   Calculation results (Section-1)

Riverbed elevations after 3 days of steady flow are compared in the plan view of the riverbed elevations in Fig. 10. The figure shows the elevation differences of the riverbeds as compared by using a constant bed slope of 0.00610. Hereafter, elevation differences are compared in the same manner.

The results shown in the figure suggest that discharge of $300\,\mathrm{m}^3\,\mathrm{s}^{-1}$ had a dominant influence on the development of the meandering flows. Moreover, what should be noted here is that the locations of the meandering channels (point bars) formed on the downstream side of the bend of the river channel and at intervals between them. Wavelengths slightly varied depending on the scale of discharge; however, the meandering channels that have wavelengths of approximately 600 m, formed at almost regular intervals. These meandering

channels are equivalent to M-1, M-2 and M-3 at the actual field site along the Otofuke River, and the locations and intervals of the meandering channels in the analysis are roughly in accordance with those at the site shown in Fig. 3 (meandering channel configurations: 1991–2005, KP18.5–KP20.3).

Next, riverbed elevations were compared after discharge with the same flow rate as that of the 1981 flood was introduced for 3 days (red line in Fig. 8), which is shown in the plan view of the riverbed elevations in Fig. 11. In the figure, the blue dotted line represents the configuration of the main stream in the recession period of the 1981 flood (blue line in Fig. 4), and the red line represents the configuration of the main stream during the low-water discharge in 1991. The analysis results show that three meandering parts (M-1, M-2 and M-3) that had wavelengths of approximately 600 m formed at regular intervals in the 1.8 km-long section. Comparison between the configuration of the flow channels shown in calculation results and those in 1981 or 1991 shows that the locations and intervals of the meandering channels are roughly in accordance with those in 1981 or 1991.

The results above suggest the possibility that the meanders formed at an actual location where the point bars were necessarily brought about due to the curvature of the normal line in the low-water channel or the planform shape of the river bank.

3.4   River channel formation process during 1981 flood

Factors that led to the formation of the above-mentioned sandbars will be discussed on the basis of the transition process of bed morphology during the flood. Figure 12 shows the calculated river channel formation during the period when discharge was lower in the 1981 flood, and the displayed time of each result corresponds to each displayed time of from (1) to (6) in Fig.13. The transition process of bed morphology during the flood can be roughly described as follows: a sign of change started to appear on the surface of the riverbed after the peak discharge, and then multiple-row bars on the riverbeds changed into double-row bars over time. For the time period from (5) to (6), differences in bed morphology are found between the section immediately upstream of the bend and the section immediately downstream of the bend in the river channel. While double-row bars still remain in the upstream section, the trend toward the development of single-row bars is already clearly seen in the downstream section. This is probably because the downstreamward migration of the sandbar was limited at the bend in the river channel, which provided conditions better than a straight river channel for the development of sandbars, and the developed sandbars eventually completely stopped moving, which then promoted the development of single-row bars. However, it is unknown whether such a phenomenon was taking place in the actual river. But in calculation, it describes the spin-up process in the numerical model as the flow transitions

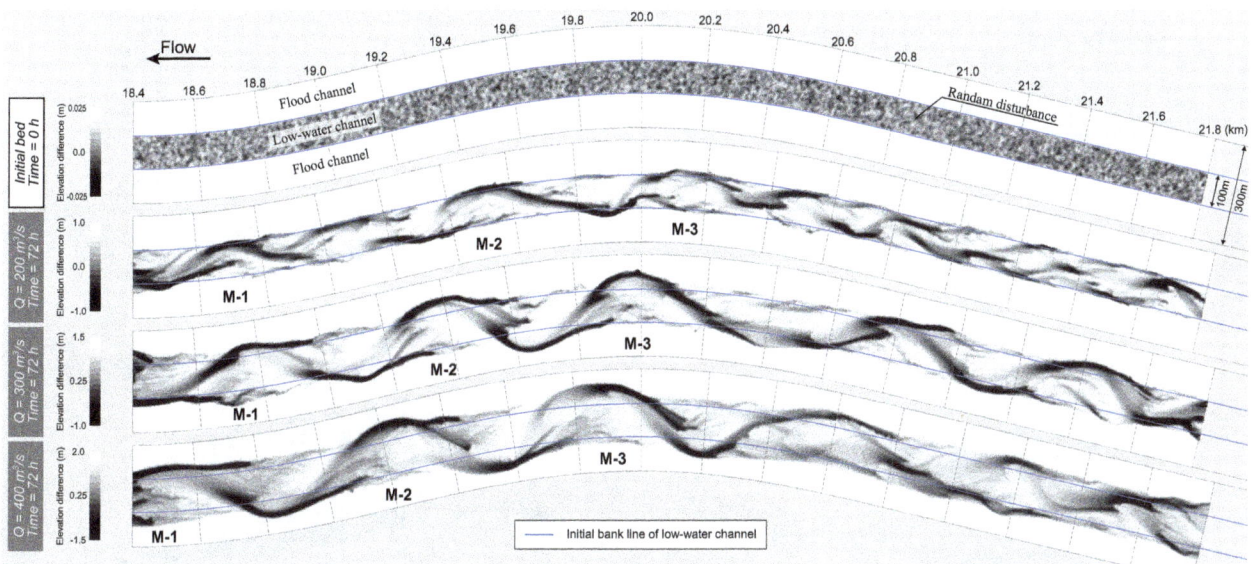

Figure 10. Analysis results of steady flow in Section-1 (riverbed elevation).

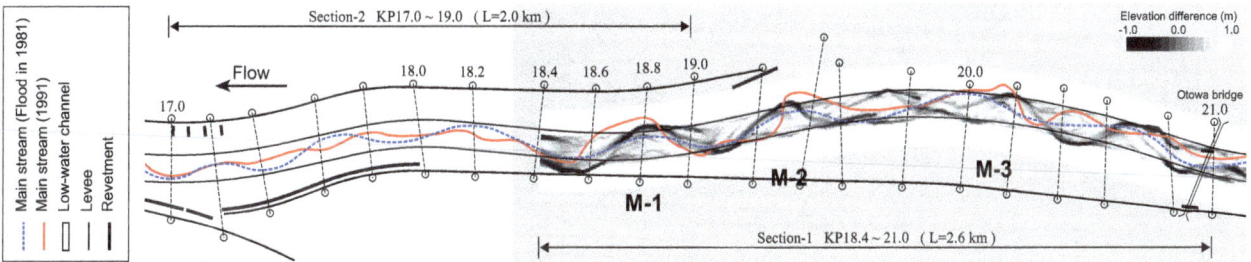

Figure 11. Analysis results of the 1981 flood in Section-1 (riverbed elevation).

from the plane bed through small-scale multiple-row bars to single-row alternate bars.

Further, it was observed in this analysis that sandbars in M-1, which formed immediately downstream of the bend, eventually triggered the development of sandbars in M-2 on the upstream side, and those sandbars then gradually increased the degree of meandering. That is, it is considered that, since point bars that form in the bend area of the river channel can even limit the migration of sandbars on the upstream side, the impact from those point bars will spread further upstream indirectly.

3.5 Influence of meandering angle on the development of meandering channels

As already mentioned, the on-site normal line in the low-water channel has a planform shape with a meandering angle of $\theta_0 = 13°$. Analysis results suggest the possibility that the planform shape of the low-water channel limits the migration of sandbars and induces the development of point bars near the river bend. From this, the next step is to evaluate how differences in the meandering angle of a curved river channel

influence the development of a meandering channel and the limiting gradient that stops the migration of sandbars. For the calculation condition, that used in the previous analysis was adopted, and the development of meandering channels after the 3 day flood of 1981 was evaluated, with only the meandering angle being changed within the range of $\theta_0 = 0 \sim 26°$.

Some examples of the calculation results are shown in Fig. 14. Additionally, as shown in Fig. 15, this calculation condition can allow the trend toward the development of single-row bars even in a straight channel (meandering angle $\theta_0 = 0°$), because single-row bars can develop even in a straight channel around the peak discharge. Therefore, it is difficult to extract only the influence that the curvature of the river channel has on the development of single-row bars; however, it is possible to evaluate the influence of the difference in meandering angle on the development of meandering channels to some degree by comparing the development of meandering channels using the river channel configurations with the meandering angle of $\theta_0 = 0°$ as a benchmark.

Figure 14 shows the general trend in which the riverbed configurations become double-row bars with decreases in

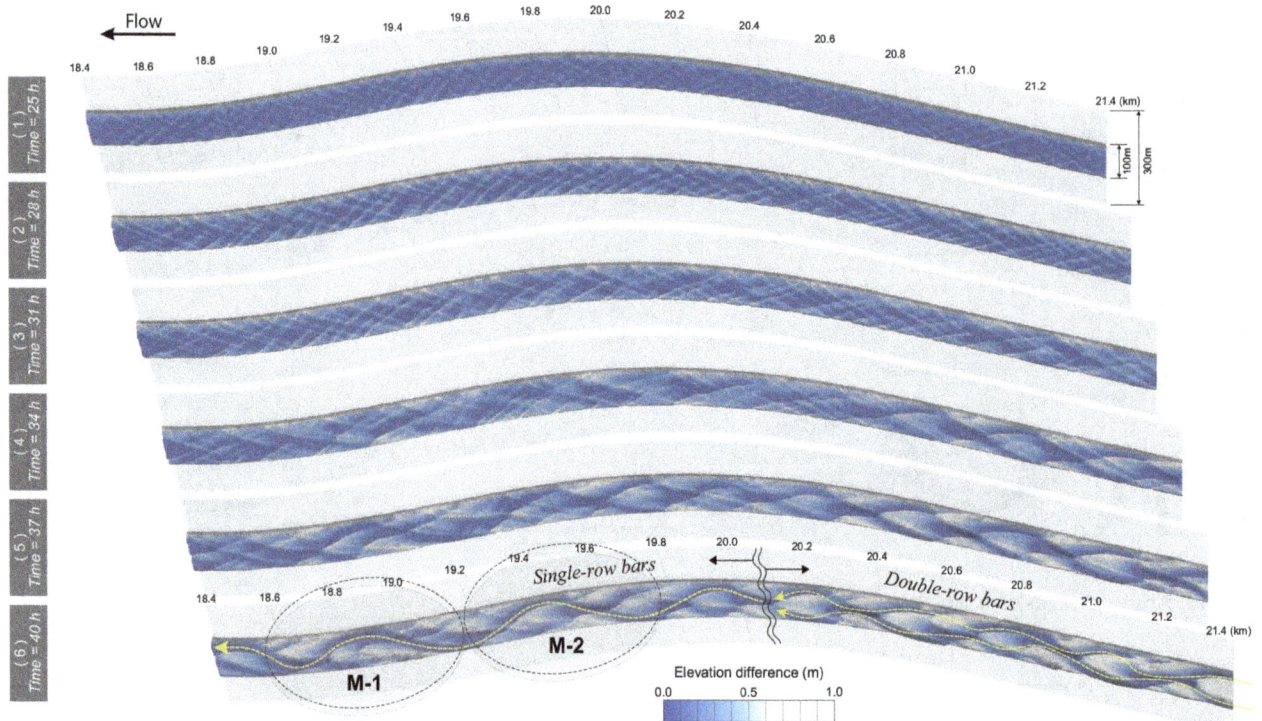

Figure 12. River channel formation process after peak discharge (1981 flood, KP18.4–KP21.4, Section-1).

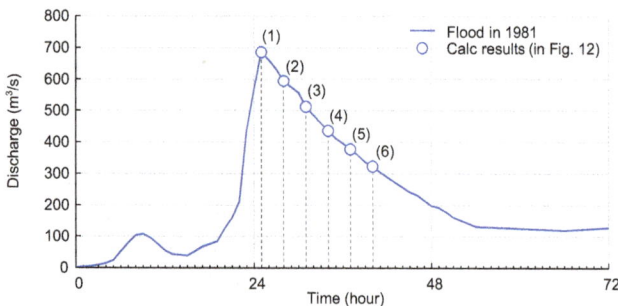

Figure 13. Display time of calculation results (1981 flood).

meandering angle, and in which the trend toward the development of single-row bars is observed to become clearer with increases in the meandering angle. In particular, in the meanders, M-1, M-2 and M-3, which are the colored parts in the figure, comparatively sharply defined point bars were formed when the meandering angle was around $\theta_0 = 10°$ or more.

4   Propagation of meandering waveforms resulting from the planform shape of the river bank

As mentioned above, the analysis of Section-1 suggests that the meanders (M-1, M-2 and M-3) that formed in the section between KP18.4 and KP21.0 are point bars inevitably resulting from the curvature of the normal line in the low-water channel. Also, from the history of the river course migra-

tion, it was found that large-scale river improvement work had never been performed in this section during the 30 year period and that these three sandbars had remained at roughly the same location since the 1981 flood, despite year after year of increases in the degree of meandering.

It is believed that a highly developed sandbar is functionally equivalent to the riverbank at the low-water channel and can strongly direct flood flow toward the right or the left river bank, and thus it promotes the development of the meandering channels. In other words, the development of sandbars in M-1 is considered to have had a significant impact on the bank disaster (near KP18.2) that occurred immediately downstream. Therefore, in the next step, data obtained from Section-2 (KP17.0–KP19.0), which includes the sandbars in M-1 and the area around the damaged location, are analyzed.

4.1   Development of point bars in M-1

First, in order to model the conditions of the river channels to be used in the analysis, the details of the development of the point bars in M-1 before the 2011 flood were confirmed. Figures 16 and 17 show the conditions of the river channel in Section-2 based on the laser profiler (LP) measurement data obtained in 2006. Figure 16 (left) is a bird's-eye view and Fig. 16 (right) is a transverse section of M-1, while Fig. 17 is a plan view of the riverbed elevation. In the transverse section of M-1, Bar-1, which formed on the left-bank side, developed to the point of completely filling up the low-water channel

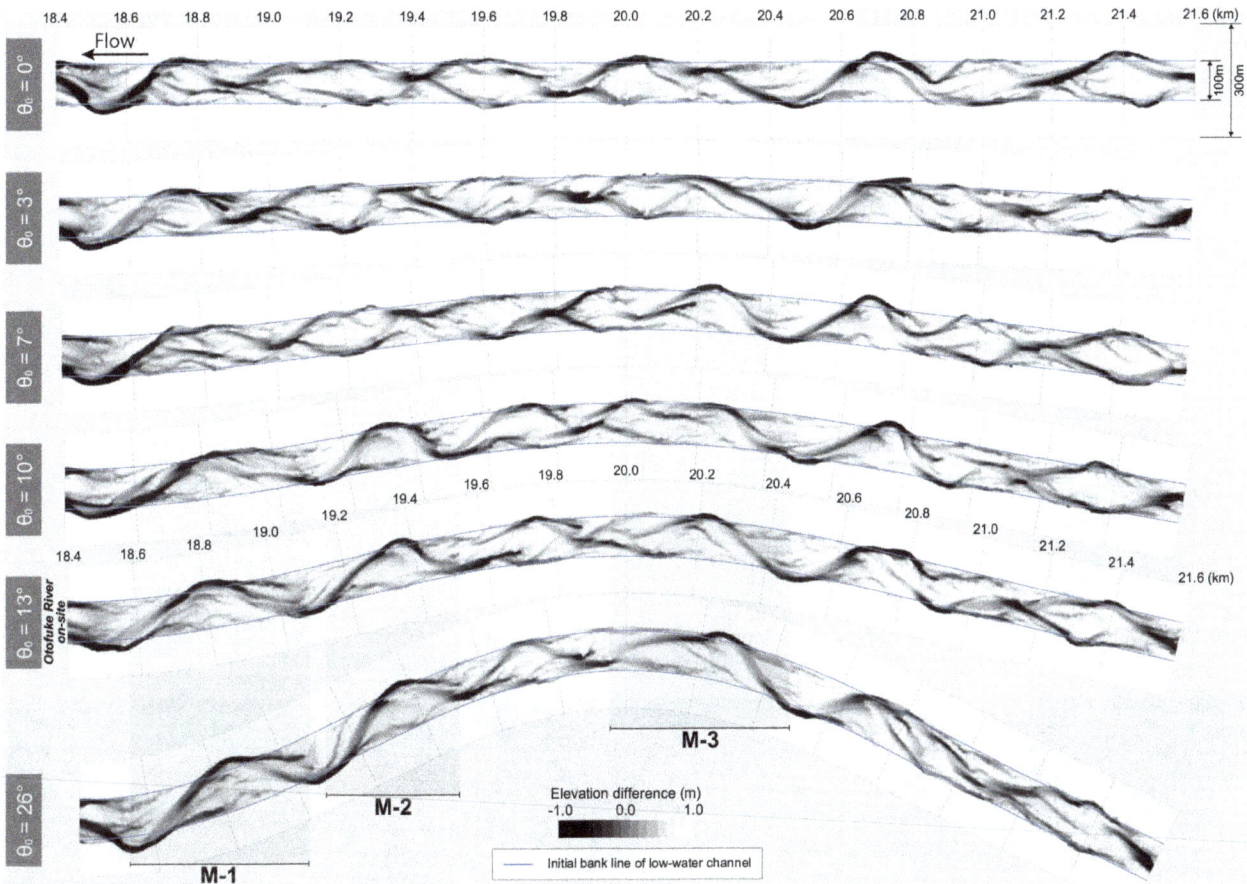

Figure 14. Development of meandering channels at various meandering angles (1981 flood, KP18.4–KP21.6).

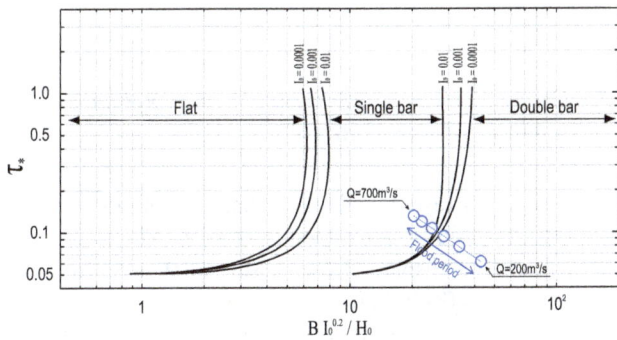

Figure 15. Mesoscale bed morphology classification (e.g., Kuroki and Kishi, 1983) during the 1981 flood.

Major points included the following: the low-water channel on the upstream side of the damaged location was curved sharply toward the right bank due to the developed point bar (M-1), and the downstream section still included a straight channel, since river improvement work was performed in 2005.

Thus, a planform-shaped riverbank that was designed to imitate the meandering path (M-1), as shown by the dotted line in Fig. 17, was formed on the upstream side of the analysis section, and a straight river channel with a 100 m-wide low-water channel was connected to the section on the downstream side. More information on the planar configuration of the riverbank is shown in the uppermost part of Fig. 21.

4.2 Calculation conditions (Section-2)

In this analysis, the 2011 flood (3 days), shown in Fig. 18, was reproduced as an external force, with the purpose of identifying the dominant factors that triggered the development of meandering flows that caused the bank disaster, in addition to the developing process of the meandering flows being examined. The 2 km-long section from KP17.0 to KP19.0 was determined as the section for analysis, and

that is indicated by the red dotted line for 1981, which shows that solid point bars, whose wave heights were as high as the crest of the riverbank in low-water channel, formed.

In this analysis, the terrain model was simplified as much as possible by extracting only major points of the on-site river channel configuration, for the purpose of identifying the dominant factors in the development of meandering flows.

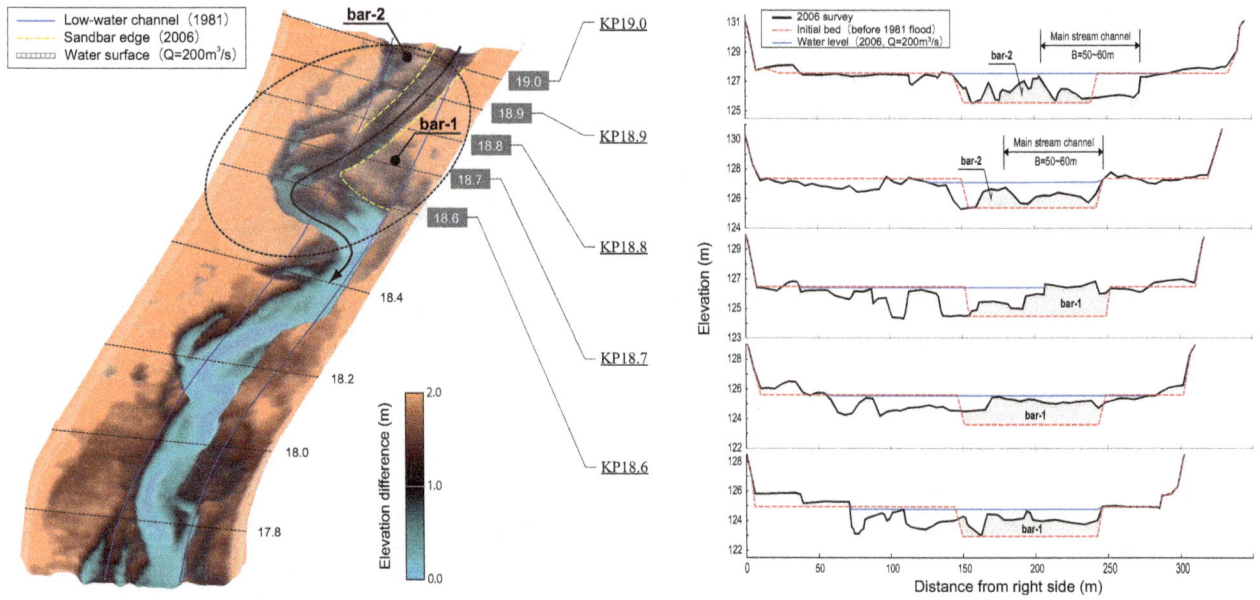

Figure 16. Development of sandbars before the 2011 flood (LP data: 2006.5, Section-2).

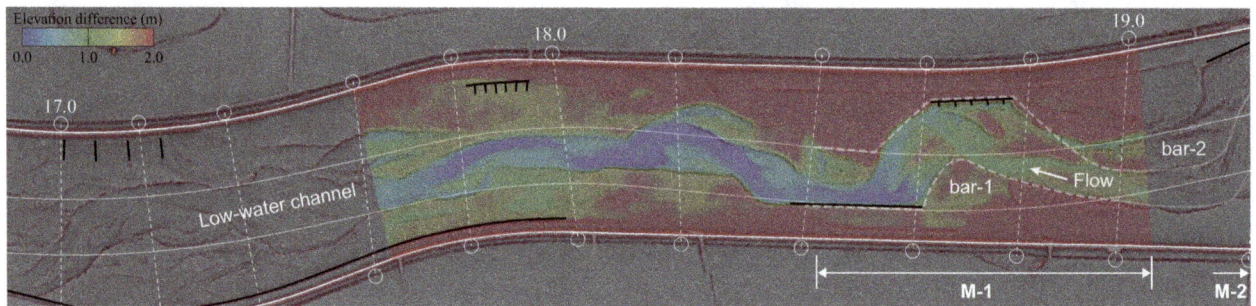

Figure 17. Observed riverbed elevation before the 2011 flood (LP data: 2006.5, Section-2).

an initial riverbed was designed to have a topography imitating the on-site meandering path as described in the previous paragraph. For the rest, the calculation condition used for the analysis of Section-2 was the same as that used for Section-1.

Here, supplementary information is given to show the validity of setting KP19.0 as the upstream end of the analysis section. In the area near KP19.0, due to the point bars (M-1 and M-2), the flow running on the left-bank side was maintained even during a flood, and no transverse direction change in watercourse was observed from that shown in Fig. 19.

In addition, at the site, natural chute cutoffs occurred in the meandering flow (M-2; Figs. 19, 20) during the flood; however, there is no major difference between the cross-sectional shape of the river channel near KP19.0 before the flood and after the flood. In this analysis, therefore, it is assumed that the cross-section profile of KP19.0 (upstream end of calculation range) was in a dynamic equilibrium during the 2011 flood.

4.3 Calculation results (Section-2) – propagation of meandering waveforms

The calculation results and the actual configuration of the riverbed measured after the flood are shown in Fig. 21. Additionally, the time display of each result corresponds to those of (1) to (6) and (8) in Fig. 18.

The calculation results confirmed that major changes started to take place in the river channel during the discharge-increase period after the discharge exceeded $200 \text{ m}^3 \text{ s}^{-1}$, and the meandering channels took their general form gradually over the period from (2) to (6), around the peak discharge. The changes that took place from (6) to (8) show that only the meandering amplitude was increasing during this period, accompanied by little change in wavelength and phase.

The blue lines show the following change that took place during the period from (2) to (6): the flood flow was restrained in the meander (M-1) due to the curvature of the low-water riverbank, which led to the meandering flow being maintained during the flood. It is considered that, since

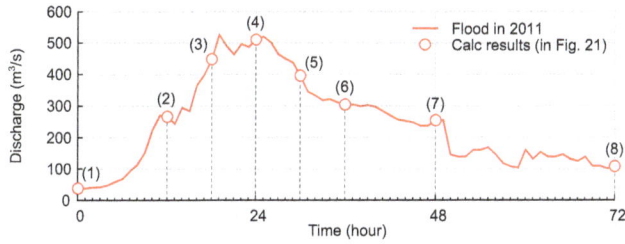

Figure 18. Time display of calculation results (2011 flood).

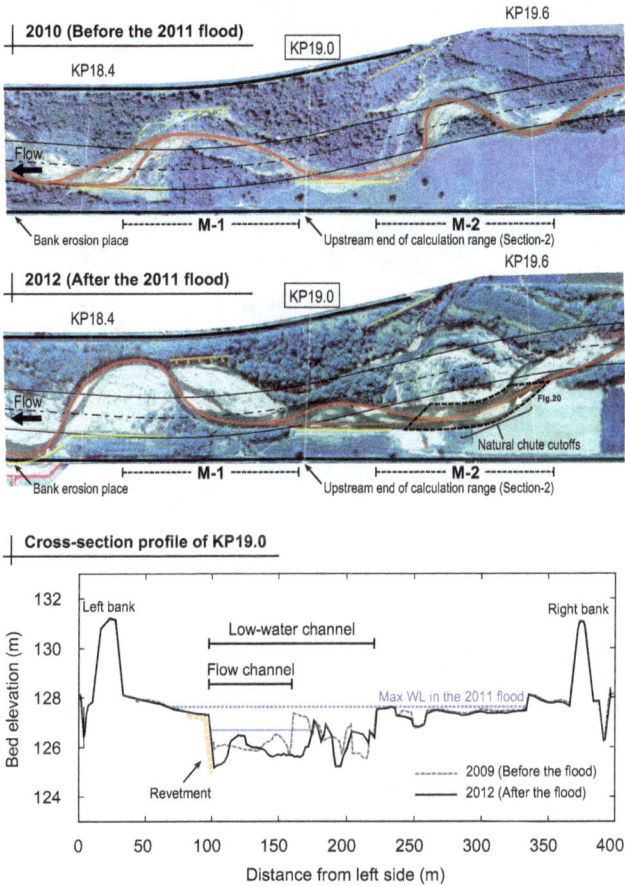

Figure 19. Planform shape of meandering channels and cross-section profiles before and after the 2011 flood (around KP19.0).

this state continued for a long time during the flood, the meandering waveforms of M-1 gradually propagated downstream, which finally led to the formation of uniform meandering channels. Further, at this time, the meandering flow slightly changed its phase to the downstream side while eroding the riverbank. As a result, the location of the riverbank with which the flood flow collided deviated from the existing revetment, which served as a decisive factor in the process leading to this bank disaster.

The contour figure in the bottom of Fig. 21 shows the actual configuration of the riverbed measured after the flood. Comparing it with the calculation results of (8) indicates

Figure 20. New flow channel caused by the chute cutoffs in M-2.

that the calculation results and the measured data are almost identical in terms of the characteristics of meandering flow (wavelength, amplitude). In particular, there are many similarities between the riverbed configuration of (8) and that of measured data, including phase shifting in M-1, traces of sandbar edge and watercourse left on the formed sandbar, which shows that the obtained calculation results are valid.

Figure 22 shows the sandbar evolution process that was seen in the upstream side of the bank disaster location during the flood. This calculation result indicates that traces of sandbar edge and watercourse formed through the process shown in this figure (Step 1~4). It is considered that meandering flow rapidly shifted towards the river dike due to the bank erosion, as a result, traces of a transverse scrolling sandbar edge is left on the formed sandbar. These characteristics of traces left on the sandbar are very similar to those of the actual place (Fig. 23).

4.4   Changes of the cross-sectional profile of a sandbar

In the development of meandering flow, how the cross-sectional profile of a sandbar changes with time is extremely important. Here, the cross-sectional profile of an alternating bar is expressed with wave height H_B and river width B (Fig. 24), and the time-series variation of a sandbar profile for representative cross sections of Section-1 and Section-2 are shown in Fig. 25. The upper part of the figure and Eq. (3) show the following: the results from the experiment on the equilibrium wave height of alternating bars that was conducted using a straight channel and the region where the equilibrium wave height could be found, which was indicated by dimensional analysis (e.g., Ikeda, 1983) and onto which the results obtained from this analysis were overwritten. In addition, the lower part of the figure shows the time-series variation of the cross slopes (H_B/B).

$$\frac{H_B}{D} = \left(\frac{B}{d}\right)^{-0.45} 9.34 \exp\left(2.53 \operatorname{erf} \frac{\log_{10}\frac{B}{D} - 1.22}{0.594}\right) \qquad (3)$$

Calculation results

(1) Time = 0 h
Initial bed

(2) Time = 12 h

(3) Time = 18 h

(4) Time = 24 h

(5) Time = 30 h

(6) Time = 36 h

(8) Time = 72 h
After the flood

Propagation of meandering waveforms

Measurement data

- Survey method: Laser profiler
- Survey section: KP17.0 ~ 19.0
- Survey date: 20 May 2012
  (The flood occured on 7 Sep. 2011)

— Revetment (constructed before the flood)
— Revetment (constructed after the flood)

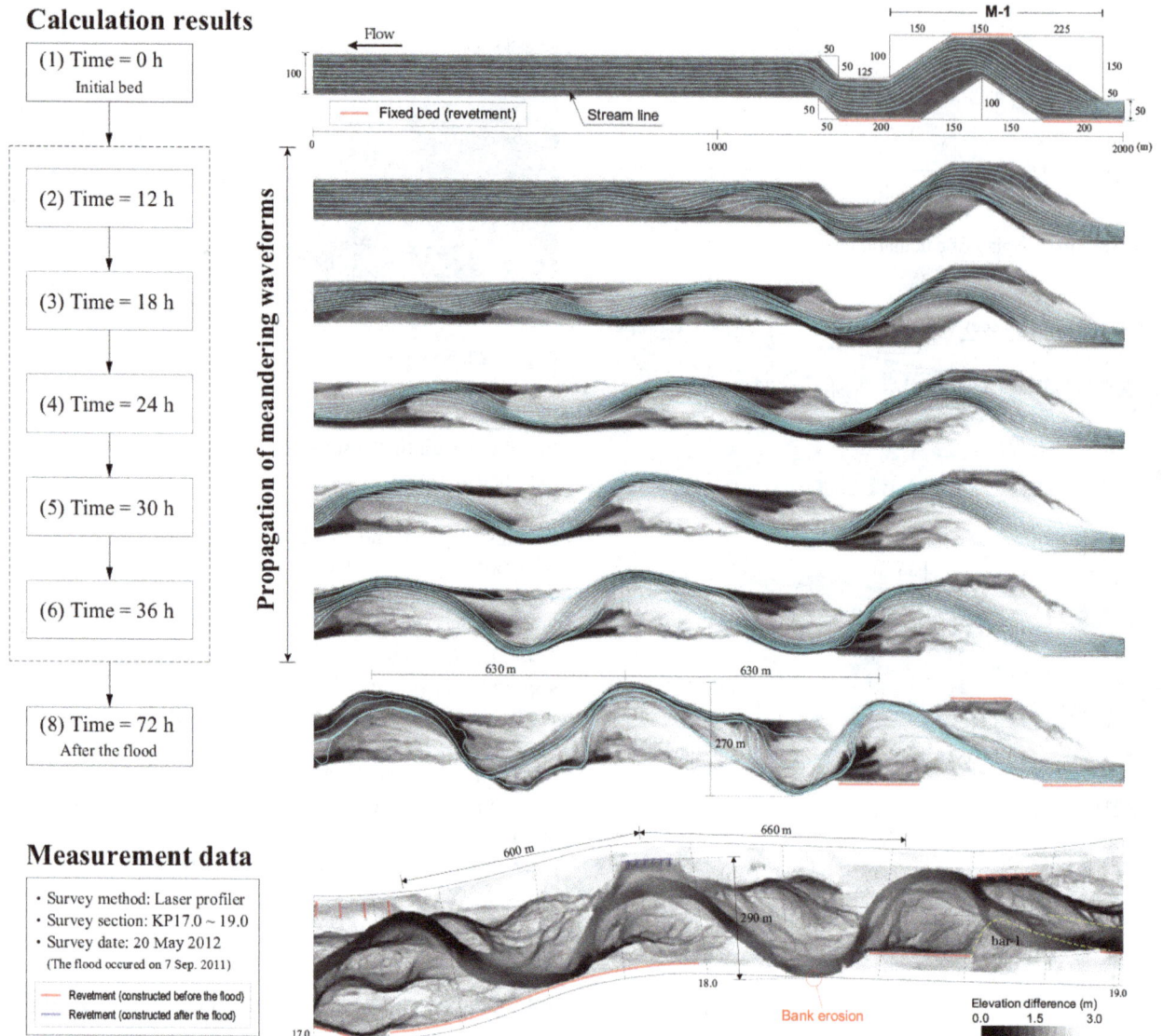

Figure 21. Calculation results and the actual configuration of the riverbed measured after the flood.

Calculation results

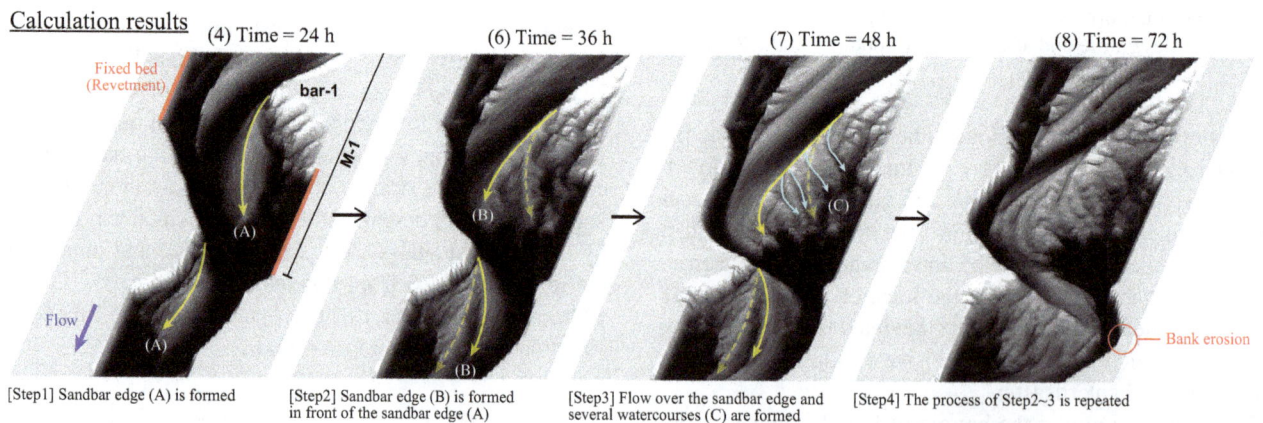

(4) Time = 24 h   (6) Time = 36 h   (7) Time = 48 h   (8) Time = 72 h

[Step1] Sandbar edge (A) is formed

[Step2] Sandbar edge (B) is formed in front of the sandbar edge (A)

[Step3] Flow over the sandbar edge and several watercourses (C) are formed

[Step4] The process of Step2~3 is repeated

Figure 22. Sandbar evolution process after the peak discharge (traces of sandbar edge and watercourse left on the formed sandbar).

Measurement data

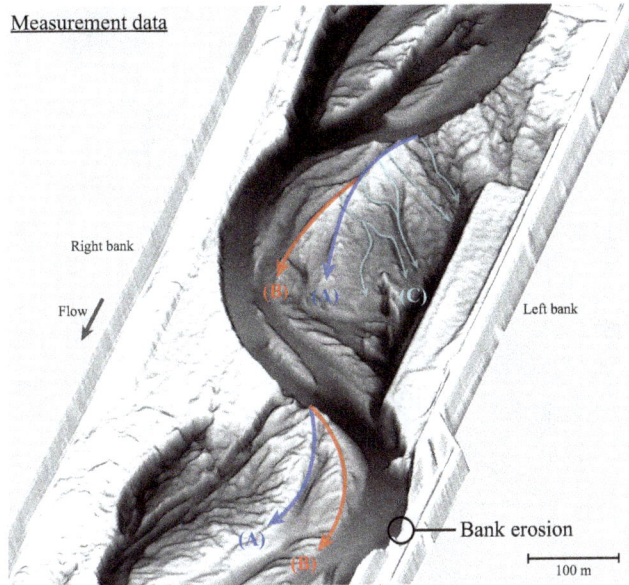

Figure 23. Actual configuration of the riverbed after the 2011 flood.

Cross-section profile

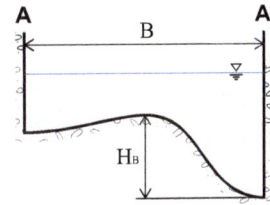

Plane view

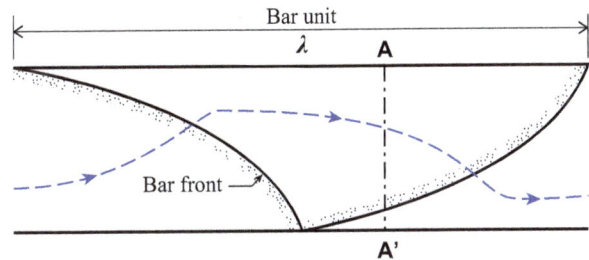

Figure 24. Definition of single-row alternate bars.

The details are as follows: H_B, sandbar wave-height (m); D, average water depth (m); B, river width (m); and d, sand grain size (m).

The two sections can be said to have the following two points in common. The first is that the sandbar wave height increases with time; however, after the sandbar wave height reaches the equilibrium wave height, the sandbar moves while maintaining the same state. The second is that the time-series variation of the sandbar cross slope reaches the maximum value when the sandbar wave height reaches equilibrium. Thus, both sections show a convex-shaped variation that has the maximum value as its peak value. This result can be roughly interpreted as follows (Fig. 26).

A. Before the wave height has fully developed, the change in the vertical direction predominates (sandbar development process).

B. When the sandbar wave height reaches the equilibrium state, the cross slope also reaches its peak ($H_B/B = 0.020$–0.025).

C. After that, the change in the transverse direction predominates, while the equilibrium height is maintained (bank erosion process).

In other words, what the sections have in common is that the planform shape of the meandering flow develops after the cross-sectional profile of sandbars has developed. The largest difference between the two sections lies in the time at which the sandbar wave height reaches equilibrium. It is possible to interpret it as follows: in Section-2, where the sandbar wave height reaches equilibrium at an early stage of the flood, bank erosion continues for a prolonged time, which

leads to the development of meandering flows that is more remarkable than that in Section-1. Additionally, the difference in the timing of when the sandbar wave height reaches the equilibrium wave height is brought about by the presence or absence of the remaining traces in the watercourse or sandbars that were produced during previous floods. Therefore, different planform shapes of river bank were given to Section-1 and Section-2 as the initial riverbeds for analysis. If a highly developed sandbar like M-1 exists in a low-water channel, it will cause strong meandering flow during a flood. It is assumed that such meandering flow increases the development rate of sandbars to a great degree (peak arrival time of H_B/B).

5 Conclusions

As stated above, this study focused on both sandbar topography and the planform shape of a low-water riverbank, and used numerical analysis to investigate the dominant factors that led to bank damage in September 2011. Based on the analysis of KP18.4–KP21.0, it was presumed that the curvature of the low-water channel had induced the formation of point bars at this section. Also, it was found from the history of the migration of the river course that those sandbars had been gradually increasing the degree of meandering over the course of about 30 years, and their wave height had finally reached the elevation of the low-water riverbank.

Furthermore, on the basis of the results obtained from the analysis on KP17.0 ~ KP19.0, it was presumed that the meandering flows were maintained for a prolonged time due to the planar configuration of the low-water riverbank (curvature) and the meandering waveforms gradually propagated downstream, which finally led to the development of the

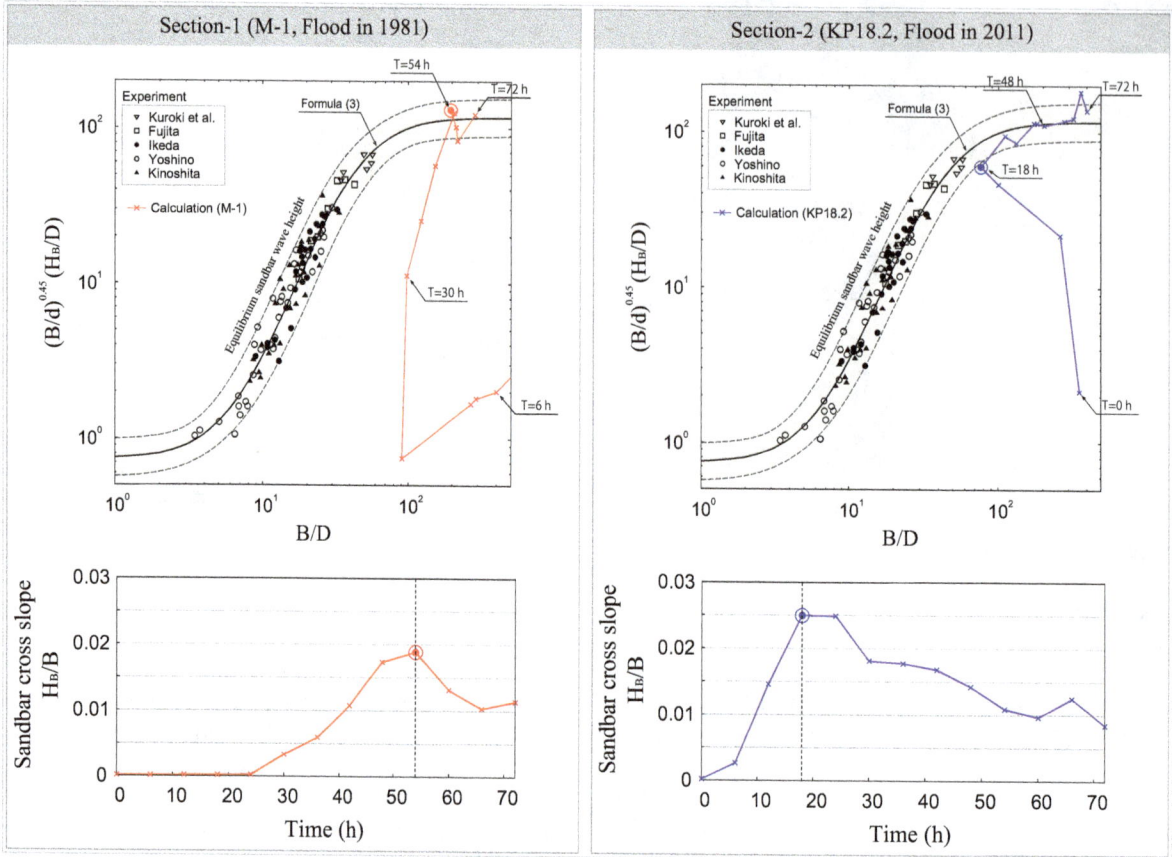

Figure 25. Sandbar wave-height development process and changes of the sandbar cross slope (Sections-1 and 2).

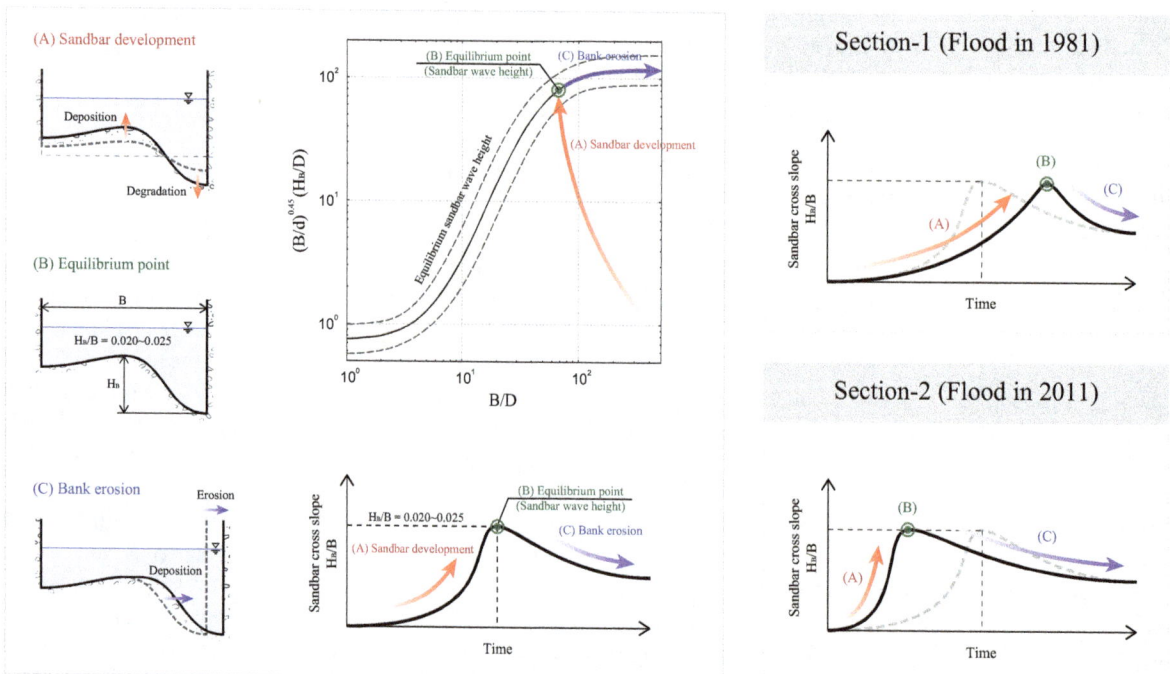

Figure 26. Schematic diagram: change of cross-sectional profile of single-row alternate bars in the flood (Sections-1 and 2).

large meandering channel that reached the river dike. The following are the major factors that caused the disaster.

1. Immobilization of the sandbars due to the curvature of the low-water channel.

2. The development of point bars, which occurred over a period of several decades.

3. Propagation of meandering waveforms due to the planform shape of the low-water riverbank.

When temporary measures are taken for the section of a low-water riverbank where protective measures have not been taken, it is considered to be effective to identify critical locations where the above-mentioned factor 2 is found, and to implement measures that lessen the influence from developed point bars. Specifically, the excavation of the upper portion of sandbars is an effective measure; however, relating to this measure, there are some factors to be carefully examined in the future, including the potential influence that it would have on the downstream side.

Acknowledgements. We are deeply grateful to anonymous reviewers whose comments and suggestions were of inestimable value for our study. We would also like to thank the members of the river engineering research team whose meticulous comments were an enormous help to us. Finally, we would like to express our gratitude to Hokkaido Regional Development Bureau who provided us the aerial photographs and many other data used in this study.

Edited by: F. Metivier

References

Blondeaux, P. and Seminara, G.: A unified bar-bend theory of river meanders, J. Fluid Mech., 157, 449–470, 1985.

Colombini, M., Seminara, G. and Tubino, M.: Finite-amplitude alternate bars, J. Fluid Mech., 181, 213–232, 1987.

Fukuoka, S. and Yamasaka, M.: Theoretical study on meander development caused by bank erosion and deposition, J. Jpn. Soc. Civ. Eng., 327, 73–85, 1981.

Hasegawa, K.: A Study on flows and bed topographies in meandering channels, J. Jpn. Soc. Civ. Eng., 338, 105–114, 1983.

Hasegawa, K. and Yamaoka, I.: The effect of plane and bed forms of channels upon the meander development, J. Jpn. Soc. Civ. Eng., 296, 143–152, 1980.

Ikeda, H.: On the origin of bars in meandering channels, Bull. ERC., Univ. Tsukuba, 1, 17–31, 1977.

Ikeda, H.: Progress of studies on the origin of bars in meandering rivers, Bull. ERC., Univ. Tsukuba, 15, 11–19, 1991.

Ikeda, H. and Ohta, A.: On the formation of stationary bars in a straight flume, Bull. ERC., Univ. Tsukuba, 10, 105–113, 1986.

Ikeda, H., Iseya, F., and Shinzawa, Y.: Cristate bed configuration in a meander bend, Bull. ERC., Univ. Tsukuba, 9, 27–42, 1985.

Ikeda, H., Lisle, T., Pizzuto, J., Iseya, F., Kodama, Y., Maita, H., Iijima, H., Wei, Y., and Suzuki, M.: Channel response to sediment wave propagation in an experimental channel, Bull. ERC., Univ. Tsukuba, 20, 1–9, 1995.

Ikeda, S.: Wavelenght and height of single row alternate bars, Annu. J. Hydraul. Eng. JSCE, 27, 689–695, 1983.

iRIC website: available at: http://i-ric.org/en/ (last access: October 2013), 2013.

Jang, C. and Shimizu, Y.: Numerical simulation of relatively wide shallow channels with erodible banks, ASCE J. Hydraul. Eng., 131, 7, 565–575, 2005.

Kinoshita, R. and Miwa, H.: The flow pass shape by which the location of the sand and gravel bar is stabilized, J. Jpn. Soc. Eros. Control Eng., 94, 12–17, 1974.

Kishi, T. and Kuroki, M.: Bed forms and resistance to flow in erodible-bed channels (1) – hydraulic relations for flow over sand waves –, Bull. Fac. Eng., Hokkaido Univ., 67, 1–23, 1972.

Kondo, Y., Shimizu, Y., Kimura, I., and Parker, G.: Numerical computation of free meandering process of rivers considering effect of slump blocks and inner bank deposition, Annu. J. Hydraul. Eng. JSCE, 53, 769–774, 2009.

Kuroki, M. and Kishi, T.: Regime criteria on bars and braids in alluvial straight channels, J. Jpn. Soc. Civ. Eng., 342, 87–96, 1983.

Kuroki, M. and Kishi, T.: Regime criteria on bed forms and flow patterns in alluvial streams, Bull. Fac. Eng., Hokkaido Univ., 118, 1–12, 1984.

Lajeunesse, E., Malverti, L. and Charru, F.: Bed load transport in turbulent flow at the grain scale: Experiments and modeling, J. Geophys. Res-earth., 115, F04001, doi:10.1029/2009JF001628, 2010.

Lanzoni, S. and Seminara, G.: On the nature of meander instability, J. Geophys. Res., 111, F04006, doi:10.1029/2005JF000416, 2006.

Motta, D., Abad, J. D., Langendoen, E. J., and García, M. H.: The effects of floodplain soil heterogeneity on meander planform shape, Water Resour. Res., 48, W09518, doi:10.1029/2011WR011601, 2012.

Nagata, T., Kakinuma, T., and Kuwamura, T.: River course change and meander characteristics of the Otofuke River, Monthly Report of Civil Engineering Research Institute for Cold Region, 706, 2–11, 2012.

Nagata, T., Watanabe, Y., Yasuda, H., and Ito, A.: Development of a meandering channel caused by a shape of the alternate bars, J. Jpn. Soc. Civ. Eng. Ser. B1, 69, 1099–1104, 2013.

Parker, G. and Johannesson, H.: Observations on several recent theories of resonance and overdeepening in meandering channels, Water Res. M., 12, 379–415, 1989.

Parker, G., Shimizu, Y., Wilkerson, G. V., Eke, E. C., Abad, J. D., Lauer, J. W., Paola, C., Dietrich, W. E., and Voller, V. R.: A new framework for modeling the migration of meandering rivers, Earth Surf. Proc. Land., 36, 70–86, 2011.

Seminara, G., Zolezzi, G., Tubino, M., and Zardi, D.: Downstream and upstream influence in river meandering, Part 2. Planimetric development, J. Fluid Mech., 438, 213–230, 2001.

Shimizu, Y.: Mutual effects of bed and bank deformation in channel plane formation, Annu. J. Hydraul. Eng. JSCE, 47, 643–648, 2003.

Shimizu, Y., Watanabe, Y., and Toyabe, T.: Finite amplitude bed topography in straight and meandering rivers, J. Jpn. Soc. Civ. Eng., 509, 67–78, 1995.

Shimizu, Y., Hirano, M., and Watanabe, Y.: Numerical calculation of bank erosion and free meandering, Annu. J. Hydraul. Eng. JSCE, 40, 921–926, 1996.

Takahashi, G., and Yasuda, H.: Study on conservation condition of central bars, J. Jpn. Soc. Civ. Eng. Ser. B1, 68, 961–966, 2012.

van Dijk, W. M., van de Lageweg, W. I., and Kleinhans, M. G.: Experimental meandering river with chute cutoffs, J. Geophys. Res., 117, F03023, doi:10.1029/2011JF002314, 2012.

van Dijk, W. M., van de Lageweg, W. I., and Kleinhans, M. G.: Formation of a cohesive floodplain in a dynamic experimental meandering river, Earth Surf. Proc. Land., 38, 1550–1565, doi:10.1002/esp.3400, 2013.

Watanabe, Y.: Weakly nonlinear analysis of bar mode reduction process, Annu. J. Hydraul. Eng. JSCE, 50, 967–972, 2006.

Watanabe, Y. and Kuwamura, T.: Experimental study on mode reduction process of double-row bars, Annu. J. Hydraul. Eng. JSCE, 48, 195–200, 2004.

Watanabe, Y. and Kuwamura, T.: Suitability of weakly nonlinear analysis to mode reduction process of double-row bars, Annu. J. Hydraul. Eng. JSCE, 49, 943–948, 2005.

Watanabe, Y., Sato, K., and Oyama, F.: Experimental study on bar formation under unsteady flow conditions, Annu. J. Hydraul. Eng. JSCE, 46, 725–730, 2002.

Wong, M. and Parker, G.: Reanalysis and correction of bed-load relation of Meyer-Peter and Muller using their own database, J. Hydraul. Eng., 132, 1159–1168, 2006.

13

Morphodynamic regime change in a reconstructed lowland stream

J. P. C. Eekhout[1], R. G. A. Fraaije[2], and A. J. F. Hoitink[1]

[1]Hydrology and Quantitative Water Management Group, Wageningen University, Wageningen, the Netherlands
[2]Ecology and Biodiversity Group, Institute of Environmental Biology, Utrecht University, Utrecht, the Netherlands

Correspondence to: J. P. C. Eekhout (joriseekhout@gmail.com)

Abstract. With the aim to establish and understand morphological changes in response to channel reconstruction, a detailed monitoring plan was implemented in a lowland stream called Lunterse Beek, located in the Netherlands. Over a period of almost 2 years, the monitoring programme included serial morphological surveys, continuous discharge and water level measurements, and riparian vegetation mapping, from photographs and field surveys. Morphological processes occurred mainly in the initial period, before riparian vegetation developed. The initial period was largely dominated by upstream sediment supply, which was associated with channel incision upstream from the study area. Herbaceous vegetation started to develop approximately 7 months after channel reconstruction. The monitoring period included two growing seasons. A clear increase of riparian vegetation cover from first to the second year was observed. Detailed morphological and hydrological data show a marked difference in morphological behaviour between the pre-vegetation and post-vegetation stage. A linear regression procedure was applied to relate morphological activity to time-averaged Shields stress. In the initial stage after channel reconstruction, with negligible riparian vegetation, channel morphology adjusted, showing only a weak response to the discharge hydrograph. In the subsequent period, morphological activity in the channel showed a clear relation with discharge variation. The two stages of morphological response to the restoration measures may be largely associated with the upstream sediment supply in the initial period. Riparian vegetation may have played a substantial role in stabilizing the channel banks and floodplain area, gradually restricting the morphological adjustments to the channel bed.

1 Introduction

Stream restoration in the Netherlands increasingly involves the development of riparian zones along reconstructed channel reaches. These riparian zones are constructed with the aim to accommodate water during flood events and to improve the connection between aquatic and terrestrial ecology. The current contribution describes the initial morphological, hydrological and ecological developments of a reconstructed lowland stream, focussing on the controls of upstream sediment supply and riparian vegetation on the initial morphologic developments after channel reconstruction.

A stream is considered to be in morphological equilibrium when, over a period of years, the channel slope is delicately adjusted to provide, with available discharge and the prevailing channel characteristics, enough flow velocity required for transportation of all of the sediment load supplied from upstream (Mackin, 1948). Leopold and Bull (1979) extended Mackin's definition with more dependent variables, including channel depth, channel width, bed roughness and planform pattern. The dependent variables describe the morphological regime resulting from the coupled system of flow, sediment transport and bed morphology, which may be developing towards a new equilibrium. In this theoretical framework, a local stream restoration effort can be seen as a local perturbation of the morphodynamic system that may trigger a regime change.

After completion of a lowland restoration project, there are two factors particularly relevant in the morphological adjustments towards a new equilibrium, viz. conditions at the boundaries and riparian vegetation development. Typically, stream restoration is applied to isolated sections of a stream, which may cause a compatibility mismatch at the boundaries of the reconstructed stream section. Longitudinal bed level profile adjustments are often initiated at these boundaries, where gradients in flow velocity and sediment transport cause erosion or deposition of the bed. In agricultural areas, such as those presented in this study, weirs located upstream often act as a sediment trap, resulting in supply-limited conditions at the upstream boundary of the reconstructed stream section. At the downstream boundary of the reconstructed stream, backwater effects may cause deceleration of the flow, resulting in deposition and flattening of the bed level slope (Eekhout et al., 2013).

The role of riparian vegetation in channel morphodynamics has been widely recognized, which is highly relevant in the context of stream restoration. Root-reinforced soils are more resistant to deformation and failure than bare soils (Abernethy and Rutherfurd, 2001), which may contribute to bank stability in reconstructed streams (e.g. Simon and Collison, 2002; Pollen-Bankhead and Simon, 2009). Hydrological processes induced by riparian vegetation may decrease bank stability, acting to control infiltration through macropores and the associated pore-water pressure variation (Simon and Collison, 2002). Laboratory experiments have shown how riparian vegetation may stabilize floodplain material, causing initial braided channel patterns to shift towards single-thread meandering channels (Gran and Paola, 2001; Tal and Paola, 2007; Braudrick et al., 2009; Tal and Paola, 2010). Model studies have shown that riparian vegetation is potentially responsible for changes in meander planform characteristics (Perucca et al., 2007). Riparian vegetation density and root structure have shown to be the most influential parameters controlling reach scale morphodynamics (Van de Wiel and Darby, 2004), which in turn may govern the morphological development towards a new equilibrium in a reconstructed stream.

Serial digital elevation models (DEMs) offer the opportunity to quantify reach-scale morphological change. Several survey techniques have been used to collect topographic data for DEM construction in fluvial environments, including total station (Fuller et al., 2003), ground-based GPS (Brasington et al., 2003), photogrammetry (Lane et al., 2010), unmanned radio-controlled platforms (Lejot et al., 2007), terrestrial laser scanning (TLS; Wheaton et al., 2013) and airborne lidar (Croke et al., 2013). Temporal morphological changes could be detected when a study reach is surveyed more than once. A DEM of difference (DoD) may quantify morphological changes when comparing two serial DEMs (Lane et al., 2003). DoDs in braided rivers have been extensively applied on a reach scale (e.g. Lane et al., 2010; Wheaton et al., 2010b, 2013), whereas DoD analysis in me-

andering rivers is often restricted to a bend scale (e.g. Gautier et al., 2010; Kasvi et al., 2013), although there are several exceptions (Fuller et al., 2003; Erwin et al., 2012; Croke et al., 2013). Until now, DoD analyses have been typically based on annual (e.g. Wheaton et al., 2013) or biannual surveys (Fuller et al., 2003; Lane et al., 2010). The low temporal resolution can cause erosional and depositional patches to overlap, complicating sediment budget estimation. In addition to sediment budget estimation, DoD analysis has been used to relate individual erosional or depositional patches to morphological and ecological processes (e.g. Grove et al., 2013; Wheaton et al., 2010a), which renders it a valuable tool for morphology monitoring after stream reconstruction.

In the context of stream restoration and beyond, there is a need for reach-scale field studies on the interaction between morphology and riparian vegetation (Camporeale et al., 2013). Here, a field study is presented based on 14 high-resolution morphological surveys over a period of almost 2 years, which places this study among the highest temporal resolution DoD analyses to date. We focus on a reach-scale study site, covering three meander bends and the adjacent floodplain. Morphological and terrestrial ecological data are combined under varying discharge conditions. The field study is performed in the context of stream restoration. Stream restoration projects are rarely subject to monitoring schemes combining morphological, hydrological and ecological surveys, although there are exceptions (e.g. Gurnell et al., 2006). The objective is to establish and understand the morphological response of a reconstructed lowland stream to riparian vegetation development and varying discharge conditions.

2 Study area

In October 2011 a stream restoration project was realized in the Lunterse Beek, representing a small lowland stream located in the central part of the Netherlands (52°4′46″ N, 5°32′30″ E) (see Fig. 1). A straightened channel was replaced by a new channel with a sinuous planform. The course of the new channel crosses the former channel at several locations (Fig. 1c). The new channel was constructed with a width of 6.5 m, a depth of 0.4 m and a longitudinal slope of $0.96 \, \mathrm{m\,km^{-1}}$. A lowered floodplain surrounded the channel, with an average width of 20 m. The bed material mainly consists of fine sand, with a median grain size of 258 μm (Eekhout, 2014).

Figure 1b shows the location of the study area in the catchment. The catchment covers an area of $63.6 \, \mathrm{km^2}$. The elevation within the catchment varies between 3 and 25 m above mean sea level. The study area is located in a lowland catchment, which implies a mild bed slope. The subsurface of the catchment mainly consists of aeolian-sand deposits. The average yearly precipitation amounts to 793 mm (KNMI, 2014). The average daily discharge amounts to

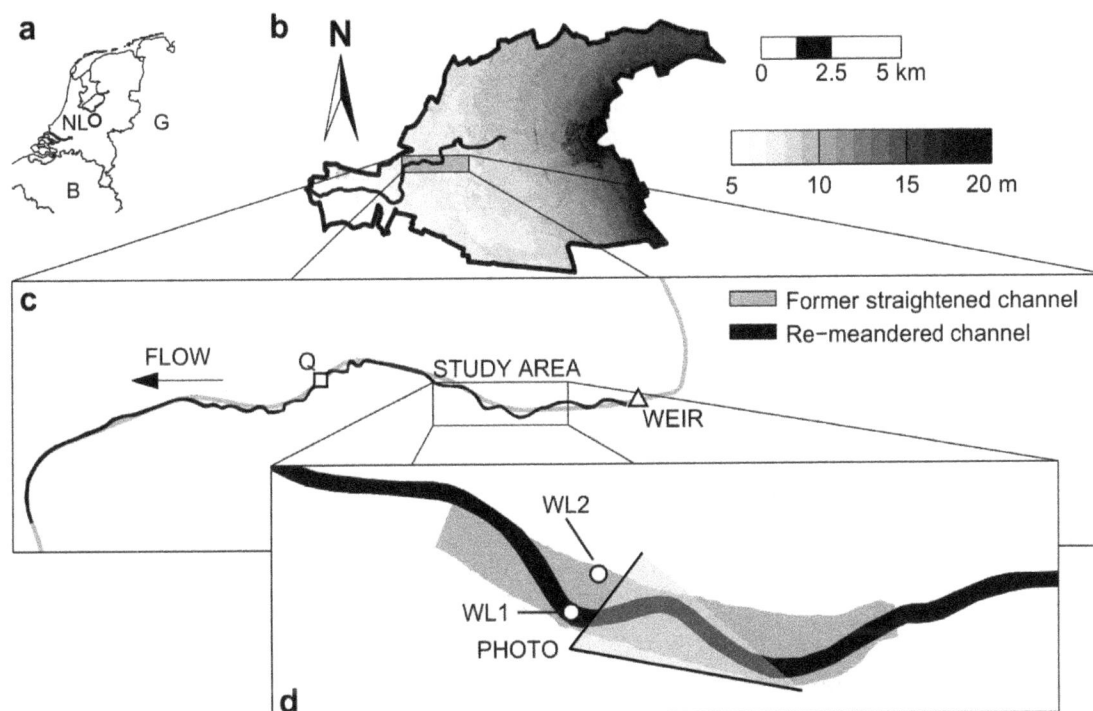

Figure 1. Overview of the study area: (**a**) location of the study area in the Netherlands; (**b**) elevation model of the catchment (Actueel Hoogtebestand Nederland, AHN; Van Heerd et al., 2000); (**c**) planform of the reconstructed reach, with the squared marker indicating the location of the discharge station (Q); and (**d**) sketch of the study area indicating the location of the water level gauges (WL1 and WL2) and the approximate extent of available terrestrial photographs (grey-shaded triangle, Fig. 9).

$0.33\,\mathrm{m^3\,s^{-1}}$ and the peak discharge during the study period was $6.46\,\mathrm{m^3\,s^{-1}}$.

A chute cutoff occurred within 3 months after realization of the stream restoration project, which is described in Eekhout (2014). Prior to the cutoff, a plug bar was deposited in the bend to be cut off. Hydrodynamic model results show that the location of the plug bar coincides with the region where flow velocity drops below the threshold of sediment motion, indicating the sediment deposition was caused by a backwater effect. Upstream from the plug bar, an embayment formed in the floodplain at a location where the former channel was located. The former channel was filled with sediment prior to channel reconstruction. The sediment fills originated from other parts of the study area, where excess sediment was available from the construction of the new channel. It is likely that the sediment at this location was less consolidated and therefore prone to erosion. The chute channel continued to incise and widen into the floodplain and, after 6 months, acted as the main channel, conveying the discharge during the majority of time.

3 Material and methods

3.1 Morphological Monitoring

The temporal evolution of the bathymetry has been monitored over a period of almost 2 years. Morphological data were collected in the area within the lowered floodplain over a length of 180 m, indicated in grey in Fig. 1d. Morphological data were collected with an average frequency of 52 days, using real-time kinematic (RTK) GPS equipment (Leica GPS 1200+ for surveys 1–13 and Leica Viva GS10 for survey 14). The RTK-GPS equipment allows for the surface elevation to be measured with an accuracy between 1 and 2 cm. The surface elevation data were collected along cross sections between the two floodplain edges. The survey strategy proposed by Milan et al. (2011) was followed, focusing on breaks of slope. Following this strategy, the point density was increased in the vicinity of steep slopes (e.g. channel banks), and decreased on flat surfaces (e.g. floodplains).

During five morphological surveys (1, 2, 4, 8 and an additional survey at day 687), bed elevation data were also obtained from 180 m upstream from the study area, over a length of approximately 350 m. These measurements were focused on the channel bed and provide information on the temporal evolution of the channel bed. The location of the crests of the channel bank in each cross section were marked during the surveys. The channel bed elevation was obtained by subtracting the hydraulic radius from the average elevation of the two opposing crests of the channel banks. The hydraulic radius is defined as the cross-sectional area divided by the wetted perimeter.

3.2   DEM construction and processing

DEMs of each of the 14 datasets were constructed. The data were transformed to (s, n) coordinates using the method described by Legleiter and Kyriakidis (2007). Since the data were collected in cross sections, an anisotropy factor was used within the interpolation routine, accounting for dispersion of the collected data in the longitudinal direction. The anisotropy factor was determined by dividing the average streamwise distance between the subsequent cross sections by the average cross-stream distance between the individual point measurements. The data were split into a channel section and a floodplain section, to account for a higher density of point measurements in the channel than in the floodplain. A separate anisotropy factor was applied for the channel and the floodplain data. Table 1 lists the point density and anisotropy factors for all individual surveys, specified for the channel and floodplain sections.

The data were projected onto a curvilinear grid, following the channel centerline for the channel data and the valley centerline for the floodplain data. A triangular irregular network (TIN) was constructed using a Delaunay triangulation routine in Matlab, following Heritage et al. (2009). Subsequently, the TIN was interpolated onto a grid using nearest neighbour interpolation, with a grid spacing of 0.25 m. After interpolation, the data were transformed back to (x, y) coordinates. The interpolated channel and floodplain were then merged to facilitate the comparison between all 14 surveys.

Deposition and erosion patterns were obtained by subtracting two subsequent morphological surveys, yielding a DoD. Real morphological change can be different from apparent morphological change, which may arise from uncertainties in the individual DEMs, e.g. from instrumental errors or errors that arise from the interpolation routine. The uncertainty can be established by determining the threshold level of detection (LoD). The method by Milan et al. (2011) was adopted to construct the spatially distributed LoD. Milan et al. (2011) account for the increase of spatially distributed error in a DEM near steep surfaces (e.g. channel bank edges), by inferring the relationship between the standard deviation of elevation errors and local topographic roughness. The method by Milan et al. (2011) was adjusted, by adopting a single linear regression model to obtain the spatial standard deviation of elevation error grids. The regression model was obtained after combining elevation errors and local topographic roughness values, from all 14 surveys.

In each grid cell the LoD was obtained according to:

$$U_{\mathrm{crit}} = t\sqrt{(\sigma_{\mathrm{e},1})^2 + (\sigma_{\mathrm{e},2})^2} \tag{1}$$

Table 1. Overview of all morphological measurements, with point density (PD, points m^{-2}) and anisotropy factor (AF) as used in the interpolation routine, specified for the channel (ch.) and floodplain (fp.) areas, respectively.

Survey no.	Survey date (day)	No. data points	PD (all)	PD (ch.)	AF (ch.)	PD (fp.)	AF (fp.)
1	0	379	0.16	0.25	9.74	0.12	4.59
2	93	956	0.32	0.44	5.04	0.24	2.35
3	133	918	0.27	0.34	3.57	0.24	2.15
4	191	742	0.20	0.31	4.24	0.15	2.04
5	231	1158	0.30	0.45	4.19	0.24	2.09
6	288	1376	0.35	0.53	4.42	0.29	2.15
7	341	1296	0.32	0.50	4.04	0.26	1.86
8	377	1655	0.42	0.62	3.56	0.35	1.68
9	426	1484	0.39	0.58	3.50	0.31	1.64
10	454	1472	0.36	0.55	4.12	0.30	2.03
11	489	1420	0.35	0.51	4.08	0.29	2.05
12	525	1462	0.37	0.56	4.66	0.29	2.15
13	558	1256	0.31	0.50	4.08	0.24	1.77
14	672	1501	0.44	0.85	10.66	0.23	2.67

where U_{crit} is the critical threshold error, t is the critical Student's value and $\sigma_{\mathrm{e},1}$ and $\sigma_{\mathrm{e},2}$ are the standard deviation of elevation error, for the first and second survey, respectively. The threshold value is based on a critical Student's t value, at a chosen confidence level. Following Milan et al. (2011), a confidence limit equal to 95 % was applied, which results in $t \geq 1.96$ (2σ). The elevation difference between two subsequent surveys at a particular grid cell is insignificant when $z_{i,j}(\mathrm{new}) - z_{i,j}(\mathrm{old}) < U_{\mathrm{crit}}$, where $z_{i,j}(\mathrm{new})$ and $z_{i,j}(\mathrm{old})$ are the elevations of the two subsequent morphological surveys at grid cell (i, j). A lower bound for the critical threshold error was adopted, to account for the accuracy of the RTK-GPS equipment. This lower bound was set to $U_{\mathrm{crit}} = 0.04$ m, which is 2 times the maximum error of the RTK-GPS equipment.

3.3   Quantifying morphological activity

For each DoD, the morphological activity was quantified for the study area as a whole and for isolated geomorphic zones. In each individual DEM, the study area was segregated into four geomorphic zones: channel bank, channel bed, floodplain and cutoff channel. Figure 2a shows an example of the distribution of each of the four geomorphic zones in the study area for the period between surveys 4 and 5, days 191–231. Segregation of each of the four geomorphic zones was accomplished as follows. The channel bank mask is defined as the zone between the channel bank lines of two subsequent surveys plus a strip of 0.5 m width on either side of the zone. The channel bed mask is defined as the zone between the two channel bank masks. The remainder of the grid domain is labelled as floodplain, with the cutoff channel considered as

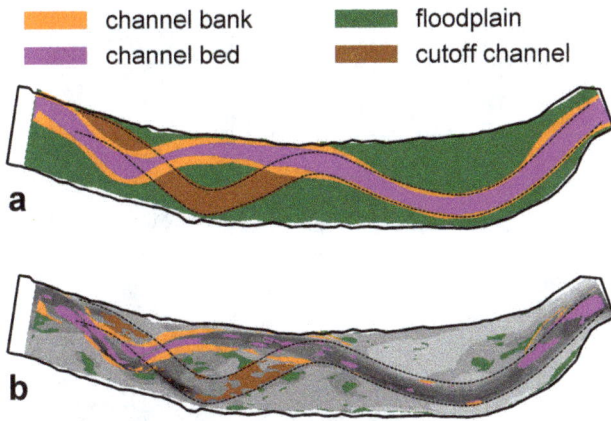

Figure 2. Segregation of geomorphic zones between surveys 4 and 5. (**a**) Masks of each of the four geomorphic zones, including channel bank (orange), channel bed (purple), floodplain (green) and cutoff channel (brown). (**b**) Resulting segregation.

a separate geomorphic zone. Finally, each of the patches of morphological change was attributed to one of the geomorphic features (Fig. 2b).

Two methods were applied to quantify morphological activity. First, the volumetric change in sediment storage was determined. Volumetric change in sediment storage was determined by integrating elevation change over area, distinguishing between gross erosion and gross deposition. Net sediment change was determined by subtracting gross erosion from gross deposition. The uncertainty associated with the volumetric change in sediment storage were estimated using the method described by Wheaton et al. (2013), who defined the uncertainty as plus or minus one standard deviation of the volumetric error. The latter standard deviation is approximated from the error surface, obtained from the LoD method (Milan et al., 2011), as the sum of squared errors.

Second, the root-mean-square elevation (RMSD, m) difference was determined:

$$\mathrm{RMSD} = \sqrt{\frac{\sum \left[z_{i,j}(\mathrm{new}) - z_{i,j}(\mathrm{old}) \right]^2}{n}}, \tag{2}$$

where $z_{i,j}$ is the elevation in grid cell (i, j) of the DEM and n is the number of grid cells. The volumetric change in sediment storage and the RMSD were expressed as a rate of change by dividing both quantities by the time between two successive surveys.

3.4 Hydrological monitoring

Discharge data were collected downstream of the study reach at a discharge measurement weir, indicated by Q in Fig. 1c. Discharge estimates were acquired with a 1 h frequency. Water level data were collected at a water level gauge in the study area, indicated by WL1 in Fig. 1d. Initially, the water level gauge was placed in the channel that became subject to

cutoff. After day 288, the water level gauge was moved to a new location, on the other side of the floodplain, indicated by WL2 in Fig. 1d. Water level data were acquired with a 1 h frequency. Due to a malfunctioning pressure sensor, water level data were unavailable in the first 103 days after channel reconstruction. Longitudinal water level profiles were measured with RTK-GPS equipment during 13 of the 14 morphological surveys.

3.5 Linear regression modelling

To investigate the degree to which morphological activity responds to discharge variation, values of the RMSD were related to time-averaged dimensionless bed shear stress (Shields stress). The Shields stress θ (–) is defined as

$$\theta(t) = \frac{S_\mathrm{w} R(t)}{d_{50} s}, \tag{3}$$

where S_w is the longitudinal water level slope (–); $R(t)$ is the hydraulic radius (m); and $s = (\rho_s - \rho)/\rho$ is the relative submerged specific gravity of the sediment (–), with $\rho = 1000\,\mathrm{kg\,m^{-3}}$ the density of water and $\rho_s = 2650\,\mathrm{kg\,m^{-3}}$ the density of sediment. The average longitudinal water level slope S_w was used, amounting to $0.49\,\mathrm{m\,km^{-1}}$. The average value was calculated from the last 11 water level profiles from the morphological surveys, excluding the two pre-cutoff water level profiles. The hydraulic radius $R(t)$ was obtained by dividing the cross-sectional flow area $A(t)$ by the wetted perimeter $P(t)$, obtained from the cross section at the location of the water level gauge. Equation 3 was applied to the discharge and water level time series and averaged between two successive morphological surveys in order to obtain an estimate of the time-averaged Shields stress per period.

3.6 Riparian vegetation

The restored stream was constructed in bare soil. Riparian vegetation started to develop halfway through the first year. The temporal development of the riparian vegetation was monitored by means of field surveys, aerial photographs and oblique terrestrial photographs.

Coverage of riparian vegetation species were estimated during two field surveys, at day 339 (September 2012) and day 645 (July 2013). The surveys were performed in three cross sections: at 17, 84, and 120 m upstream from the study area. In each cross section, species cover was mapped visually by estimating the vertical projection of a species' shoot area to the soil surface. Species cover was mapped in five rectangular plots (25 cm × 50 cm), located on the channel bed (1 plot), at the channel bank toe (1 plot), in the floodplain (2 plots) and on the floodplain edge (1 plots). Data from two plots per cross section were used to estimate the dominant riparian vegetation species. The two plots were located in the floodplain, at 2.7–5.5 and 7.9–11.7 m from the channel banks.

The spatial distribution of the riparian vegetation was obtained from three aerial photographs, taken at day 188 (April 2012), day 289 (July 2012) and at day 636 (July 2013). The aerial photographs were taken with a 10 cm (day 188) and 25 cm (days 289 and 636) resolution. Figure 3a shows an example of the aerial photograph taken at day 289. The last two aerial surveys included colourized infrared (CIR) images (Fig. 3b), which contain data from the near-infrared (NIR) wavelengths (0.78–3 μm). With these data the riparian vegetation development at these two dates was quantified with the normalized difference vegetation index (NDVI):

$$NDVI = \frac{NIR - VIS}{NIR + VIS},\tag{4}$$

where NIR is the intensity of the NIR wavelengths and VIS is the intensity of the visible red wavelengths. Equation 4 was applied to the entire image, from which the spatial variation of the NDVI within the study area was obtained. NDVI varies between −1.0 and 1.0. Positive values generally correspond to vegetation, whereas negative values correspond to water or other media that adsorb the infrared wavelengths (Clevers, 1988).

Figure 3c shows the NDVI from the second aerial survey for the study area and surrounding agricultural fields and roads. The panel clearly shows positive values for the pasture area on the north side of the study area. Negative values are obtained inside the channel and on the road, on the south side of the study area. The obtained NDVI values within the study area are in a range between 0 and 0.2, which is associated with bare soil (Holben, 1986). Nevertheless, the aerial photo and field observations show that riparian vegetation was present in the study area, where higher NDVI values corresponded to more abundant riparian vegetation. Further analysis of the aerial photographs only considers the areas where positive NDVI values are obtained.

Oblique terrestrial photographs were taken during the majority of the morphological surveys. The series of photographs provide a qualitative view on the temporal development of the riparian vegetation in the study area.

4 Results

4.1 Morphodynamics

Figures 4 and 5 show, respectively, the DEMs and DoDs of the 14 morphological surveys. The first three surveys (days 0, 93 and 133) show the sequence of morphological changes that capture the chute cutoff event. The first DoD (day 0–93) shows deposition of sediment on the channel bed, which is associated with the formation of a plug bar (Eekhout, 2014). An initial embayment had formed upstream from the bend to be cutoff. The actual cutoff occurred in the second period (day 93–133), where a new channel incised into the flood-

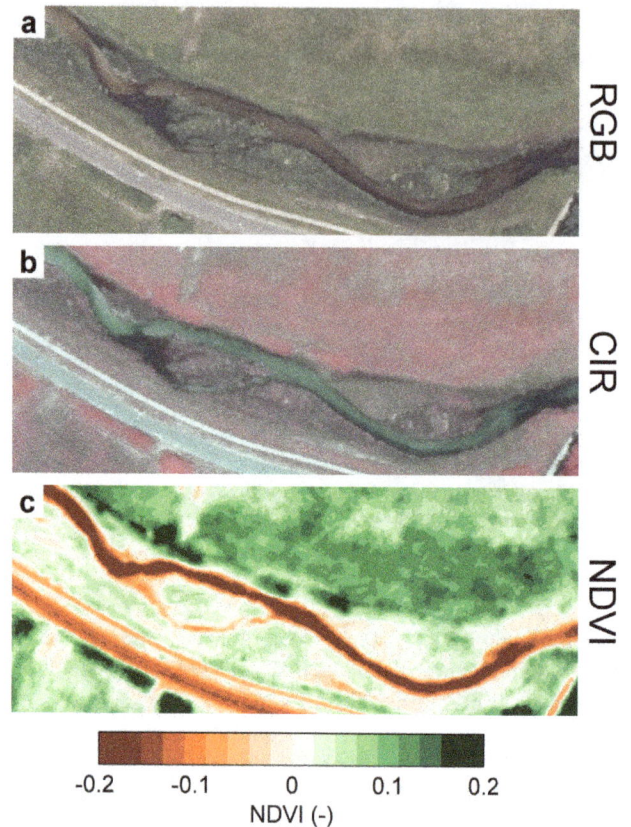

Figure 3. An example of the second aerial photograph (day 289) from which the NDVI was determined, showing **(a)** the RGB image (red, green, blue), **(b)** the CIR image (colourized infrared) and **(c)** the NDVI. In **(c)**, the green-coloured areas indicate positive NDVI values, and orange-coloured areas indicate negative NDVI values. All images have a 25 cm resolution.

plain. In the subsequent period (day 133–231), the morphological changes occurred mainly in the bend at the downstream end of the study area. During this period, both erosion of the outer bank and accretion in the inner bank were observed in this bend. Subsequently, there was a period of little morphological change (day 231–288). In the last period (day 341–672), morphological changes were restricted to the channel bed and banks. At that time, little morphological change was observed in the floodplain.

Overall, most morphological developments were observed in the downstream half of the study area. Figure 4 clearly illustrates the channel bed incision of the bend at the downstream end of the study reach in the period following the chute cutoff. The upstream half of the study reach did not show pronounced morphological changes. Apart from occasional changes in the channel bed, no structural bank erosion or accretion was observed.

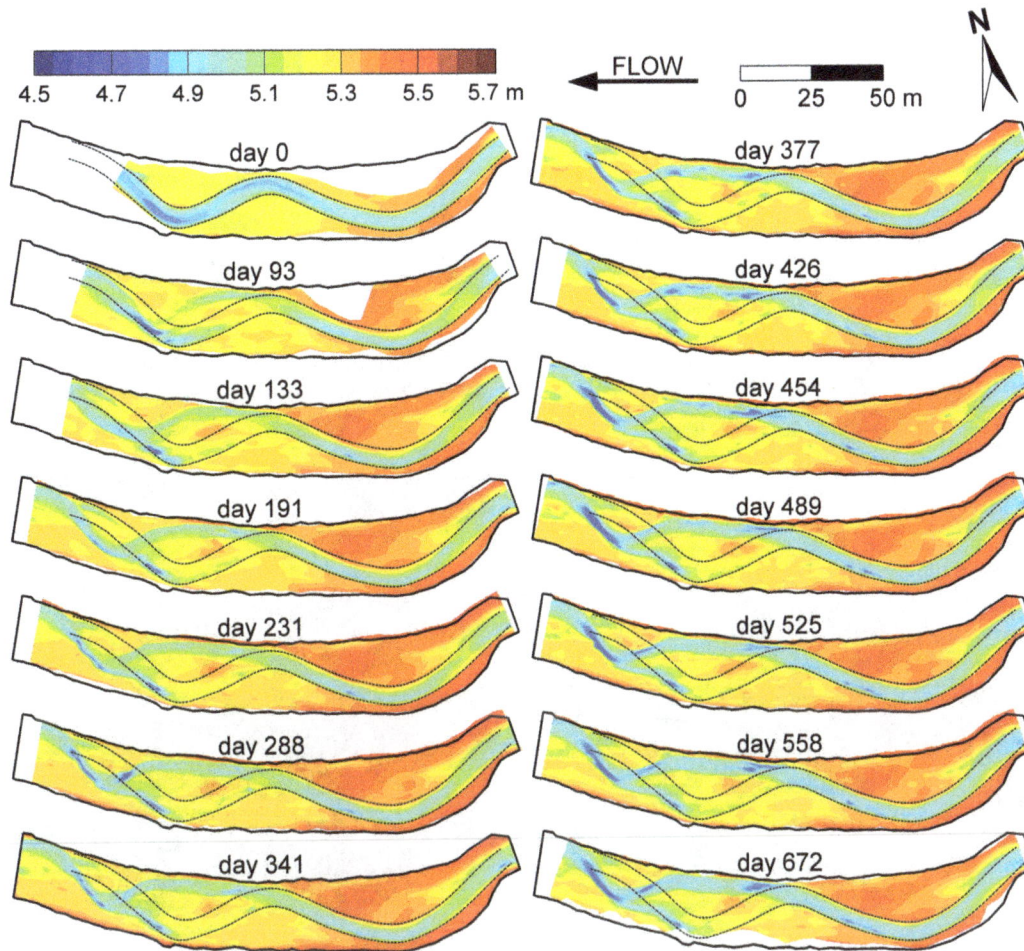

Figure 4. DEMs of all 14 morphological surveys. The number of days indicate the time since channel reconstruction. The dashed black lines indicate the location of the channel banks of the reconstructed channel. The solid black line surrounding the DEMs indicates the extent of the seventh morphological survey (day 341). Elevation is indicated in metres above mean sea level.

Figure 6 and Table 2 show the sediment budgets, derived from the DoDs (Fig. 5). Figure 6a shows volumetric change over the study area as a whole, aggregated for regions of erosion and deposition as well as net change. Figure 6b shows the volumetric change subdivided per geomorphic zone. Considering the study area as a whole (Fig. 6a), a shift from net deposition to net erosion is observed. In the first period (between surveys 1 and 2), deposition is mainly caused by the deposition of sediment in the channel bed and in the cutoff channel. This is followed by a period of limited net change between surveys 2 and 4, where bank erosion is balanced by net deposition in the rest of the study area. From the fifth survey onwards, the sediment balance shifts towards net erosion. Channel bed processes (channel bed incision) are the dominant contributors to the net erosion of sediment in the study area.

Figure 6a shows that net deposition occurred in the periods between surveys 1 and 2 as well as surveys 4 and 5. This implies that sediment was transported towards the

study area from upstream. This observation is confirmed by Fig. 7, which shows longitudinal channel bed elevation data obtained over an area ranging from a distance of 180 m upstream from the study area to the downstream end of the study reach. The figure shows that between days 0 and 93, upstream channel bed incision led to channel bed aggradation in the downstream part. In the upstream area only minor changes in channel bed elevation occurred in the period following day 93. The deposited sediment in the study area gradually eroded, which explains the negative sediment balance in several periods between surveys 5 and 14 (Fig. 6a).

4.2 Riparian vegetation development

Table 3 lists the dominant (average cover > 10 %) riparian vegetation species and their characteristics obtained from two field surveys. The dominant species in 2012 were *Juncus articulatus* (jointed rush) and *Juncus bufonius* (toad rush). *Juncus articulatus* was also among the dominant species in 2013, together with *Juncus effusus* (soft rush) and *Trifolium*

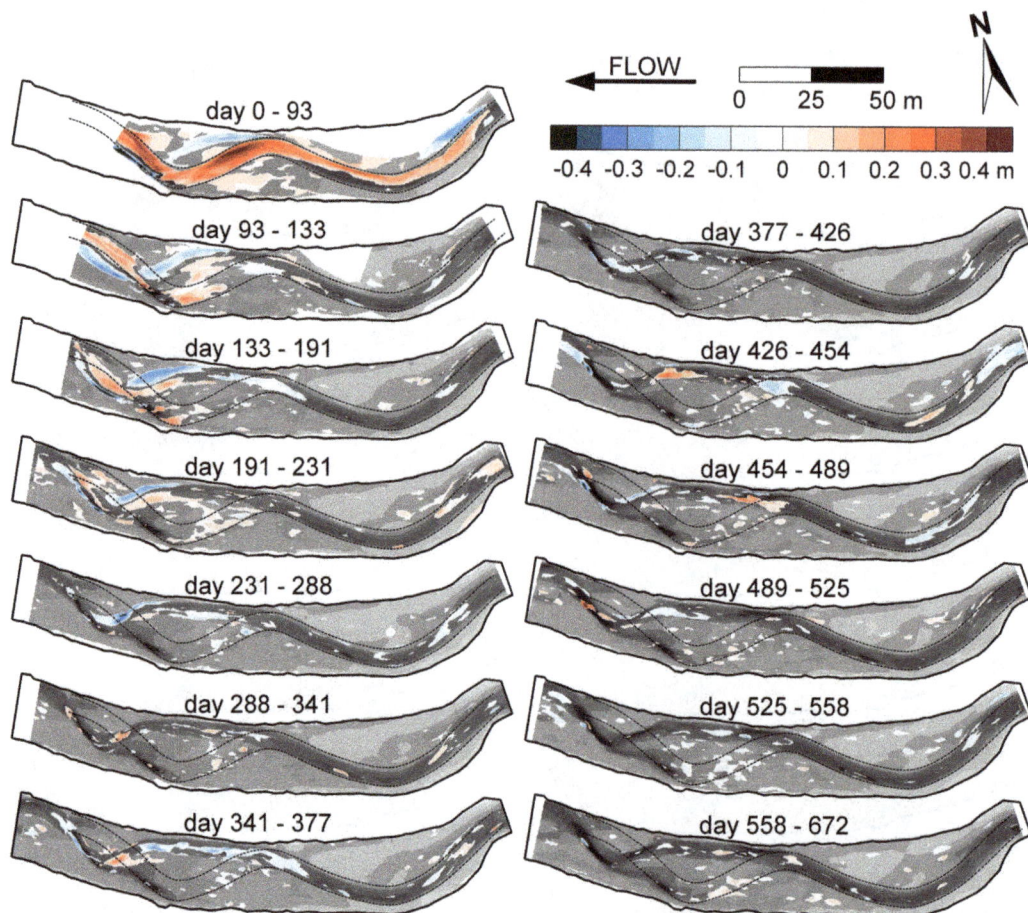

Figure 5. DoDs of all 13 periods between the 14 morphological surveys. The number of days indicates the time since channel reconstruction. The dashed black lines indicate the location of the channel banks of the reconstructed channel. The solid black line surrounding the DEMs indicates the extent of the seventh morphological survey (day 341). Erosion is indicated in blue and deposition in red. The DEM of the first of the two DEMs is shown in greyscale.

repens (white clover). All species are classified as herbaceous vegetation and are perennial. The growing season of the dominant species is roughly from June to September, which explains the maximum coverage in September as observed in Fig. 9 (days 341 and 672). Maximum root depth varies between less than 10 cm and 100 cm. The presented root depth values are valid for fully grown specimens. It is likely that the root depths for the observed riparian vegetation in this case study were shorter.

Figure 8 shows the three aerial photographs and the derived positive NDVI values for the last two aerial surveys. Panel a shows that riparian vegetation was absent in the study area at day 188. The second aerial photograph (day 289, panels b and d) shows that a patchy pattern of riparian vegetation started to emerge in the floodplain area. At the time of taking the second aerial photograph, riparian vegetation cover (observed as a positive NDVI value) amounted to 56 % of the study area. Specified per geomorphic zone, riparian vegetation covered 33 % of the channel bank zone, 3 % of the chan-

nel bed zone, 80 % of the floodplain zone and 57 % of the cutoff channel zone. Riparian vegetation in the cutoff channel did not develop as abundantly as in the rest of the floodplain. There is also a clear distinction between channel (bed and bank) and floodplain. The third aerial photograph (day 636, panels c and e) clearly shows that riparian vegetation cover increased from the first to the second year, when riparian vegetation cover amounted to 77 % of the study area. Riparian vegetation cover increased in all geomorphic zones, amounting to 84 % of the channel bank zone, 22 % of the channel bed zone, 97 % of the floodplain zone and 92 % of the cutoff channel zone.

Figure 9 shows the series of 14 oblique terrestrial photographs taken from the location indicated in the lower-right corner of the figure. Just after channel reconstruction had finished at day 0 (first photo on row one), riparian vegetation was visible on the left channel bank. Overall, riparian vegetation was absent in the floodplain. In the subsequent period, until day 161 (third photo on row one), no change in

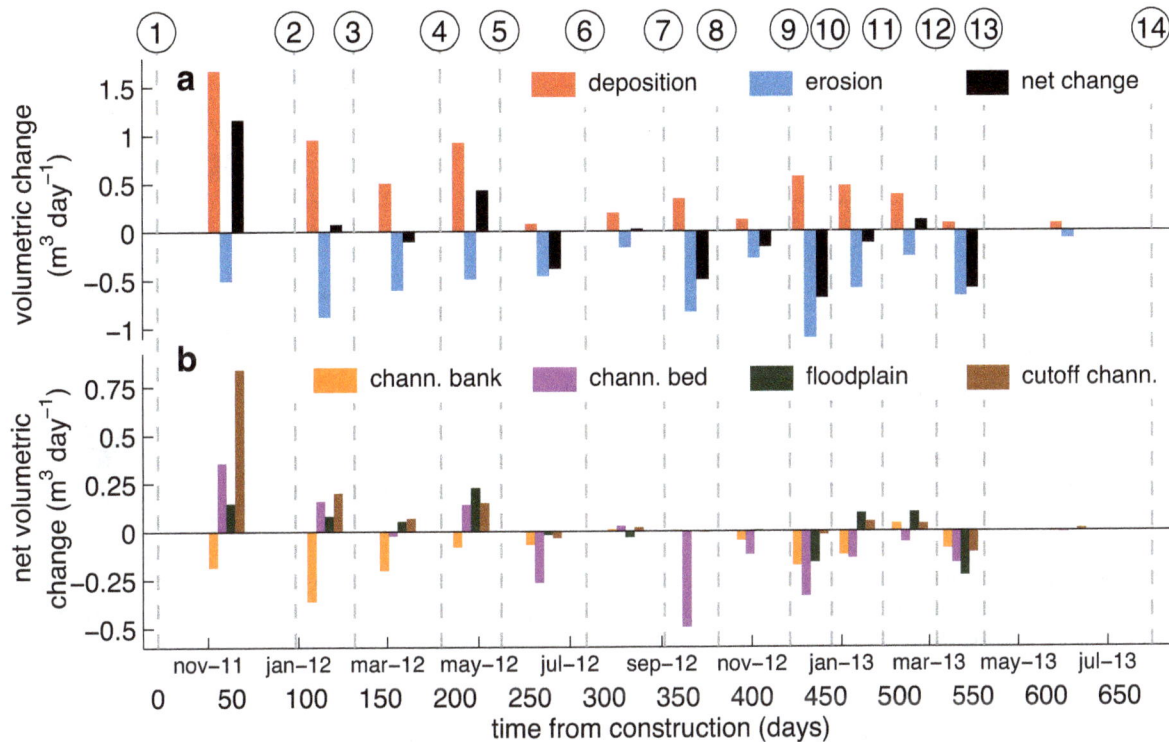

Figure 6. Temporal evolution of the volumetric change in sediment storage. (**a**) Volumetric change in sediment storage (m³ day⁻¹) for regions of deposition (red), erosion (blue) and net change (black). (**b**) Volumetric change in sediment storage (m³ day⁻¹), specified per geomorphic feature, with channel bank (orange), channel bed (purple), floodplain (green) and cutoff channel (brown). The dashed vertical lines indicate the surveying moments, and the numbers at the top of the figure correspond to the numbers in Table 1.

Table 2. Summary of volumetric change in storage by deposition, erosion and net change (deposition minus erosion). Volumetric change is specified in gross change ($\sum$ volume) and the standard deviation (SD).

Period (days)	Deposition $\sum$ volume $\pm$ SD (m³)	Erosion $\sum$ volume $\pm$ SD (m³)	Net change $\sum$ volume $\pm$ SD (m³)
0–93	155.1 ± 23.5	-47.2 ± 7.9	107.9 ± 31.4
93–133	38.1 ± 8.0	-35.2 ± 6.4	2.9 ± 14.4
133–191	29.1 ± 7.1	-35.2 ± 6.7	-6.0 ± 13.8
191–231	36.8 ± 10.9	-19.7 ± 4.3	17.2 ± 15.2
231–288	4.4 ± 1.5	-26.4 ± 7.3	-22.0 ± 8.8
288–341	10.1 ± 3.2	-8.9 ± 2.9	1.2 ± 6.1
341–377	12.2 ± 3.4	-30.0 ± 8.8	-17.9 ± 12.2
377–426	5.8 ± 2.1	-13.7 ± 4.6	-7.9 ± 6.7
426–454	15.9 ± 4.6	-35.2 ± 9.5	-19.3 ± 14.1
454–489	16.4 ± 5.1	-20.7 ± 6.7	-4.3 ± 11.8
489–525	13.6 ± 4.7	-9.4 ± 3.3	4.2 ± 8.0
525–558	2.7 ± 1.0	-22.2 ± 7.7	-19.5 ± 8.7
558–672	9.0 ± 2.9	-8.6 ± 3.2	0.3 ± 6.2

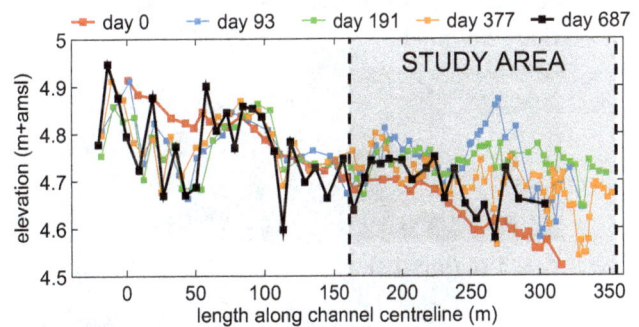

Figure 7. Temporal evolution of the longitudinal bed elevation.

riparian vegetation coverage was observed. Riparian vegetation started to develop around day 231 (last photo on row one). Some patches of riparian vegetation were emerging in the floodplain and on the channel banks. The development of riparian vegetation continued in the following period and a maximum riparian vegetation coverage was observed at day 341 (second photo on row two). At that time, the floodplain was almost entirely covered with riparian vegetation. Only in the cutoff channel did riparian vegetation develop less abundantly (see also Fig. 8d). In the period until the end of the study period, from day 377 (third photo on row two) until day

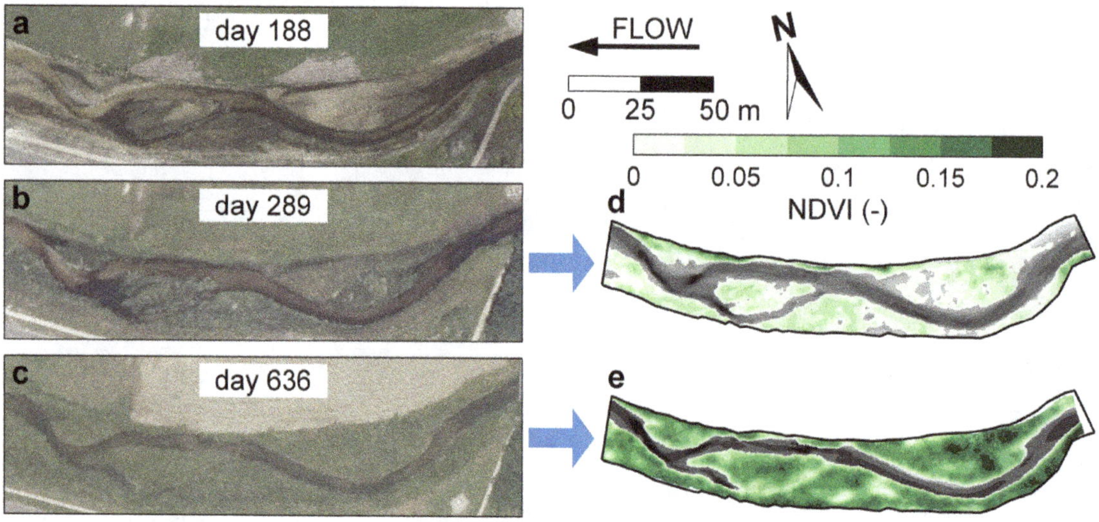

Figure 8. Three aerial photographs taken on (**a**) day 188 (April 2012), (**b**) day 289 (July 2012) and (**c**) day 636 (July 2013); (**d**) and (**e**) show positive NDVI values derived from the aerial photographs from (**b**) and (**c**), respectively. The DEMs from survey 6 (**d**) and survey 13 (**e**) are shown in greyscale as a reference.

Figure 9. Series of 14 oblique terrestrial photographs taken from the location indicated in Fig. 1d, except for the third photo (day 161), which was taken from the other side of the floodplain.

Table 3. Characteristics of the most dominant riparian vegetation species found in the floodplain, including average cover (%), lifetime, growing season and maximum root depth. The presented characteristics were obtained from http://www.wilde-planten.nl/ (*Wilde planten in Nederland en België* – Wild plants in the Netherlands and Belgium).

Scientific name	Cover (%)	Lifetime	Growing season	Root depth
September 2012				
Juncus articulatus	19	Perennial	Jun–Sep	10–20 cm
Juncus bufonius	22	Annual	Jun–Sep	< 10 cm
July 2013				
Juncus articulatus	14	Perennial	Jun–Sep	10–20 cm
Juncus effusus	14	Perennial	Jun–Aug	< 100 cm
Trifolium repens	28	Perennial	May–Oct	10–50 cm

558 (last photo on row three), the riparian vegetation cover started to decrease to a level approximately similar to the situation between days 231 and 288. The riparian vegetation cover started to increase again from day 558 onwards. This resulted in the maximum riparian vegetation cover at the end of the survey period, at day 672 (second photo on row four).

4.3 Morphodynamic regime change

Figure 10 combines the data obtained from the morphological surveys, oblique terrestrial photographs and discharge measurements. Figure 10a shows morphological change (RMSD, m year^{-1}), which has been derived from Fig. 5, with Eq. 2. Figure 10b shows the discharge hydrograph, obtained from the measurement weir located 360 m downstream from the study area (Fig. 1c). The temporal evolution of the morphological change (black diamonds in Fig. 10a) shows that the RMSD metric of morphological change was relatively high in the first two periods, between surveys 1 and 3. In the subsequent period, morphological changes show a decreasing trend until the period between surveys 6 and 7. In the final period, incidental peaks are observed during periods of increased discharges, i.e. in periods between surveys 7 and 8 as well as surveys 9 and 10.

From Fig. 10a it appears that the study period can be divided into two stages. The first stage can be considered an apparent morphological disequilibrium. The interval between surveys 1 and 3 is largely dominated by the chute cutoff event, which is followed by an interval until survey 5 that is dominated by channel bank processes. During the second stage, the morphology within the study reach can be considered to tend towards an equilibrium, in the sense that the reach-scale morphology has largely stabilized. Both channel bank and channel bed processes dominate morphological change in the latter stage. During the whole study period, morphological changes in the floodplain contributed only slightly to the overall changes in the study area.

The temporal development of riparian vegetation, indicated by the green shading in Fig. 10, shows reach-scale riparian vegetation development from day 231 onwards. An initial maximum riparian vegetation coverage is observed in the period between surveys 7 and 8, which corresponds to the photo taken on day 341 (Fig. 9, second photo on row two). In the subsequent period, until survey 13, a decrease of riparian vegetation cover is observed. A second maximum riparian vegetation coverage is observed in the period following survey 14, which corresponds to the photo taken on day 672 (Fig. 9, second photo on row four). Figure 10b shows that two periods of extremely high discharges occurred, i.e. in the periods between surveys 1 and 2 and surveys 9 and 10, featuring discharge peaks with a return period of 120 days and 180 days per year^{-1}, respectively.

Figure 10 shows an initial stage of morphological disequilibrium, when rates of morphological change are relatively high without showing a clear response to discharge variation. In the subsequent near-equilibrium stage, a clear response to the varying discharge is evident. A linear regression model was established, relating the RMSD to time-averaged Shields stress for the study area as a whole. The linear regression model was established for the period after riparian vegetation emerged (between surveys 5 and 14) only. The evaluation was based on the hypothesis that morphological activity in the period prior to riparian vegetation growth was significantly different from the period after riparian vegetation emerged. Additionally, the 95 % confidence interval was determined based on the Student t distribution, with the degrees of freedom corresponding to the number of periods after riparian vegetation emerged.

Figure 11 shows the results from the linear regression between RMSD and time-averaged Shields stress. The RMSD in a certain period is significantly different from predicted values when the value is outside the 95 % confidence interval. Considering the study area as a whole, the period prior to riparian vegetation growth is outside the 95 % confidence interval. These results indicate that the morphodynamic behaviour during the period prior to riparian vegetation growth is significantly different from the period after riparian vegetation emerged.

5 Discussion

Figures 4 and 5 show that active meander processes occurred in the initial period prior to riparian vegetation growth, including a chute cutoff and meander migration. In the subsequent period, when riparian vegetation started to emerge, the active meander processes were less pronounced, although localized bank erosion was observed. From Fig. 6 and Table 2 it appears that the study period can be subdivided into an initial period of net deposition (prior to survey 5) and a subsequent period of net erosion (after survey 5).

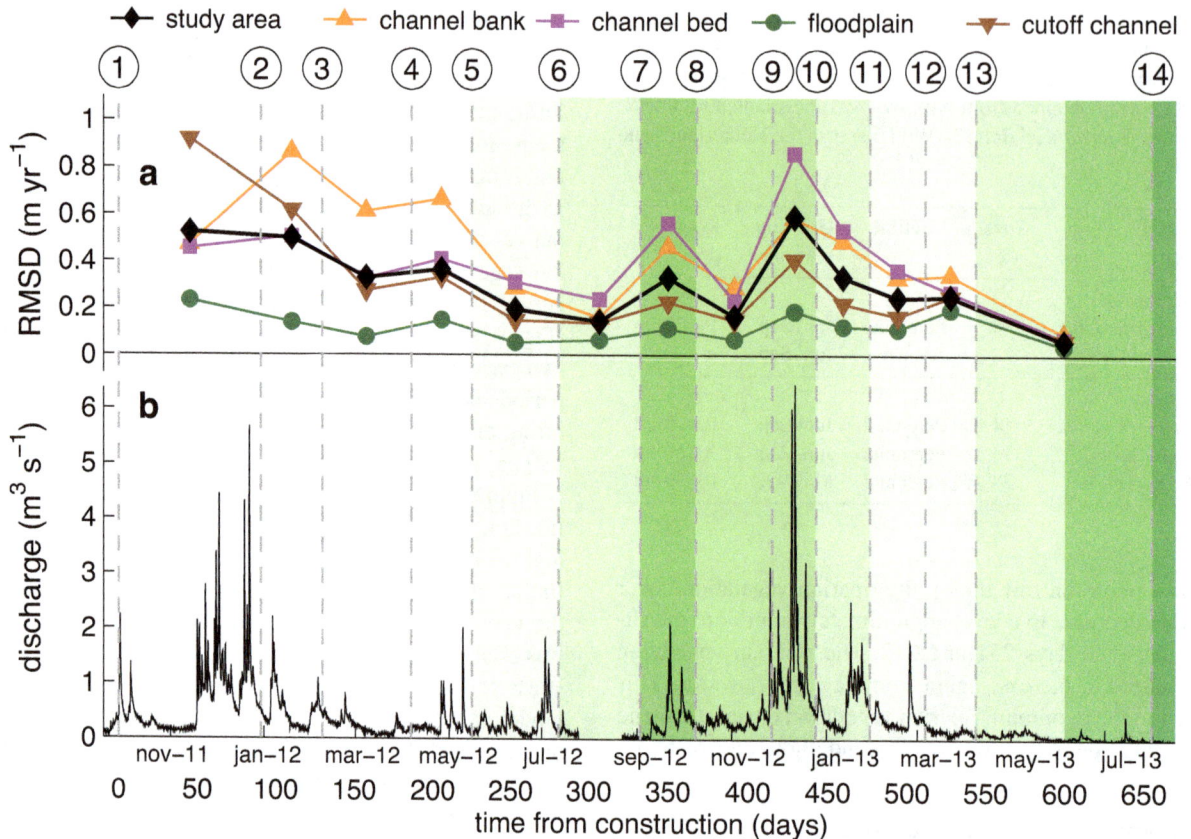

Figure 10. Temporal evolution of (**a**) the rate of morphological change (m year^{-1}) for the study area as a whole (black diamonds) and specified per morphological feature, with channel bank (orange triangles), channel bed (purple squares), floodplain (green circles) and cutoff channel (brown triangles), as well as (**b**) discharge hydrograph (m^3 s^{-1}). The green shading indicates the riparian vegetation coverage, estimated from Fig. 9. The dashed vertical lines indicate the surveying moments, and the numbers at the top of the figure correspond to the numbers in Table 1.

Figure 7 shows that in the period between surveys 1 and 2 upstream channel bed incision caused channel bed aggradation in the study area. The active meandering processes, as observed in the period prior to riparian vegetation growth, may be related to the associated net sediment accumulation. Variation in sediment supply has been previously related to differences in morphological regime. An increased sediment supply in a river may result in braided or actively meandering planforms, including chute cutoffs, whereas reduced sediment supply may result in laterally stable, sinuous planforms (Church, 2006; Kleinhans, 2010). In the present case study, the regime difference between the two periods may partially be explained by the difference in sediment supply. In the initial period, the channel may have acted as an actively meandering river, as a consequence of excessive sediment supply. After the sediment supply reduced, the channel may have acted as a laterally stable, sinuous river. The marked difference between the two periods of morphological behaviour may also relate to the growing season of the observed riparian vegetation. It is infeasible to isolate the role of reduced upstream sediment supply from the effect of riparian vegeta-

tion development; both factors may assert a substantial influence on the morphological regime change.

The dominant riparian vegetation species are all classified as herbaceous vegetation. The root structure of herbaceous vegetation differs from other vegetation types, such as shrubby and woody vegetation. Herbaceous vegetation shows a more graminoid-shaped root structure, including fibrous roots (Wynn et al., 2004; Burylo et al., 2011), as opposed to shrubby and woody vegetation, which include a taplike root system without fibrous roots (Burylo et al., 2011). These differences in root structure cause differences in erosional resistance. Recent studies have shown that herbaceous vegetation provides more soil reinforcement and is more efficient in improving aggregate stability than shrubby and woody vegetation (Burylo et al., 2011; Fattet et al., 2011). The advantages of erosion resistance for herbaceous vegetation are most effective in the top soil layer (< 30 cm; Wynn et al., 2004). Herbaceous vegetation types are among the first to (re-)colonize the substrate after environmental disturbance or land restoration (Cammeraat et al., 2005; Burylo et al., 2007). Although most of the studies supporting this

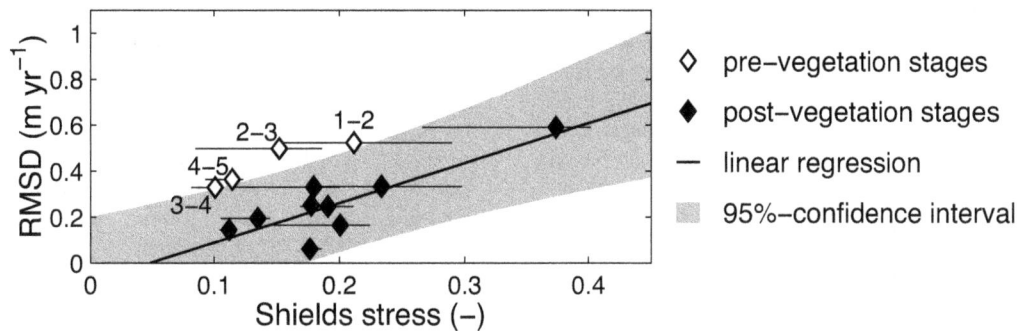

Figure 11. Linear regression between time-averaged Shields stress (–) and the root-mean-square elevation difference (RMSD, m year^{-1}) for the study area as a whole. The open and filled markers correspond to the pre-vegetation and post-vegetation stages, respectively. The solid black line denotes the linear regression curve for the period after riparian vegetation emerged (between surveys 5 and 14). The grey-shaded area indicates the extent of the 95 % confidence interval of the regression model. The extent of the black horizontal lines indicate the 25 % and 75 % quantiles.

were performed in different geographical regions and within different climates, these results show that herbaceous vegetation may have a significant impact on soil stability in the study area. The influence of herbaceous vegetation on bank stability may be particularly relevant for lowland streams due to their low channel depths. The temporally high rate at which riparian vegetation has colonized the lowered floodplains confirms a previous study by Gurnell et al. (2006).

For future research, it may be worthwhile better quantifying the rate at which riparian vegetation colonizes the floodplain and channel banks after stream restoration measures are implemented. If riparian vegetation indeed effectively fixes the channel banks, which cannot be clearly demonstrated from this study, riparian vegetation may exert an influence on hydraulic geometry relations in the equilibrium state. Hydraulic geometry theory quantifies scale relations between channel depth, width, mean flow velocity and bed slope in alluvial rivers and streams, which is often used in stream restoration practice (Rosgen, 1994; Copeland et al., 2001; Shields et al., 2003). For rivers, such scaling relations generally do not account for riparian vegetation (Wilkerson and Parker, 2011). For streams with channel depths similar to root depths, such as lowland streams, hydraulic geometry relations may be much dependent on riparian vegetation species, and on the timing of the grow season relative to the discharge hydrograph. It is not unlikely that details in the first year after reconstruction of a stream are crucial for the terminal channel geometry.

6 Conclusions

A detailed monitoring plan was implemented to monitor a lowland stream for a period of almost 2 years after channel reconstruction. Morphological, hydrological and ecological data were combined to establish interactions between reach-scale morphodynamics, discharge dynamics and riparian vegetation development. In the initial stage after chan-

nel reconstruction, when negligible riparian vegetation was present, channel morphology adjusted rapidly towards an alternative, complex topography. This period can be characterized by the occurrence of a chute cutoff and large-scale bank erosion. Upstream sediment supply may have played a substantial role in the initial morphological development in the study area. Riparian vegetation emerged in the subsequent stage. Channel bed incision and localized bank erosion dominate the morphodynamic developments after riparian vegetation had developed. Linear regression analysis shows that the morphological response to Shields stress variation is significantly different between the pre-vegetation and post-vegetation stages, and hence a morphological regime change had occurred.

The two stages of morphological adjustments reveal the rate at which a lowland stream adjusts towards a new morphological equilibrium. The two stages of morphological response may be largely associated with upstream sediment supply, caused by channel incision upstream from the study area. From the obtained field evidence, the role of riparian vegetation in the morphological regime change cannot be isolated from the effect of reduced sediment supply. The characteristics of the herbaceous vegetation and channel dimensions of lowland streams suggest that the initial riparian vegetation may exert a significant control on the stability of the channel banks. It may be worthwhile minimizing the duration of the initial stage of morphological adaptations after a lowland stream restoration project is realized. To some extent, this duration can be manipulated by changing the season in which the new channel is planned to be reconstructed. This has an impact on the duration of the pre-vegetation period, when the channel-forming discharges may cause significant morphological changes. In streams where channel depths are similar to root depths, the developments of riparian vegetation growth in the first years may be decisive for the terminal morphological equilibrium.

Acknowledgements. This study is part of a research project funded by the STOWA, the Foundation for Applied Water Research (project code 443209), and the research project Beekdalbreed Hermeanderen funded by Agency NL (project code KRW 09023) and co-funded by STOWA. We thank Philip Wenting (Wageningen University) and Andrés Vargas-Luna (Deltares) for their contribution to the fieldwork campaign, and Paul Torfs (Wageningen University) for his help in post-processing of field data. We thank CycloMedia Technology B.V. and Slagboom en Peeters Luchtfotografie B.V. for providing the aerial photographs. Also, we thank Remko Uijlenhoet (Wageningen University), Piet Verdonschot (Alterra), Christian Huising (Waterschap Vallei en Veluwe), Simon Dixon (University of Birmingham) and two anonymous reviewers for their comments on the manuscript. The morphological and hydrological data are made available through doi:10.6084/m9.figshare.960038.

Edited by: F. Metivier

References

Abernethy, B. and Rutherfurd, I. D.: The distribution and strength of riparian tree roots in relation to riverbank reinforcement, Hydrol. Process., 15, 63–79, 2001.

Brasington, J., Langham, J., and Rumsby, B.: Methodological sensitivity of morphometric estimates of coarse fluvial sediment transport, Geomorphology, 53, 299–316, doi:10.1016/S0169-555X(02)00320-3, 2003.

Braudrick, C. A., Dietrich, W. E., Leverich, G. T., and Sklar, L. S.: Experimental evidence for the conditions necessary to sustain meandering in coarse-bedded rivers, P. Nat. Acad. Sci. USA, 106, 16936–16941, doi:10.1073/pnas.0909417106, 2009.

Burylo, M., Rey, F., and Delcros, P.: Abiotic and biotic factors influencing the early stages of vegetation colonization in restored marly gullies (Southern Alps, France), Ecol. Eng., 30, 231–239, doi:10.1016/j.ecoleng.2007.01.004, 2007.

Burylo, M., Hudek, C., and Rey, F.: Soil reinforcement by the roots of six dominant species on eroded mountainous marly slopes (Southern Alps, France), Catena, 84, 70–78, doi:10.1016/j.catena.2010.09.007, 2011.

Cammeraat, E., van Beek, R., and Kooijman, A.: Vegetation succession and its consequences for slope stability in SE Spain, Plant Soil, 278, 135–147, doi:10.1007/s11104-005-5893-1, 2005.

Camporeale, C., Perucca, E., Ridolfi, L., and Gurnell, A. M.: Modeling the interactions between river morphodynamics and riparian vegetation, Rev. Geophys, 51, 379–414, doi:10.1002/rog.20014, 2013.

Church, M.: Bed material transport and the morphology of alluvial river channels, Annu. Rev. Earth Pl. Sc., 34, 325–354, doi:10.1146/annurev.earth.33.092203.122721, 2006.

Clevers, J. G. P. W.: The derivation of a simplified reflectance model for the estimation of leaf area index, Remote Sens. Environ., 25, 53–69, 1988.

Copeland, R. R., McComas, D. N., Thorne, C. R., Soar, P. J., Jonas, M. M., and Fripp, J. B.: Hydraulic Design of Stream Restoration Projects, Tech. rep., US Army Corps of Engineers, Washington DC, 2001.

Croke, J., Todd, P., Thompson, C., Watson, F., Denham, R., and Khanal, G.: The use of multi temporal LiDAR to assess basin-scale erosion and deposition following the catastrophic January 2011 Lockyer flood, SE Queensland, Australia, Geomorphology, 184, 111–126, doi:10.1016/j.geomorph.2012.11.023, 2013.

Eekhout, J. P. C.: Morphological Processes in Lowland Streams. Implications for Stream Restoration, Ph.D. thesis, Wageningen University, the Netherlands, 178 pp., 2014.

Eekhout, J. P. C., Hoitink, A. J. F., and Mosselman, E.: Field experiment on alternate bar development in a straight sand-bed stream, Water Resour. Res., 49, 8357–8369, doi:10.1002/2013WR014259, 2013.

Erwin, S. O., Schmidt, J. C., Wheaton, J. M., and Wilcock, P. R.: Closing a sediment budget for a reconfigured reach of the Provo River, Utah, United States, Water Resour. Res., 48, W10512, doi:10.1029/2011WR011035, 2012.

Fattet, M., Fu, Y., Ghestem, M., Ma, W., Foulonneau, M., Nespoulous, J., Le Bissonnais, Y., and Stokes, A.: Effects of vegetation type on soil resistance to erosion: Relationship between aggregate stability and shear strength, Catena, 87, 60–69, doi:10.1016/j.catena.2011.05.006, 2011.

Fuller, I. C., Large, A. R. G., and Milan, D. J.: Quantifying channel development and sediment transfer following chute cutoff in a wandering gravel-bed river, Geomorphology, 54, 307–323, doi:10.1016/S0169-555X(02)00374-4, 2003.

Gautier, E., Brunstain, D., Vauchel, P., Jouanneau, J., Roulet, M., Garcia, C., Guyol, J., and Castro, M.: Channel and floodplain sediment dynamics in a reach of the tropical meandering Rio Beni (Bolivian Amazonia), Earth Surf. Proc. Land., 35, 1838–1853, doi:10.1002/esp.2065, 2010.

Gran, K. and Paola, C.: Riparian vegetation controls on braided stream dynamics, Water Resour. Res., 37, 3275–3284, doi:10.1029/2000WR000203, 2001.

Grove, J., Croke, J., and Thompson, C.: Quantifying different riverbank erosion processes during an extreme flood event, Earth Surf. Proc. Land., 38, 1393–1406, doi:10.1002/esp.3386, 2013.

Gurnell, A. M., Morrissey, I. P., Boitsidis, A. J., Bark, T., Clifford, N. J., Petts, G. E., and Thompson, K.: Initial Adjustments Within a New River Channel: Interactions Between Fluvial Processes, Colonizing Vegetation, and Bank Profile Development, Environ. Manage., 38, 580–596, doi:10.1007/s00267-005-0190-6, 2006.

Heritage, G. L., Milan, D. J., Large, A. R. G., and Fuller, I. C.: Influence of survey strategy and interpolation model on DEM quality, Geomorphology, 112, 334–344, doi:10.1016/j.geomorph.2009.06.024, 2009.

Holben, B. N.: Characteristics of maximum-value composite images from temporal AVHRR data, Int. J. Remote Sens., 7, 1417–1434, doi:10.1080/01431168608948945, 1986.

Kasvi, E., Vaaja, M., Alho, P., Hyyppä, J., Kaartinen, H., and Kukko, A.: Morphological changes on meander point bars associated with flow structure at different discharges, Earth Surf. Proc. Land., 38, 577–590, doi:10.1002/esp.3303, 2013.

Kleinhans, M. G.: Sorting out river channel patterns, Prog. Phys. Geog., 34, 287–326, doi:10.1177/0309133310365300, 2010.

KNMI: Langjarige gemiddelden en extremen, tijdvak 1971–2000, http://www.knmi.nl/klimatologie/normalen1971-2000/index.html (last access: 26 January 2014), 2014.

Lane, S. N., Westaway, R. M., and Hicks, D. M.: Estimation of erosion and deposition volumes in a large, gravel-bed, braided river using synoptic remote sensing, Earth Surf. Proc. Land., 28, 249–271, doi:10.1002/esp.483, 2003.

Lane, S. N., Widdison, P. E., Thomas, R. E. ans Ashworth, P. J., Best, J. L., Lunt, I. A., Sambrook Smith, G. H., and Simpson, C. J.: Quantification of braided river channel change using archival digital image analysis, Earth Surf. Proc. Land., 35, 971–985, doi:10.1002/esp.2015, 2010.

Legleiter, C. J. and Kyriakidis, P. C.: Forward and inverse transformations between cartesian and channel-fitted coordinate systems for meandering rivers, Math. Geol., 38, 927–958, doi:10.1007/s11004-006-9056-6, 2007.

Lejot, J., Delacourt, C., Piégay, H., Fournier, T., Trémélo, M.-L., and Allemand, P.: Very high spatial resolution imagery for channel bathymetry and topography from an unmanned mapping controlled platform, Earth Surf. Proc. Land., 32, 1705–1725, doi:10.1002/esp.1595, 2007.

Leopold, L. B. and Bull, W. B.: Base level, aggradation, and grade, P. Am. Philos. Soc., 123, 168–202, 1979.

Mackin, J. H.: Concept of the Graded River, Geol. Soc. Am. Bull., 59, 463–512, doi:10.1130/0016-7606(1948)59[463:COTGR]2.0.CO;2, 1948.

Milan, D. J., Heritage, G. L., Large, A. R. G., and Fuller, I. C.: Filtering spatial error from DEMs: Implications for morphological change estimation, Geomorphology, 125, 160–171, doi:10.1016/j.geomorph.2010.09.012, 2011.

Perucca, E., Camporeale, C., and Ridolfi, L.: Significance of the riparian vegetation dynamics on meandering river morphodynamics, Water Resour. Res., 43, W03430, doi:10.1029/2006WR005234, 2007.

Pollen-Bankhead, N. and Simon, A.: Enhanced application of root-reinforcement algorithms for bank-stability modeling, Earth Surf. Proc. Land., 34, 471–480, doi:10.1002/esp.1690, 2009.

Rosgen, D. L.: A classification of natural rivers, Catena, 22, 169–199, 1994.

Shields, F. D., Copeland, R. R., Klingeman, P. C., Doyle, M. W., and Simon, A.: Design for stream restoration, J. Hydraul. Eng. ASCE, 129, 575–584, doi:10.1061/(ASCE)0733-9429(2003)129:8(575), 2003.

Simon, A. and Collison, A. J. C.: Quantifying the mechanical and hydrologic effects of riparian vegetation on streambank stability, Earth Surf. Proc. Land., 27, 527–546, doi:10.1002/esp.325, 2002.

Tal, M. and Paola, C.: Dynamic single-thread channels maintained by the interaction of flow and vegetation, Geology, 35, 347–350, doi:10.1130/G23260A.1, 2007.

Tal, M. and Paola, C.: Effects of vegetation on channel morphodynamics: Results and insights from laboratory experiments, Earth Surf. Proc. Land., 35, 1014–1028, doi:10.1002/esp.1908, 2010.

Van de Wiel, M. J. and Darby, S. E.: Riparian Vegetation and Fluvial Geomorphology, chap. Numerical modeling of bed topography and bank erosion along tree-lined meandering rivers, pp. 267–282, Water Science and Application Series 8, American Geophysical Union, Washington, DC, 2004.

Van Heerd, R. M., Kuijlaars, E. A. C., Teeuw, M. P., and Van 't Zand, R. J.: Productspecificatie AHN 2000, Tech. Rep. MDTGM 2000.13, Rijkswaterstaat, Adviesdienst Geo-informatie en ICT, Delft, 2000.

Wheaton, J. M., Brasington, J., Darby, S. E., Merz, J., Pasternack, G. B., Sear, D., and Vericat, D.: Linking geomorphic changes to salmond habitat at a scale relevant to fish, River Res. Appl., 26, 469–486, doi:10.1002/rra.1305, 2010a.

Wheaton, J. M., Brasington, J., Darby, S. E., and Sear, D. A.: Accounting for uncertainty in DEMs from repeat topographic surveys: Improved sediment budgets, Earth Surf. Proc. Land., 35, 136–156, doi:10.1002/esp.1886, 2010b.

Wheaton, J. M., Brasington, J., Darby, S. E., Kasprak, A., Sear, D., and Vericat, D.: Morphodynamic signatures of braiding mechanisms as expressed through change in sediment storage in a gravel-bed river, J. Geophys. Res., 118, 759–779, doi:10.1002/jgrf.20060, 2013.

Wilkerson, G. V. and Parker, G.: Physical basis for quasi-universal relationships describing bankfull hydraulic geometry of sand-bed rivers, J. Hydraul. Eng. ASCE, 137, 739–753, doi:10.1061/(ASCE)HY.1943-7900.0000352, 2011.

Wynn, T. M., Mostaghimi, S., Burger, J. A., Harpold, A. A., Henderson, M. B., and Henry, L. A.: Variation in Root Density along Stream Banks, J. Environ. Qual., 33, 2030–2039, 2004.

14

Morphological coupling in multiple sandbar systems – a review

T. D. Price[1,*], B. G. Ruessink[1], and B. Castelle[2]

[1]Department of Physical Geography, Faculty of Geosciences, Utrecht University, Utrecht, the Netherlands
[2]UMR EPOC, Université de Bordeaux 1, Bordeaux, France
[*]now at: Department of Ecological Science, Faculty of Earth and Life Sciences, VU University Amsterdam, Amsterdam, the Netherlands

Correspondence to: T. D. Price (t.d.price@vu.nl)

Abstract. Subtidal sandbars often exhibit alongshore variable patterns, such as crescentic plan shapes and rip channels. While the initial formation of these patterns is reasonably well understood, the morphodynamic mechanisms underlying their subsequent finite-amplitude behaviour have been examined far less extensively. This behaviour concerns, among other aspects, the coupling of alongshore variable patterns in an inner bar to similar patterns in a more seaward bar, and the destruction of crescentic patterns. This review aims to present the current state of knowledge on the finite-amplitude behaviour of crescentic sandbars, with a focus on morphological coupling in double sandbar systems. In this context we include results from our recent study, based on a combination of remote-sensing observations, numerical modelling and data–model integration. Morphological coupling is an inherent property of double sandbar systems, where the inner bar may attain a type of morphology not found in single bar systems. Coupling is governed by water depth variability along the outer-bar crest and by various wave characteristics, including the offshore wave height and angle of incidence. In recent research, the role of the angle of wave incidence for sandbar morphodynamics has received more attention. Numerical modelling results have demonstrated that the angle of wave incidence is crucial to the flow pattern, sediment transport, and thus the emerging morphology of the coupled inner bar. Moreover, crescentic patterns predominantly vanish under high-angle wave conditions, highlighting the role of alongshore currents in straightening sandbars and challenging the traditional conception that crescentic patterns vanish under high-energy, erosive wave conditions only.

1 Introduction

Subtidal sandbars are shore-parallel ridges of sand in less than 10 m water depth fringing wave-dominated coasts along great lakes, semi-enclosed seas and open oceans (e.g. Evans, 1940; Saylor and Hands, 1970; Greenwood and Davidson-Arnott, 1975; Lippmann et al., 1993; Ruessink and Kroon, 1994; Shand et al., 1999; Almar et al., 2010; Kuriyama, 2002; Ruessink et al., 2003; Wijnberg and Terwindt, 1995, and references therein). Sandbars often have multi-annual lifetimes and can occur as a single feature, or as a multiple bar (most often two, sometimes up to five) system. Intriguingly, sandbars often exhibit quasi-regular undulations in their height and cross-shore position (Fig. 1). These so-called crescentic sandbars can be viewed as a more-or-less rhythmic sequence of shallow horns (shoals) and deep bays (cross-shore troughs) alternating shoreward and seaward of an imaginary line parallel to the coast. In addition, crescentic sandbars are often associated with similar rhythmic perturbations in onshore morphology, such as the shoreline (e.g. Sonu, 1973; Van de Lageweg et al., 2013) or a more landward located inner sandbar (e.g. Ruessink et al., 2007a). Depending on the wave conditions and the currents they induce in the nearshore zone, these sandbar patterns continuously change, vanish or reappear. It is this perpetual variability of nearshore sandbars that continues to draw the attention of nearshore researchers, just as it has done over the past decades.

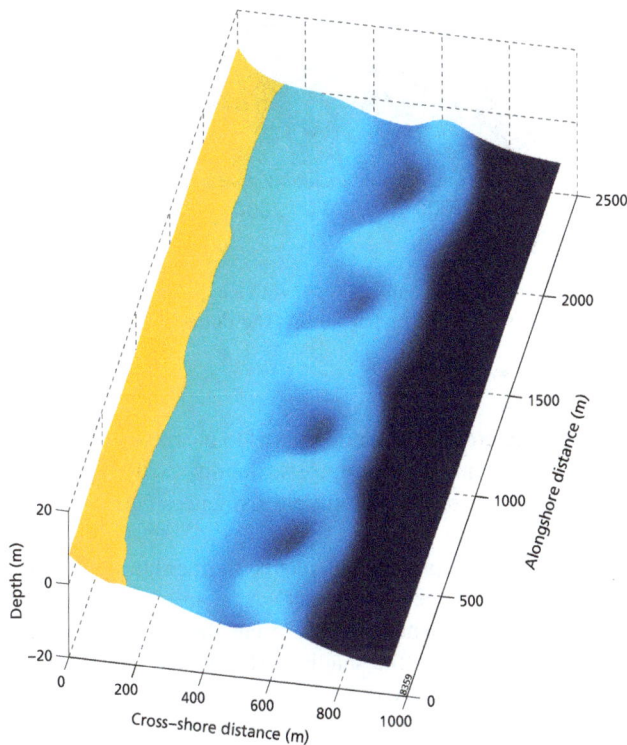

Figure 1. Bathymetry of a beach with a crescentic sandbar. This bathymetry was measured during the ECORS-Truc Vert 2008 field experiment (see Almar et al. 2010).

Besides their intriguing morphological appearance and evolution, sandbars are also of significant societal importance by forming a natural barrier between the hinterland and the ocean. Sandbars safeguard beaches by dissipating storm waves before they impact the shore. Morphological coupling, for instance, can lead to alongshore variations in wave dissipation, resulting in localised beach and dune erosion and subsequent property loss during storms (Thornton et al., 2007). Many present-day soft engineering measures to improve coastal safety, such as shoreface nourishments, involve direct or indirect modifications to sandbars (e.g. Grunnet and Ruessink, 2005; Ojeda and Guillén, 2008). A comprehensive understanding of the processes that govern sandbar behaviour and the development of the capability to predict this behaviour are thus of significant importance when it comes to minimising human and economic losses.

A key element to the understanding of morphological sandbar behaviour is frequent (daily), and long-term (~ years) monitoring of the nearshore zone, and the subsequent investigation of patterns and regularities in behaviour emerging from this observational data. Numerous observations and long-term monitoring of the nearshore zone have revealed the wide range of shapes that nearshore sandbars may attain (e.g. Wright and Short, 1984; Lippmann and Holman, 1990; Van Enckevort et al., 2004; Ranasinghe et al., 2004). Despite each observed sandbar configuration being

unique, and the continuous change in shape under the influence of waves and currents, a certain regularity in sandbar morphology has been observed. For single-barred beaches, Wright and Short (1984) developed the most widely accepted and applied beach state classification model, based on observations of beaches with contrasting environmental conditions over a period of 3 years. Such an aggregation facilitates answers as to when certain behaviour, such as morphological coupling, actually happens. Whereas the Wright and Short (1984) classification model is essentially applicable to single-barred beaches only, Short and Aagaard (1993) devised a multi-bar state model where each bar can go through the same states as in the single bar model. The sandbars are essentially treated as independent features and the role of coupling between the bars for the behaviour of the composite double sandbar system is thus disregarded.

Another key approach to the understanding of sandbar dynamics is the development, use, and validation of simplified exploratory and detailed simulation models. Model studies first explained alongshore sandbar variability from a *hydrodynamic* template in the water motion (Bowen and Inman, 1971; Holman and Bowen, 1982); present-day models rely on the principle of self-organisation (Hino, 1975; Sonu, 1972; Falqués et al., 2000; Coco and Murray, 2007), in which a crescentic sandbar forms spontaneously through the positive feedback between the flow, sediment processes and the evolving morphology. The genesis of crescentic patterns in single sandbar systems is thus reasonably well understood. In a double sandbar system, with a more landward inner bar and a more seaward outer bar, the distinction between a forcing template and self-organisation becomes blurred (Castelle et al., 2010a, b). In this case, the crescentic outer-bar morphology acts as a morphological template for the inshore flow patterns through the breaking and focussing of waves across the outer bar. It is obvious that this morphological coupling no longer relates to the initial formation of patterns, but relates to finite-amplitude behaviour instead. Here, the morphological template of the crescentic outer bar may suppress local self-organisation mechanisms at the inner bar and hence govern the shape of the inner bar.

Although morphological coupling has been observed and finite-amplitude behaviour of sandbars has been shown to be one of the largest sources of nearshore morphodynamic variability, it is not understood when and why morphological variations in an outer bar impact the geometry of an inner bar. The increasing availability of high-resolution (daily), long-term (many years) time series of nearshore video imagery (Holman and Stanley, 2007), together with advances in the non-linear modelling of nearshore morphodynamics and in data–model integration techniques, have recently advanced our knowledge of the finite-amplitude behaviour of alongshore sandbar variability considerably.

This review aims to present the current state of knowledge on the finite-amplitude behaviour of crescentic sandbars, with a focus on morphological coupling in double sandbar

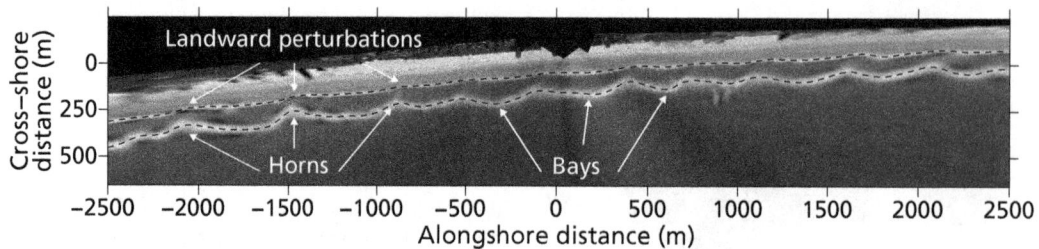

Figure 2. Example of a time-exposure image from the Gold Coast, Australia, showing the Idt coupling type, with a crescentic outer bar and a terraced inner bar with landward perturbations coupled to the alongshore positions of the outer-bar horns. The dotted lines indicate the video-derived inner and outer barlines. Source: Price et al. (2013).

systems. In this context we include results from our recent study, based on an approximately 9.3-year long data set of low-tide time-exposure video images of the double-barred Surfers Paradise, Gold Coast, Queensland, Australia, a swell-dominated site where the waves are usually obliquely incident. Measurements from nearby wave buoys provided concurrent wave data, i.e. root-mean-square wave height H_{rms}, peak wave period T_p and angle of wave incidence with respect to shore-normal in 15 m depth θ. First, the morphodynamic states that characterise a double sandbar system are described in Sect. 2, followed by a discussion of observations and modelling efforts of morphological coupling in double sandbar systems in Sect. 3. In Sect. 4, we conclude with a brief synthesis and perspectives for future research.

2   Alongshore sandbar variability

Although considerable research has been devoted to the state dynamics of a double-barred system, observations were mostly based on data which were either temporally limited to a single accretionary/erosional sequence (e.g. Van Enckevort et al., 2004; Ruessink et al., 2007a), spatially limited to (an alongshore transect of) the inner bar (e.g. Lippmann and Holman, 1990; Shand et al., 2003; Sénéchal et al., 2009) or based on data acquired at different locations or at irregular intervals (Short and Aagaard, 1993; Castelle et al., 2007). Furthermore, the large relaxation times of outer bars, in relation to the offshore wave forcing, have often prevented an abundance of state transitions of the outer bar to occur during the studied periods (see e.g. Goldsmith et al., 1982; Ferrer et al., 2009). While these observations each provide a varying amount of insight into sandbar behaviour, the use of scarce data or the selective use of data carries the risk of assuming the identified behaviour to be representative of the characteristic system dynamics. With the rise of video monitoring of the nearshore zone over the last 3 decades (Holman and Stanley, 2007), the trend towards frequent, long-term monitoring is increasing. The sequential behaviour of the bar states of a double-barred system at a single site, however, had not been studied under a wide range of wave conditions. Accordingly, an important first step in our study of the finite-amplitude

behaviour of crescentic sandbars was to characterise the typical development of alongshore variability within a double sandbar system, based on multiple sequences (see Price and Ruessink, 2011).

The most conspicuous elements in the low-tide time-exposure images mentioned in Sect. 1 are the alongshore continuous white bands that represent the foam created by wave breaking above the sandbars (Lippmann and Holman, 1989; Fig. 2). We tracked this optical breaker line (hereafter referred to as the barline) of both the inner and outer bar in all available (2995) low-tide images, allowing us to quantify the alongshore variability of both bars (see Price and Ruessink, 2011). During the 9.3 years studied, the outer bar was predominantly (two thirds of the time) alongshore variable, whereas the inner bar existed as a shore-attached terrace with a rhythmic terrace edge almost half of the time (shown in Fig. 2). This alongshore rhythmicity of the inner terrace contrasts with shore-attached terraces in single bar systems, which are mostly alongshore-uniform. For more alongshore-uniform outer-bar shapes (a third of the time), an inner terrace was less common and, instead, rip channels dominated the inner-bar morphology.

As mentioned in Sect. 1, the development of crescentic sandbars has been attributed to self-organisation processes, with the traditional conception that the wave energy alone governs their evolution. The development of crescentic sandbars has been found to develop during low-energy, accretive wave conditions (e.g. Ranasinghe et al., 2004; Van Enckevort et al., 2004); a so-called downstate sequence (Wright and Short, 1984). Their alongshore variability is associated with wave-driven circulation patterns that consist of weak onshore flow over the horns and strong offshore flow through the bays. Under continuing low waves the horns of the crescentic bar weld to the shore, causing the initially alongshore continuous trough to disappear and the bays to evolve into distinct cross-shore troughs (rip channels) with strong currents (up to 2 m s^{-1}) (e.g. Brander, 1999; Houser et al., 2013). On the other hand, the straightening of an alongshore variable sandbar, called an upstate sequence (Wright and Short, 1984) or morphological reset, has traditionally been associated with

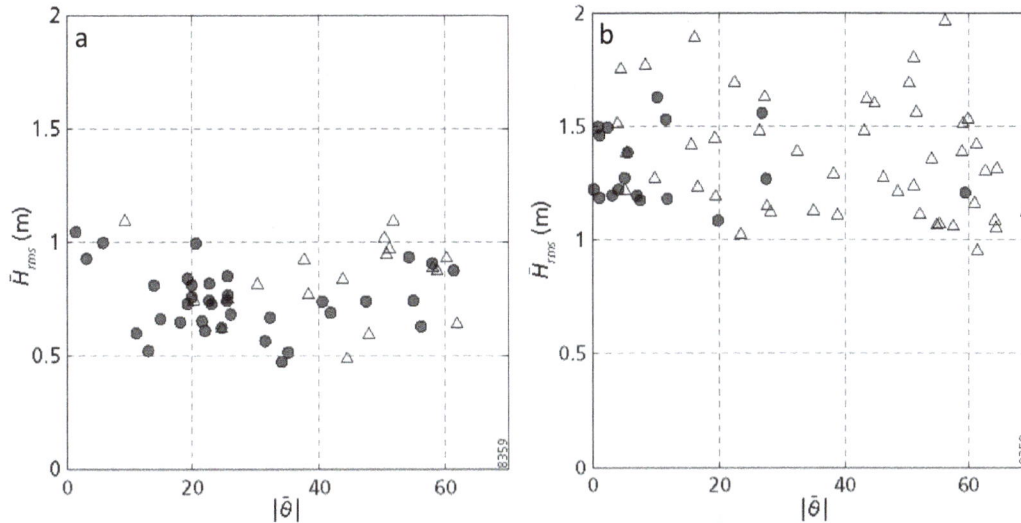

Figure 3. Mean wave conditions during (**a**) low-energetic and (**b**) moderately energetic downstate (circles) and upstate (triangles) transitions of the outer bar, showing $\bar{H}_{rms}$ versus $|\bar{\theta}|$. A downstate transition corresponds to the further development of rip channels, an upstate transition to a sandbar straightening. Adapted from: Price and Ruessink (2011).

high-energy, erosive-wave conditions, without an actual account of which processes lead to straightening.

Observations from the Gold Coast video data set challenge the need for high-energy wave conditions for the straightening of an alongshore variable sandbar; instead, they stress the effect of wave obliquity in morphological evolution. For example, Fig. 3a illustrates that low-energetic wave conditions ($\bar{H}_{rms} = 0.5-1$ m) generally resulted in the further development of rip channels in the outer bar, especially when θ is small (say, less than 30°), while the same waves with a larger angle of incidence ($\theta > 20°$) were observed to cause a reset. Similarly, Fig. 3b illustrates that moderately energetic wave conditions ($\bar{H}_{rms} = 1-2$ m) generally led to sandbar straightening, while the further development of rip channels was observed during smaller angles of wave incidence ($\theta < 30°$). The straightening of a shore-attached crescentic sandbar to a shore-parallel linear bar by obliquely incident waves generally happened gradually (1–5 days). During this transition, the barline straightened and the rip channels became obliquely oriented, leading to a characteristic sandbar morphology (Fig. 4).

Whereas the morphodynamics of the outer bar at the Gold Coast could be related to offshore wave conditions (as a single bar system), two types of inner-bar morphodynamics were distinguished, governed by the outer-bar state: the inner bar mostly existed as an alongshore variable terrace for alongshore variable outer-bar states. As more wave energy reached the inner bar during alongshore-uniform outer-bar states, the inner-bar behaviour resembled more that of a single-barred system, with its frequent separation from the shoreline and the persistent development of rip channels. This interaction implies that sandbars in a double-barred system should not be studied as independent features, but that

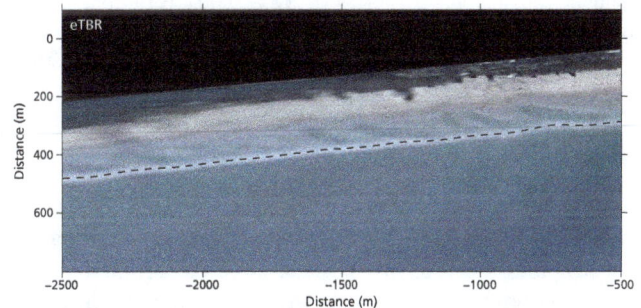

Figure 4. Example of the characteristic sandbar morphology during the straightening of a shore-attached crescentic sandbar to a shore-parallel linear bar by obliquely incident waves. Here, the barline (dashed line) straightens and the rip channels become obliquely oriented. Adapted from: Price and Ruessink (2011).

the behaviour of the composite sandbar system should be taken into account.

3 Sandbar coupling

3.1 Observations

Various observations indicate that the inner bar may possess remarkably smaller and often more variable alongshore scales than the outer bar (e.g. Bowman and Goldsmith, 1983; Van Enckevort et al., 2004). This has long been interpreted as self-organisation at the scale of the individual bar and the absence of interaction between sandbars. Other observations, summarised in Castelle et al. (2010a), demonstrate that inner-bar patterns can also couple to those in the outer bar, indicative of a type of interaction that Castelle et al. (2010a) termed

Figure 5. Examples of coupled morphology, showing **(a)** out-of-phase (180°) coupled sandbars, **(b)** out-of-phase coupling between sandbar and shoreline (courtesy of A.D. Short), **(c)** in-phase (0°) coupled sandbars (taken from Bowman and Goldsmith, 1983), and **(d)** two inner-bar rip channels for each outer-bar bay (taken from Castelle et al., 2007).

morphological coupling. Ruessink et al. (2007a), for example, found that the inner bar increasingly coupled to the outer-bar shape as the outer bar became more crescentic and migrated onshore, i.e. during a downstate transition of the outer bar (Wright and Short, 1984; Price and Ruessink, 2011). Coupling examples (Fig. 5) include the systematic occurrence of two inner-bar rip channels within one outer-bar crescent (Castelle et al., 2007; Fig. 5d), that of seaward perturbations in the inner bar facing outer-bar horns (a 180°, or out-of-phase relationship; Van Enckevort and Wijnberg, 1999; Fig. 5a), and that of shoreward perturbations in the inner bar facing outer-bar horns (a 0°, or in-phase relationship; Bowman and Goldsmith, 1983; Castelle et al., 2007; Fig. 5c). The out-of-phase relationship is reminiscent of the commonly observed relationship between inner-bar patterns and shoreline rhythms (Sonu, 1973; Orzech et al., 2011; Fig. 5b). Additionally, Ruessink et al. (2007a) and Quartel (2009) found coupled sandbar patterns with gradual phase changes (ranging from 0 to 180°), thought to be related to the persistent non-zero angle of wave incidence and larger alongshore migration rates of the subtidal bar with respect to the inner bar, respectively.

The aforementioned field observations of sandbar coupling were either based on sporadic observations (e.g. Bowman and Goldsmith, 1983; Castelle et al., 2007) or a short single event (e.g. Ruessink et al., 2007a). Although this previous work has provided clear examples of the phenomenon

of sandbar coupling, the frequency or predominance of either of the coupling patterns remained unclear. As a first step towards understanding when and how often certain coupling types develop, we addressed the representativeness of these findings using the barlines derived from the low-tide time-exposure video images. Cross-correlation of the barlines allowed detecting coupled inner- and outer-bar morphology (Price and Ruessink, 2013). Intriguingly, 40 % of all observations were found to have statistically significant (at the 98 % confidence level) coupling. The images unveiled five characteristic coupling types (Fig. 6). The bars either coupled in-phase, with an outer-bar horn facing a shoreward perturbation of the inner barline, or out-of-phase, where the outer-bar horn coincided with a seaward bulge in the inner barline. Four of the five observed coupling types coincided with a downstate sequence of the outer bar. The morphology of the inner bar was found to be either terraced (with no trough or channels intersecting the bar) or characterised by the presence of rip channels. These properties were used to give abbreviated names to the coupling types (Fig. 6): I or O (in-phase or out-of-phase), d or u (downstate or upstate) and t or r (terraced or with rips). By far the most common coupling type at the Gold Coast was, however, the Idt type, with a wavy terraced inner bar showing landward perturbations displaced slightly ($\approx$ 100 m) alongshore with respect to the outer-bar horns (Fig. 6a and Fig. 2). This coupling type corresponds to the coupled morphology observed by Ruessink et al. (2007a) at the same site.

Using a numerical model with synthetic wave-input conditions and bathymetries, Castelle et al. (2010a) demonstrated that, under shore-normal waves, coupling processes arise because of alongshore variability in wave height, and associated flow patterns over the inner bar that are induced by the water depth variability along the outer-bar crest. As summarised in Fig. 7, a large fraction of wave breaking over the outer bar leads to out-of-phase coupled sandbars (Fig. 7a). For a small fraction of wave breaking, wave focusing by refraction over the outer-bar horns overwhelms the effect of wave breaking, leading to in-phase coupled sandbars (Fig. 7b). Figure 8 summarises the Gold Coast observations in a conceptual model, in which the type of coupling is governed by the offshore wave height, the angle of wave incidence and the depth variation along the outer bar. The two coupling types explored in Castelle et al. (2010a), under shore-normal wave incidence, correspond to Odr (Fig. 7a) and Idr (Fig. 7b). The predominance of the Idt coupling type is related to the fairly large waves that persistently arrive with a large angle of incidence (30°). We hypothesised that such wave conditions drive a meandering alongshore current (Sonu, 1972; MacMahan et al., 2010) that prevents the outer-bar horns from welding to the inner bar and leads to downdrift-positioned landward perturbations in the inner terrace. When the meandering current is less strong (smaller wave height or more shore-normal incidence), the outer-bar horns can weld ashore and lead to the Odt coupling type.

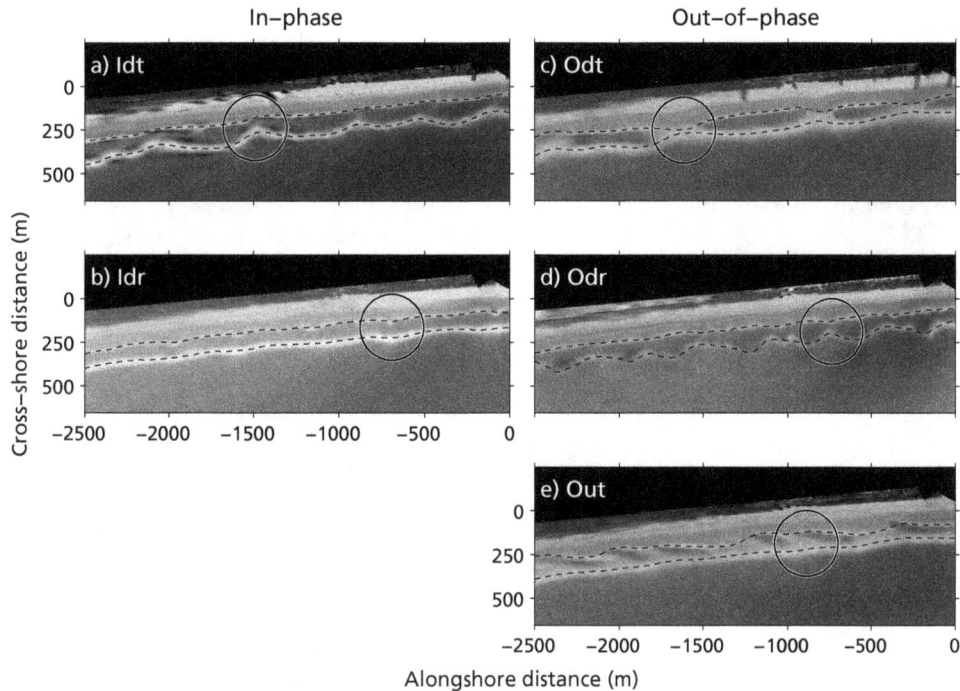

Figure 6. Examples of observed types of morphological coupling between the inner and outer barlines; low-tide time-exposure plan-view images of in-phase images with (**a**) an inner terrace and (**b**) inner rips, out-of-phase coupling with (**c**) an inner terrace and (**d**) inner rips in downstate direction, and (**e**) out-of-phase coupling with a clear alongshore offset between the inner- and outer-barline features in upstate direction. The dotted lines indicate the detected barlines, and the circles indicate a characteristic coupling feature for each coupling type. Source: Price and Ruessink (2013).

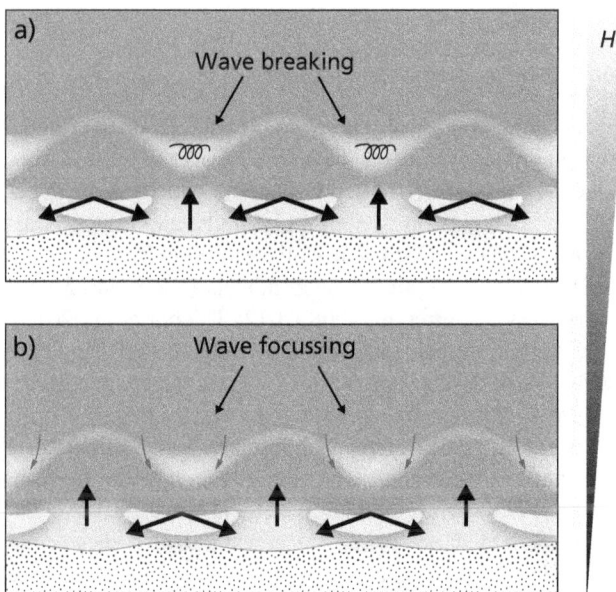

Figure 7. Coupling patterns found by Castelle et al. (2010a), showing (**a**) out-of-phase coupling and (**b**) in-phase coupling, depending on the wave height H. The thick black arrows indicate the associate flow patterns, whereas the gray arrows indicate wave refraction.

When the waves are highly energetic and obliquely incident, the outer bar becomes more alongshore-uniform (see also Sect. 2); the outer-bar horns separate from the outer bar to become part of the inner bar (similar to Almar et al., 2010), resulting in an alongshore variable inner terrace, the upstate coupling type Out. If the straightening persists, both bars become alongshore-uniform with alongshore continuous troughs. A sudden change toward the end of this straightening, however, leads to the Idr coupling type. Now, the small remaining depth variations along the outer bar cause wave focussing through refraction, driving a weak cell-circulation pattern over the inner bar (see also Fig. 7b).

Although the alongshore variability in the inner bar is coupled to that in the outer bar for some 40 % of the time at the Gold Coast, it remains unknown to what extent these observations represent the behaviour of other double-barred beaches. Similar to the observed behaviour at the Gold Coast (also see Ruessink et al., 2007a), observations from Duck Beach (North Carolina, USA) show the formation of an Idt coupling type, following a period of obliquely incident, moderately energetic waves (Fig. 9).

In a follow-up study, Castelle et al. (2010b) demonstrated that self-organisation and coupling processes can co-exist on an inner bar. In fact, their modelling suggests that the combination of both processes leads to stronger variability in the alongshore inner-bar scales, rather than

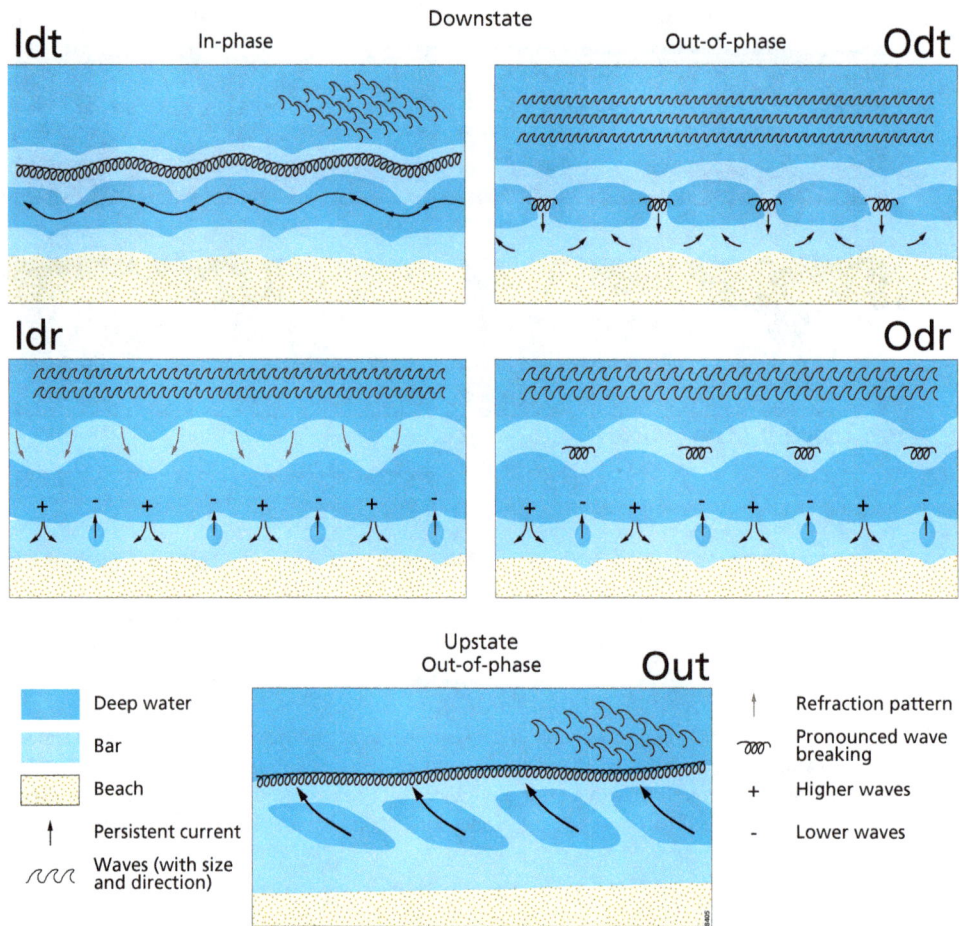

Figure 8. Conceptual model of the development of different coupling types.

self-organisational processes alone, as in single bar systems. They further demonstrated that the relative importance of self-organisation and morphological coupling changes in favour of the latter with an increase in water depth variability along the outer-bar crest. An analysis of an event during which an Idt coupling type formed, however, indicated that, under oblique wave incidence, it was not necessarily the alongshore depth variation but the alongshore shape of the outer bar which is important for altering the wave and current fields at the inner bar (Price and Ruessink, 2013). In the next section, we further discuss the role of the angle of wave incidence for the development of different coupling types.

3.2 Modelling

Although video observations provide a high-frequency long-term data set of coupled sandbar morphology, numerical models are often used to shed light on the processes underlying the observed morphodynamics. So far, numerical studies of sandbar morphology have largely focussed on single-barred beaches (e.g. Ranasinghe et al., 2004; Reniers et al., 2004; Garnier et al., 2006; Tiessen et al., 2011). The

few existing numerical studies of double sandbar systems have mainly focussed on the initial development and subsequent evolution of crescentic patterns, either using linear stability analysis (e.g. Klein and Schuttelaars, 2006; Garnier et al., 2008; Coco and Calvete, 2009; Brivois et al., 2012), nonlinear depth-averaged models (Klein and Schuttelaars, 2006; Smit et al., 2008, 2012; Thiébot et al., 2012), or quasi-three-dimensional models (Drønen and Deigaard, 2007). Whereas the simulations of Castelle et al. (2010a, b) were performed for shore-normal wave incidence only, Thiébot et al. (2012) performed numerical simulations for a large range of wave angles over initially alongshore-uniform sandbars. For slightly obliquely incident waves (10 and 15° with respect to shore normal at 8 m water depth), they found that initially the inner bar did not develop any alongshore variability due to the large alongshore current. However, when the outer bar started to develop alongshore variability, the alongshore current and the incoming wave field at the inner bar became perturbed, leading to the development of inner-bar features with an alongshore spacing similar to that of the outer-bar horns.

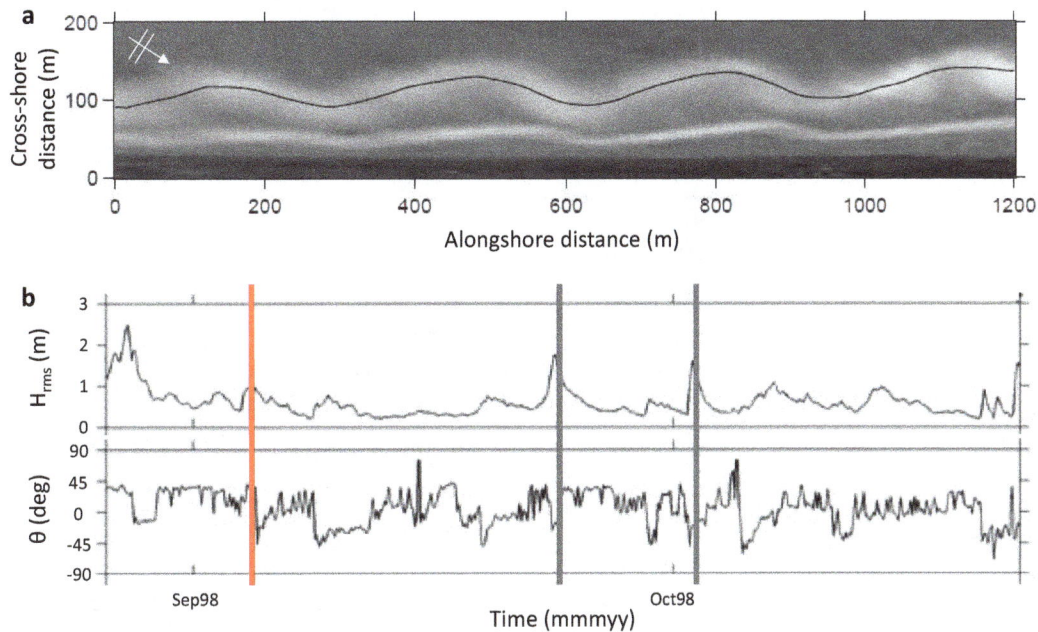

Figure 9. Example of the Idt coupling type observed at Duck, NC, USA. The video image in (**a**) is from 4 September 1998, indicated by the solid red line in the time series in (**b**), showing the offshore root-mean-square wave height H_{rms} (top) and angle of wave incidence with respect to shore normal θ (bottom). The solid grey lines correspond to moments when an existing crescentic pattern was wiped out, see Van Enckevort et al. (2004). Adapted from Van Enckevort and Ruessink (2003) (**a**) and Van Enckevort et al. (2004) (**b**).

Building upon the hypotheses from Castelle et al. (2010a) for shore-normal wave incidence and the video observations from the Gold Coast, Price et al. (2013) applied the non-linear 2DH (two horizontal dimensions) numerical model of Castelle et al. (2010a) to explore why different angles of wave incidence lead to the development of different coupling types. Modelling the finite-amplitude behaviour of nearshore bars, however, requires correct estimates of the initial bathymetric state. As bathymetric surveys of crescentic sandbar systems are scarce, they used the assimilation model of Van Dongeren et al. (2008) to estimate depth variations from the video images. This contrasts with earlier modelling efforts of double-barred systems, which used synthetic or highly idealised bathymetries. The boundary conditions for the simulations were extracted from a representative 4-day period during which the development of an Idt coupling type was observed in time-exposure video images. Subsequently, the model was run with time-invariant forcing (offshore significant wave height and period of 1.1 m and 9 s, respectively) for angles of wave incidence θ ranging from 0 to 20°, with an initially crescentic outer bar ((see Price et al., 2013, for details)).

Figure 10 shows the flow pattern along the inner bar for all θ (in 15 m depth) simulations after 2 days of simulation. Here, the grey scaling indicates the strength of the rotational nature of the flow, termed the swirling strength, over the inner bar. It can be seen that the flow is rotational (i.e. contains cell-circulation patterns) for angles of wave incidence of up to $\approx 10°$. As θ approaches 10°, the feeder current directly downdrift of the rip channel becomes weaker and eventually disappears as it becomes overridden by the alongshore current. Now, the flow field above the inner bar is dominated by a meandering alongshore current. Figure 11 shows the depth perturbations along the inner bar after 2 days of simulation. The most pronounced depth perturbations are found for the simulations with $\theta = 7°$, which are relatively deep and narrow. As the flow is still rotational (see Fig. 10), these negative perturbations correspond to rip channels. For larger angles, the negative depth perturbations decrease and become increasingly wider. Toward $\theta = 20°$, the depth perturbations have hardly developed at all. When we examine the simulations for $\theta = 10-20°$ in more detail, we find that the meandering alongshore current erodes the inner terrace downstream of the outer-bar horns, where more onshore-directed flow and accretion turn to more offshore-directed flow and erosion. This results in a landward perturbation in the terrace edge, consistent with the observations of the Idt coupling type. As such, the landward perturbations in the inner terrace for the Idt coupling type are erosional features. For $\theta < 10°$, cell-circulation patterns govern the flow at the inner bar, with offshore flow and the development of rip channels in the inner bar at the locations of the outer-bar horns, the Odr coupling type also found by Castelle et al. (2010a). On the whole, Figs. 10 and 11 confirm that the angle of wave incidence is crucial to the flow pattern, sediment transport, and thus the emerging coupling type at the inner bar.

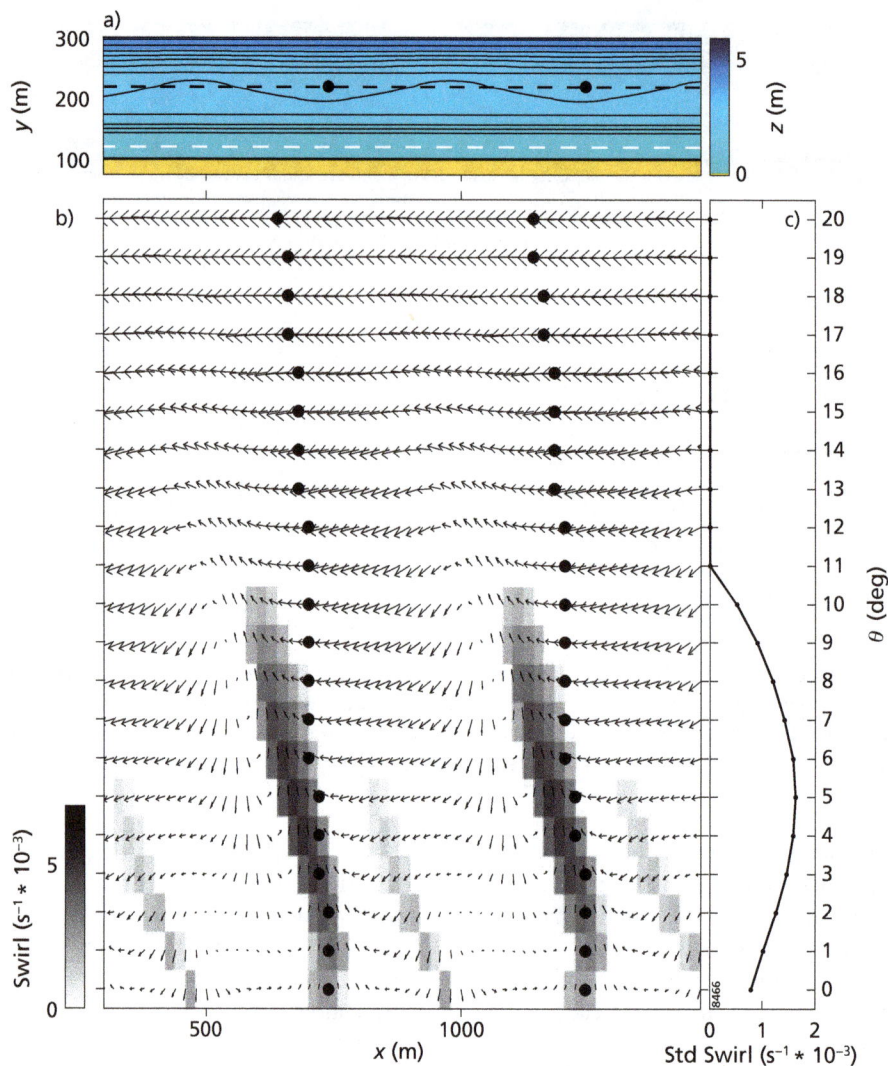

Figure 10. Model results, showing **(a)** the initial bathymetry, with isobaths (0.5 m intervals) contoured in the background, **(b)** flow velocity U (arrows) and swirling strength (shaded) along the inner bar at $y = 120$ m for all simulations after 2 days of simulation, and **(c)** the corresponding standard deviation of the swirling strength along the inner bar at $y = 120$ m. The black dots in **(a)** and **(b)** indicate the alongshore positions of the outer-bar horns along $y = 220$ m. The swirling strength is a measure of the rotational nature of the flow. Non-zero values imply the presence of cell-circulation patterns. Source: Price et al. (2013).

It is somewhat surprising that the most pronounced rip channels are found for the simulations with θ around 7° (Fig. 11), as previous modelling exercises of single bar systems (e.g. Castelle and Ruessink, 2011) found that rip channels were more pronounced when formed during shore-normal wave incidence. Also notice that the depth perturbations are located further to the left (downdrift) for larger angles of wave incidence. These findings may both be explained through the combination of the increased magnitude of the alongshore current on the one hand, and the alongshore migration and evolution of the outer bar (the morphological template for the inner bar) on the other hand (indicated by the black dots in Figs. 10 and 11a, b). Figure 11c shows that for small angles of wave incidence (up to $\theta = 7°$), the alongshore

variability of the outer bar increases with respect to the initial alongshore variability within the 2-day simulation period, whereas the outer bar becomes more alongshore-uniform for larger angles of wave incidence ($\theta > 7°$), corresponding with our observations (see Sect. 2; Price and Ruessink, 2011). Although the inner-bar depth perturbations follow the alongshore migration of the outer-bar horns at first, the straightening of the outer bar reduces the effect of the outer-bar morphological template on the inner-bar flow pattern, inhibiting the further development of inner-bar features as the flow pattern becomes alongshore-uniform. The numerical model study of Garnier et al. (2013) also stresses the effect of wave obliquity and the associated meandering current pattern in bar straightening. Their results indicated that the rip currents

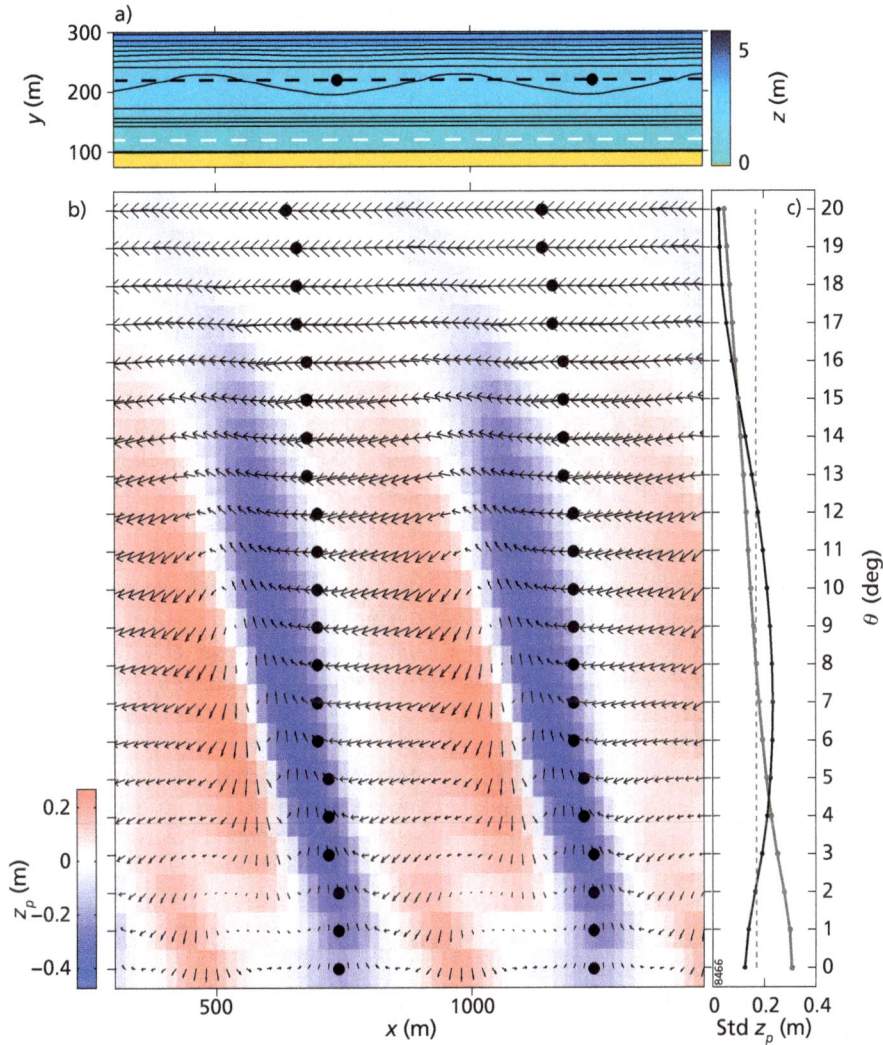

Figure 11. Model results, showing (**a**) the initial bathymetry, with isobaths (0.5 m intervals) contoured in the background, (**b**) flow velocity U (arrows) and depth perturbations z_p (colour) along the inner bar at $y = 120$ m (white dashed line in (**a**) for all simulations after 2 days of simulation). (**c**) depicts the alongshore standard deviation of z_p along the inner bar at $y = 120$ m (black) and the outer bar at $y = 220$ m (grey), and the initial standard deviation of z_p along the outer bar at $y = 220$ m (dashed). The black dots in (**a**) and (**b**) indicate the alongshore positions of the outer-bar horns along $y = 220$ m. Source: Price et al. (2013).

through the bays weakened in intensity with an increase in θ and that, at the same time, the strongest current shifted to a location downstream of the deepest part of the bay. As in Fig. 11, this shift causes the rip channels to migrate and decay. Interestingly, the transition from rip growth to rip decay at the outer bar takes place at substantially lower θ (say, 5–10°) than in the observations (Fig. 3, $\theta \approx 30°$). Similarly, Thiébot et al. (2012) showed that under slightly oblique wave incidence inner-bar perturbations did not develop before the growth of outer-bar perturbations.

4   Conclusions and perspectives

To summarise, the individual sandbars in a double-barred system should not be studied as independent features, but, instead, the interaction within the composite sandbar system should be taken into account. Morphological coupling is an inherent property of double sandbar systems. Accordingly, in double-barred systems, the inner bar may attain a type of morphology not found in single bar systems. Coupling is predominant when the outer bar is alongshore variable, both in position and depth, except for excessively large offshore angles of incidence or wave heights, leading to outer-bar straightening and sandbar de-coupling. From various observations of double sandbar systems, characteristic coupling types are distinguished. In addition to offshore wave height and depth variation along the outer bar, the offshore angle of wave incidence is crucial to the type of coupling that emerges. It strongly controls the type of flow pattern over the inner bar, with a change from cell-circulation patterns for approximately shore-normal waves to an alongshore meandering current as the angle increases.

Further work is necessary to test the generality of the findings from our Gold Coast study. The obtained results and the developed and applied methodology provide a framework for studying and describing similar data sets of multiple sandbar systems. In general, we expect intersite variability to arise from differences in sandbar mobility, which, in turn, is ascribed to sandbar volume, grain size, bottom slope, tidal range, and wave climate (e.g. see Wright and Short, 1984; Masselink and Short, 1993; Shand et al., 1999). More generally, as also suggested by Pape et al. (2010), intersite differences in sandbar behaviour are expected to depend on the ratio between the response time of a sandbar and the variability of the wave climate. Besides identifying the role of these potential variables through intersite comparison, numerical modelling becomes essential in testing the concepts formed. For example, a numerical model with different initial inner-bar morphologies, and time-variant wave forcing could shed light on this aspect of morphological coupling behaviour (see also Drønen and Deigaard, 2007; Garnier et al., 2008; Castelle and Ruessink, 2011; Tiessen et al., 2011; Smit et al., 2012). Moreover, from this, it would be interesting to assess changes in the ratio between self-organisation processes and outer-bar forced development of the inner bar (see also Castelle et al., 2010b).

In Sect. 3.2, the assimilation model of Van Dongeren et al. (2008) provided the initial bathymetric state for the numerical modelling, based on time-exposure images. Although previous work has been devoted to unravel how the observed foam relates to, for example, the roller dissipation (Aarninkhof and Ruessink, 2004; Alexander and Holman, 2004), further investigation into the relation between the observed foam and the measured wave properties on a natural beach would likely benefit the use of this assimilation technique at other sites with scarce amounts of data (see e.g. Birrien et al., 2013). Moreover, it is expected that the inclusion of multiple proxies for the bathymetry, such as wave celerity (e.g. Wilson et al., 2010), wave height (Almar et al., 2012; Gal et al., 2014), and cross-shore wave height profiles from terrestrial laser scanners (Belmont et al., 2007; Blenkinsopp et al., 2012), will enhance the assimilation results (Van Dongeren et al., 2008), and ultimately improve our understanding of finite-amplitude sandbar behaviour.

Although this review focussed on the alongshore variability of a double sandbar system, the observed effect of the outer bar on the inner-bar morphodynamics implicitly includes a cross-shore aspect. In fact, recent research (Plant et al., 2006; Splinter et al., 2011) has indicated that alongshore variations in bar crest position affect the alongshore-uniform behaviour. It was found that the horizontal cell-

circulation coinciding with the growth of alongshore variability facilitates onshore migration under low-energetic conditions. Analogously, a decrease in three-dimensionality in the outer bar coincides with offshore migration of the outer bar. Although this offshore migration has been suggested to be driven by the increased undertow over the bar during high-energetic events, it remains unknown whether undertow leads to the straightening of the bar. Both observations (Sect. 2) and modelling results (Sect. 3.2) show that sandbars do not necessarily straighten during storms, with large wave heights, but that obliquely incident waves play a crucial role in the straightening of the bar through the generation of an alongshore current. Process-based models that focus on cross-shore migration (e.g. Hoefel and Elgar, 2003; Ruessink et al., 2007b, 2012) or on alongshore variability (e.g. Reniers et al., 2004; Calvete et al., 2005; Drønen and Deigaard, 2007; Castelle and Coco, 2012) alone, have become quite mature. The key challenge will be to integrate both model concepts into a single model that can adequately simulate the complete dynamics of double sandbar systems. As such, understanding the alongshore variable sandbar behaviour will also lead to improved understanding of cross-shore behaviour.

Acknowledgements. We wish to thank Giovanni Coco and two anonymous referees for their constructive comments, which significantly improved this review. We acknowledge financial support by the Netherlands Organisation for Scientific Research (NWO) under contract 818.01.009 and by the French BARBEC project under contract ANR N2010JCJC60201. We thank Ian Turner for the use of the Gold Coast video imagery.

Edited by: G. Coco

References

Aarninkhof, S. G. J. and Ruessink, B. G.: Video observations and model predictions of depth-induced wave dissipation, IEEE Trans. Geosci. Remote Sens., 42, 2612–2622, 2004.

Alexander, P. S. and Holman, R. A.: Quantification of nearshore morphology based on video imaging, Mar. Geol., 208, 101–111, 2004.

Almar, R., Castelle, B., Ruessink, B. G., Sénéchal, N., Bonneton, P., and Marieu, V.: Two- and three-dimensional double sandbar system behaviour under intense wave forcing and a meso-macro tidal range, Cont. Shelf Res., 30, 781–792, 2010.

Almar, R., Cienfuegos, R., Catalán, P. A., Michallet, H., Castelle, B., Bonneton, P., and Marieu, V.: A new breaking wave height direct estimator from video imagery, Coast. Engin., 61, 42–48, 2012.

Belmont, M. R., Horwood, J. M. K., Thurley, R. W. F., and Baker, J.: Shallow Angle Wave Profiling Lidar, J. Atmos. Ocean. Technol., 24, 1150–1156, 2007.

Birrien, F., Castelle, B., Marieu, V., and Dubarbier, B.: On a data-model assimilation method to inverse wave-dominated beach bathymetry using heterogeneous video-derived observations, Ocean Engin., 73, 126–138, 2013.

Blenkinsopp, C. E., Turner, I. L., Allis, M. J., Peirson, W. L., and Garden, L. E.: Application of LiDAR technology for measurement of time-varying free-surface profiles in a laboratory wave flume, Coast. Engin., 68, 1–5, 2012.

Bowman, D. and Goldsmith, V.: Bar morphology of dissipative beaches: An empirical model, Mar. Geol., 51, 15–33, 1983.

Bowen, A. J. and Inman, D. L.: Edge Waves and Crescentic Bars, J. Geophys. Res., 76, 8662–8671, 1971.

Brander, R. W.: Field observations on the morphodynamic evolution of a low-energy rip current system, Mar. Geol., 157, 199–217, 1999.

Brivois, O., Idier, D., Thiébot, J., Castelle, B., Cozannet, G. L., and Calvete, D.: On the use of linear stability model to characterize the morphological behaviour of a double bar system. Application to Truc Vert beach (France), Comptes Rendus Geosci., 344, 277–287, 2012.

Calvete, D., Dodd, N., Falqués, A., and van Leeuwen, S. M.: Morphological development of rip channel systems: Normal and near-normal wave incidence, J. Geophys. Res., 110, C10006, doi:10.1029/2004JC002803, 2005.

Castelle, B. and Coco, G.: The morphodynamics of rip channels on embayed beaches, Conti. Shelf Res., 43, 10–23, 2012.

Castelle, B. and Ruessink, B. G.: Modeling formation and subsequent nonlinear evolution of rip channels: time-varying versus time-invariant wave forcing, J. Geophys. Res., 116, F04008, doi:10.1029/2011JF001997, 2011.

Castelle, B., Bonneton, P., Dupuis, H., and Sénéchal, N.: Double bar beach dynamics on the high-energy meso-macrotidal French Aquitanian Coast: A review, Mar. Geol., 245, 141–159, 2007.

Castelle, B., Ruessink, B. G., Bonneton, P., Marieu, V., Bruneau, N., and Price, T. D.: Coupling mechanisms in double sandbar systems, Part 1: Patterns and physical explanation, Earth Surf. Proc. Landforms, 35, 476–486, 2010a.

Castelle, B., Ruessink, B. G., Bonneton, P., Marieu, V., Bruneau, N., and Price, T. D.: Coupling mechanisms in double sandbar systems, Part 2: Impact on alongshore variability of inner-bar rip channels, Earth Surf. Proc. Landforms, 35, 771–781, 2010b.

Coco, G. and Calvete, D.: The use of linear stability analysis to characterize the variability of multiple sandbar systems, in: Proceedings of Coastal Dynamics 2009, Tokyo, Japan, paper no. 53, doi:10.1142/9789814282475_0055, 2009.

Coco, G. and Murray, A. B.: Patterns in the sand: From forcing templates to self-organization, Geomorphology, 91, 271–290, 2007.

Drønen, N. and Deigaard, R.: Quasi-three-dimensional modelling of the morphology of longshore bars, Coast. Engin., 54, 197–215, 2007.

Evans, O. F.: The low and ball of the eastern shore of lake michigan, The J. Geol., 48, 476–511, 1940.

Falqués, A., Coco, G., and Huntley, D.: A mechanism for the generation of wave-driven rhythmic patterns in the surf zone, J. Geophys. Res., 105, 24071–24087, doi:10.1029/2000JC900100, 2000.

Ferrer, P., Certain, R., Barusseau, J. P., and Gervais, M.: Conceptual modelling of a double crescentic barred coast (Leucate beach, France), in: Proceedings Coastal Dynamics 2009, Tokyo, Japan, 2009.

Gal, Y., Browne, M., and Lane, C.: Long-term automated monitoring of nearshore wave height from digital video, IEEE Trans. Geosci. Remote Sens., 52, 3412–3420, 2014.

Garnier, R., Calvete, D., Falques, A., and Caballeria, M.: Generation and nonlinear evolution of shore-oblique-transverse sand bars, J. Fluid Mechan., 567, 327–360, 2006.

Garnier, R., Calvete, D., Dodd, N., and Falqués, A.: Modelling the interaction between transverse and crescentic bar systems, in: River, Coastal and Estuarine Morphodynamics: RCEM 2007, edited by: Dohmen-Jansen, C. M. and Hulscher, S. J. M. H., vol. 2, Taylor and Francis Group, London, 931–937, 2008.

Garnier, R., Falqués, A., Calvete, D., Thiébot, J., and Ribas, F.: A mechanism for sand bar straightening by oblique wave incidence, Geophys. Res. Lett., 40, 2726–2730, 2013.

Goldsmith, V., Bowman, D., and Kiley, K.: Sequential stage development of crescentic bars: HaHoterim beach, southeastern Mediterranean, J. Sediment. Petrol., 52, 233–249, 1982.

Greenwood, B. and Davidson-Arnott, R. G. D.: Marine bars and nearshore sedimentary processes, Kouchibouguac Bay, New Brunswick, Canada, Nearshore sediment dynamics and sedimentation, 16, 123–150, 1975.

Grunnet, N. M. and Ruessink, B.: Morphodynamic response of nearshore bars to a shoreface nourishment, Coast. Engin., 52, 119–137, 2005.

Hino, M.: Theory on formation of rip-current and cuspidal coast, in: Proc. 14th Conf. Coastal Eng., Copenhagen, Denmark, Am. Soc. Civ. Eng., 901–919, 1975.

Hoefel, F. and Elgar, S.: Wave-Induced Sediment Transport and Sandbar Migration, Science, 299, 1885–1887, 2003.

Holman, R. A. and Bowen, A. J.: Bars, bumps, and holes: Models for the generation of complex beach topography, J. Geophys. Res., 87, 457–468, 1982.

Holman, R. A. and Stanley, J.: The history and technical capabilities of Argus, Coast. Engin., 54, 477–491, 2007.

Houser, C., Arnott, R., Ulzhöfer, S., and Barrett, G.: Nearshore circulation over transverse bar and rip morphology with oblique wave forcing, Earth Surf. Proc. Landforms, 38, 1269–1279, 2013.

Klein, M. D. and Schuttelaars, H. M.: Morphodynamic evolution of double-barred beaches, J. Geophys. Res., 111, C06017, doi:10.1029/2005JC003155, 2006.

Kuriyama, Y.: Medium-term bar behavior and associated sediment transport at Hasaki, Japan, J. Geophys. Res., 107, 3132, doi:10.1029/2001JC000899, 2002.

Lippmann, T. and Holman, R.: Quantification of sand bar morphology: a video technique based on wave dissipation, J. Geophys. Res., 94, 995–1011, 1989.

Lippmann, T. C. and Holman, R. A.: The spatial and temporal variability of sand bar morphology, J. Geophys. Res., 95, 11575–11590, 1990.

Lippmann, T. C., Holman, R. A., and Hathaway, K. K.: Episodic, nonstationary behavior of a double bar system at Duck, North Carolina, U.S.A., 1986-1991, J. Coast. Res., SI15, 49–75, 1993.

MacMahan, J., Brown, J., Brown, J., Thornton, E., Reniers, A., Stanton, T., Henriquez, M., Gallagher, E., Morrison, J., Austin, M. J., Scott, T. M., and Senechal, N.: Mean Lagrangian flow behavior on an open coast rip-channeled beach: A new perspective, Mar. Geol., 268, 1–15, 2010.

Masselink, G. and Short, A. D.: The effect of tide range on beach morphodynamics and morphology: A conceptual beach model, J. Coast. Res., 9, 785–800, 1993.

Ojeda, E. and Guillén, J.: Shoreline dynamics and beach rotation of artificial embayed beaches, Mar. Geol., 253, 51–62, 2008.

Orzech, M. D., Reniers, A. J. H. M., Thornton, E. B., and MacMahan, J. H.: Megacusps on rip channel bathymetry: Observations and modeling, Coast. Engin., 58, 890–907, 2011.

Pape, L., Plant, N. G., and Ruessink, B. G.: On cross-shore migration and equilibrium states of nearshore sandbars, J. Geophys. Res., 115, F03008, doi:10.1029/2009JF001501, 2010.

Plant, N. G., Holland, K. T., and Holman, R. A.: A dynamical attractor governs beach response to storms, Geophys. Res. Lett., 33, L17607, doi:10.1029/2006GL027105, 2006.

Price, T. D. and Ruessink, B. G.: State dynamics of a double sandbar system, Continen. Shelf Res., 31, 659–674, 2011.

Price, T. D. and Ruessink, B. G.: Observations and conceptual modelling of morphological coupling in a double sandbar system, Earth Surf. Proc. Landforms, 38, 477–489, 2013.

Price, T. D., Castelle, B., Ranasinghe, R., and Ruessink, B. G.: Coupled sandbar patterns and obliquely incident waves, J. Geophys. Res., 118, 1677–1692, 2013.

Quartel, S.: Temporal and spatial behaviour of rip channels in a multiple-barred coastal system, Earth Surf. Proc. Landforms, 34, 163–176, 2009.

Ranasinghe, R., Symonds, G., Black, K., and Holman, R.: Morphodynamics of intermediate beaches: a video imaging and numerical modelling study, Coast. Engin., 51, 629–655, 2004.

Reniers, A. J. H. M., Roelvink, J. A., and Thornton, E. B.: Morphodynamic modeling of an embayed beach under wave group forcing, J. Geophys. Res., 109, C01030, doi:10.1029/2002JC001586, 2004.

Ruessink, B., Ramaekers, G., and van Rijn, L.: On the parameterization of the free-stream non-linear wave orbital motion in nearshore morphodynamic models, Coast. Engin., 65, 56–63, doi:10.1016/j.coastaleng.2012.03.006, 2012.

Ruessink, B. G. and Kroon, A.: The behaviour of a multiple bar system in the nearshore zone of Terschelling, the Netherlands: 1965–1993, Mar. Geol., 121, 187–197, 1994.

Ruessink, B. G., Wijnberg, K. M., Holman, R. A., Kuriyama, Y., and van Enckevort, I. M. J.: Intersite comparison of interannual nearshore bar behavior, J. Geophys. Res., 108, 3249, doi:10.1029/2002JC001505, 2003.

Ruessink, B. G., Coco, G., Ranasinghe, R., and Turner, I. L.: Coupled and noncoupled behavior of three-dimensional morphological patterns in a double sandbar system, J. Geophys. Res., 112, C07002, doi:10.1029/2006JC003799, 2007a.

Ruessink, B. G., Kuriyama, Y., Reniers, A. J. H. M., Roelvink, J. A., and Walstra, D. J. R.: Modeling cross-shore sandbar behavior on the timescale of weeks, J. Geophys. Res., 112, F03010, doi:10.1029/2006JF000730, 2007b.

Saylor, J. and Hands, E.: Properties of longshore bars in the Great Lakes, in: Proceedings of the 12th International Conference on Coastal Engineering, 839–853, ASCE, Washington DC, 1970.

Sénéchal, N., Gouriou, T., Castelle, B., Parisot, J. P., Capo, S., Bujan, S., and Howa, H.: Morphodynamic response of a meso- to macro-tidal intermediate beach based on a long-term data set, Geomorphology, 107, 263–274, 2009.

Shand, R. D., Bailey, D. G., and Shepherd, M. J.: An inter-site comparison of net offshore bar migration characteristics and environmental conditinos, J. Coast. Res., 15, 750–765, 1999.

Shand, R. D., Bailey, D. G., Hesp, P. A., and Shepherd, M. J.: Conceptual beach-state model for the inner bar of a storm-dominated, micro/meso tidal range coast at Wanganui, New Zealand, in: Proceedings of Coastal Sediments 2003, Miami, USA, 2003.

Short, A. D. and Aagaard, T.: Single and multi-bar beach change models, Journal of Coastal Research, SI15, 141–157, 1993.

Smit, M., Reniers, A., and Stive, M.: Role of morphological variability in the evolution of nearshore sandbars, Coast. Engin., 69, 19–28, 2012.

Smit, M. W. J., Reniers, A. J. H. M., Ruessink, B. G., and Roelvink, J. A.: The morphological response of a nearshore double sandbar system to constant wave forcing, Coast. Engin., 55, 761–770, 2008.

Sonu, C. J.: Field observation of nearshore circulation and meandering currents, J. Geophys. Res., 77, 3232–3247, 1972.

Sonu, C. J.: Three-dimensional beach changes, J. Geol., 81, 42–64, 1973.

Splinter, K. D., Holman, R. A., and Plant, N. G.: A behavior-oriented dynamic model for sandbar migration and 2DH evolution, J. Geophys. Res., 116, C01020, doi:10.1029/2010JC006382, 2011.

Thiébot, J., Idier, D., Garnier, R., Falqués, A., and Ruessink, B. G.: The influence of wave direction on the morphological response of a double sandbar system, Contin. Shelf Res., 32, 71–85, 2012.

Thornton, E., MacMahan, J., and Sallenger, Jr., A.: Rip currents, mega-cusps, and eroding dunes, Mar. Geol., 240, 151–167, 2007.

Tiessen, M. C. H., Dodd, N., and Garnier, R.: Development of crescentic bars for a periodically perturbed initial bathymetry, J. Geophys. Res., 116, F04016, doi:10.1029/2011JF002069, 2011.

Van de Lageweg, W. I., Bryan, K. R., Coco, G., and Ruessink, B. G.: Observations of shoreline sandbar coupling on an embayed beach, Mar. Geol., 344, 101–114, 2013.

Van Dongeren, A., Plant, N., Cohen, A., Roelvink, J. A., Haller, M. C., and Catalán, P.: Beach Wizard: Nearshore bathymetry estimation through assimilation of model computations and remote observations, Coast. Engin., 55, 1016–1027, 2008.

Van Enckevort, I. M. J. and Ruessink, B. G.: Video observations of nearshore bar behaviour. Part 2: alongshore non-uniform variability, Contin. Shelf Res., 23, 513–532, 2003.

Van Enckevort, I. M. J. and Wijnberg, K. M.: Intra-annual changes in bar plan shape in a triple bar system, in: Proceedings Coastal Sediments 1999, ASCE, New York, 1094–1108, 1999.

Van Enckevort, I. M. J., Ruessink, B. G., Coco, G., Suzuki, K., Turner, I. L., Plant, N. G., and Holman, R. A.: Observations of nearshore crescentic sandbars, J. Geophys. Res., 109, C06028, doi:10.1029/2003JC002214, 2004.

Wijnberg, K. M. and Terwindt, J. H. J.: Extracting decadal morphological behaviour from high-resolution, long-term bathymetric surveys along the Holland coast using eigenfunction analysis, Mar. Geol., 126, 301–330, 1995.

Wilson, G., Özkan Haller, H., and Holman, R.: Data assimilation and bathymetric inversion in a two-dimensional horizontal surf zone model, J. Geophys. Res., 115, C12057, doi:10.1029/2010JC006286, 2010.

Wright, L. D. and Short, A. D.: Morphodynamic variability of surf zones and beaches: a synthesis, Mar. Geol., 56, 93–118, 1984.

Permissions

The contributors of this book come from diverse backgrounds, making this book a truly international effort. This book will bring forth new frontiers with its revolutionizing research information and detailed analysis of the nascent developments around the world.

We would like to thank all the contributing authors for lending their expertise to make the book truly unique. They have played a crucial role in the development of this book. Without their invaluable contributions this book wouldn't have been possible. They have made vital efforts to compile up to date information on the varied aspects of this subject to make this book a valuable addition to the collection of many professionals and students.

This book was conceptualized with the vision of imparting up-to-date information and advanced data in this field. To ensure the same, a matchless editorial board was set up. Every individual on the board went through rigorous rounds of assessment to prove their worth. After which they invested a large part of their time researching and compiling the most relevant data for our readers.

The editorial board has been involved in producing this book since its inception. They have spent rigorous hours researching and exploring the diverse topics which have resulted in the successful publishing of this book. They have passed on their knowledge of decades through this book. To expedite this challenging task, the publisher supported the team at every step. A small team of assistant editors was also appointed to further simplify the editing procedure and attain best results for the readers.

Apart from the editorial board, the designing team has also invested a significant amount of their time in understanding the subject and creating the most relevant covers. They scrutinized every image to scout for the most suitable representation of the subject and create an appropriate cover for the book.

The publishing team has been an ardent support to the editorial, designing and production team. Their endless efforts to recruit the best for this project, has resulted in the accomplishment of this book. They are a veteran in the field of academics and their pool of knowledge is as vast as their experience in printing. Their expertise and guidance has proved useful at every step. Their uncompromising quality standards have made this book an exceptional effort. Their encouragement from time to time has been an inspiration for everyone.

The publisher and the editorial board hope that this book will prove to be a valuable piece of knowledge for researchers, students, practitioners and scholars across the globe.

List of Contributors

B. W. Goodfellow
Department of Physical Geography and Quaternary Geology, and Bolin Centre for Climate Research, Stockholm University, 10691 Stockholm, Sweden
Department of Geology, Lund University, 22362 Lund, Sweden

A. P. Stroeven
Department of Physical Geography and Quaternary Geology, and Bolin Centre for Climate Research, Stockholm University, 10691 Stockholm, Sweden

D. Fabel
Department of Geographical and Earth Sciences, East Quadrangle, University Avenue, University of Glasgow, lasgow G12 8QQ, UK

O. Fredin
Department of Geography, Norwegian University of Science and Technology (NTNU),7491, Trondheim, Norway

M.-H. Derron
Geological Survey of Norway, Leiv Eirikssons vei 39, 7491 Trondheim, Norway
Institute of Geomatics and Risk Analysis, University of Lausanne, 1015 Lausanne, Switzerland

R. Bintanja
Royal Netherlands Meteorological Institute, Wilhelminalaan 10, 3732 GK De Bilt, the Netherlands

M. W. Caffee
Department of Physics, Purdue University, West Lafayette, Indiana, USA

T. A. Tran
Graduate School for Creative Cities, Osaka City University, Osaka, Japan

V. Raghavan
Graduate School for Creative Cities, Osaka City University, Osaka, Japan

S. Masumoto
Graduate School of Science, Osaka City University, Osaka, Japan

P. Vinayaraj
Graduate School for Creative Cities, Osaka City University, Osaka, Japan

G. Yonezawa
Graduate School for Creative Cities, Osaka City University, Osaka, Japan

A. Burtin
GeoForschungsZentrum, Helmholtz Centre Potsdam, Potsdam, Germany

N. Hovius
GeoForschungsZentrum, Helmholtz Centre Potsdam, Potsdam, Germany

B. W. McArdell
Swiss Federal Institute for Forest, Snow and Landscape Research WSL, Birmensdorf, Switzerland

J. M. Turowski
GeoForschungsZentrum, Helmholtz Centre Potsdam, Potsdam, Germany

J. Vergne
École et Observatoire des Sciences de la Terre, CNRS UMR7516, Strasbourg, France

S. Zhao
Department of Surveying and Mapping, College of Mining Technology, Taiyuan University of Technology, Taiyuan 030024, China

W. Cheng
State Key Laboratory of Resources and Environmental Information System, Institute of Geographic Sciences and Natural Resources Research, CAS, Beijing 100101, China

R. O. Tinoco
Environmental Hydraulics Institute IH Cantabria, University of Cantabria, Santander, Spain

G. Coco
Environmental Hydraulics Institute IH Cantabria, University of Cantabria, Santander, Spain

S. Hergarten
Universität Freiburg i. Br., Institut für Geo- und Umweltnaturwissenschaften, Freiburg, Germany

J. Robl
Universität Salzburg, Institut für Geographie und Geologie, Salzburg, Austria

K. Stüwe
Universität Graz, Institut für Erdwissenschaften, Graz, Austria

M. T. Melis
Department of Chemical and Geological Sciences, University of Cagliari, Cagliari, Italy

F. Mundula
Department of Chemical and Geological Sciences, University of Cagliari, Cagliari, Italy

F. Dessì
Department of Chemical and Geological Sciences, University of Cagliari, Cagliari, Italy

R. Cioni
Department of Earth Sciences, University of Firenze, Firenze, Italy

A. Funedda
Department of Chemical and Geological Sciences, University of Cagliari, Cagliari, Italy

L. Zhang
State Key Laboratory of Hydroscience and Engineering, Tsinghua University, Beijing, China

G. Parker
Department of Civil&Environmental Engineering and Department of Geology, Hydrosystems Laboratory, University of Illinois, Urbana, IL, USA

C. P. Stark
Lamont-Doherty Earth Observatory, Columbia University, Palisades, NY, USA

T. Inoue
Civil Engineering Research Institute for Cold Regions, Hiragishi Sapporo, Japan

E. Viparelli
Dept. of Civil&Environmental Engineering, University of South Carolina, Columbia, SC, USA

X. Fu
State Key Laboratory of Hydroscience and Engineering, Tsinghua University, Beijing, China

N. Izumi
Faculty of Engineering, Hokkaido University, Sapporo, Japan

C. H. Mohr
Institute of Earth and Environmental Science, University of Potsdam, Potsdam, Germany

A. Zimmermann
Institute of Earth and Environmental Science, University of Potsdam, Potsdam, Germany

O. Korup
Institute of Earth and Environmental Science, University of Potsdam, Potsdam, Germany

A. Iroumé
Faculty of Forest Sciences and Natural Resources, Universidad Austral de Chile, Valdivia, Chile

T. Francke
Institute of Earth and Environmental Science, University of Potsdam, Potsdam, Germany

A. Bronstert
Institute of Earth and Environmental Science, University of Potsdam, Potsdam, Germany

M. Rogerson
Department of Geography, Environment and Earth Sciences, University of Hull, Cottingham Road, Hull, HU6 7RX, UK

H. M. Pedley
Department of Geography, Environment and Earth Sciences, University of Hull, Cottingham Road, Hull, HU6 7RX, UK

A. Kelham
Department of Geography, Environment and Earth Sciences, University of Hull, Cottingham Road, Hull, HU6 7RX, UK

J. D Wadhawan
Department of Geography, Environment and Earth Sciences, University of Hull, Cottingham Road, Hull, HU6 7RX, UK

A. Pelosi
Department of Civil Engineering, Università degli Studi di Salerno (UNISA), Via Giovanni Paolo II – 84084 Fisciano (Salerno), Italy

G. Parker
Department of Civil & Environmental Engineering and Department of Geology, Hydrosystems Laboratory, University of Illinois, 301 N. Mathews Ave., Urbana, IL 61801, USA

T. Nagata
Civil Engineering Research Institute for Cold Region, Public Works Research Institute, Japan

Y. Watanabe
Department of Civil and Environmental Engineering, Kitami Institute of Technology, Japan

H. Yasuda
Research Center for Natural Hazard & Disaster Recovery,
Niigata University, Japan

A. Ito
Civil Engineering Research Institute for Cold Region,
Public Works Research Institute, Japan

J. P. C. Eekhout
Hydrology and Quantitative Water Management Group,
Wageningen University, Wageningen, the Netherlands

R. G. A. Fraaije
Ecology and Biodiversity Group, Institute of
Environmental Biology, Utrecht University, Utrecht, the
Netherlands

A. J. F. Hoitink
Hydrology and Quantitative Water Management Group,
Wageningen University, Wageningen, the Netherlands

T. D. Price
Department of Physical Geography, Faculty of
Geosciences, Utrecht University, Utrecht, the Netherlands

B. G. Ruessink
Department of Physical Geography, Faculty of
Geosciences, Utrecht University, Utrecht, the Netherlands

B. Castelle
UMR EPOC, Université de Bordeaux 1, Bordeaux, France